Liquid Crystalline Polymers

Synthesis, Properties and Applications

Related Titles

Biopolymers Based Advanced Materials
ISBN: 978-0-6482205-4-1 (e-book)
ISBN: 978-0-6482205-5-8 (hardcover)

Functional Polymer Blends and Nanocomposites
ISBN: 978-0-6482205-6-5 (e-book)
ISBN: 978-0-6482205-7-2 (hardcover)

Polymers in Oil and Gas Industry
ISBN: 978-0-6482205-0-3 (e-book)
ISBN: 978-0-6482205-1-0 (softcover)

Functional Nanomaterials and Nanotechnologies: Applications for Energy & Environment
ISBN: 978-0-6482205-2-7 (e-book)
ISBN: 978-0-6482205-3-4 (softcover)

Technology Management in Business
ISBN: 978-1-925823-02-8 (softcover)

Advances in Polymer Technology: Material Development, Properties and Performance Evaluation
ISBN: 978-1-925823-00-4 (e-book)
ISBN: 978-1-925823-01-1 (hardcover)

Polymer Nanomaterials for Specialty Applications
ISBN: 978-1-925823-03-5 (e-book)
ISBN: 978-1-925823-04-2 (hardcover)

Advanced Materials
ISBN: 978-1-925823-05-9 (e-book)
ISBN: 978-1-925823-06-6 (hardcover)

Engineering Chemistry
ISBN: 978-1-925823-10-3 (softcover)

Biofuels
ISBN: 978-1-925823-12-7 (e-book)
ISBN: 978-1-925823-13-4 (hardcover)

Liquid Crystalline Polymers

Synthesis, Properties and Applications

Dr. Vikas Mittal
Editor

CWP
Central West Publishing

This edition has been published by Central West Publishing, Australia

For more information about the books published by Central West Publishing, please visit https://centralwestpublishing.com

Disclaimer

Every effort has been made by the publisher, editor and authors while preparing this book, however, no warranties are made regarding the accuracy and completeness of the content. The publisher, editor and authors disclaim without any limitation all warranties as well as any implied warranties about sales, along with fitness of the content for a particular purpose. Citation of any website and other information sources does not mean any endorsement from the publisher and authors. For ascertaining the suitability of the contents contained herein for a particular lab or commercial use, consultation with the subject expert is needed. In addition, while using the information and methods contained herein, the practitioners and researchers need to be mindful for their own safety, along with the safety of others, including the professional parties and premises for whom they have professional responsibility. To the fullest extent of law, the publisher, editor and authors are not liable in all circumstances (special, incidental, and consequential) for any injury and/or damage to persons and property, along with any potential loss of profit and other commercial damages due to the use of any methods, products, guidelines, procedures contained in the material herein.

NATIONAL LIBRARY OF AUSTRALIA

A catalogue record for this book is available from the National Library of Australia

ISBN (print): 978-1-925823-17-2
ISBN (e-book): 978-1-925823-16-5

Contents

Preface

Liquid crystalline polymers exhibit functional properties owing to both polymeric structure and liquid crystalline mesophase. Over the last few decades, a large number of such polymer systems have been developed with tuned structure-property correlations, which drive their uses in a wide variety of application areas. Development of composites of these polymers with various nano-reinforcements further enhances the stiffness, strength, heat and chemical resistance, barrier properties as well as dimensional stability. In this respect, the book aims to provide in-depth insights into the recent developments and future outlook of various liquid crystalline polymer systems, so as to further develop the fundamental understanding for widening their application potential.

Chapter 1 summarizes photoresponsive liquid crystalline polymers with azobenzene chromophores. As a typical photoactive molecule which can photo-isomerize, along with its rigid rod shape of the thermally stable *trans* configuration, an azobenzene moiety is often introduced in photoresponsive liquid crystalline polymers as the photoactive mesogens, endowing the obtained polymers with photocontrollable features. Chapter 2 focuses on the various strategies to incorporate different types of nanoparticles, e.g. metallic or metal oxides, in liquid crystalline polymers. In Chapter 3, the applications of liquid crystalline polymers in the electronics sector have been discussed within the context of their fundamental physical origins and chemical properties. Photoresponsive main chain liquid crystalline polymers have been reviewed in detail in Chapter 4. Chapter 5 is an experimental study presenting the synthesis and liquid crystalline properties of 1,3,4-oxadiazole derivatives with a terminal (R)-1-methylheptyloxy group. In another experimental study, Chapter 6 reports a facile synthetic route to generate novel thermotropic liquid crystalline polymers by bulk reaction or reactive extrusion, along with the characterization of their mesophases. Chapter 7 reviews the synthesis and characterization of segmented liquid crystalline copolyesters, delving into the parameters required for phase separation and the impact of the domain morphology. Finally, the smectic phases of rod-like helical polymers have been analyzed in Chapter 8.

I am indebted to the chapter authors for providing their valuable insights about liquid crystalline polymers. The book would not have been successfully accomplished without their support. The book is dedicated to my family for unswerving support and constant motivation.

Vikas MITTAL

Chapter 1

Photoresponsive Liquid Crystalline Polymers with Azobenzene Chromophores

Shudeng Ma, Shuai Huang and Haifeng Yu*

Department of Materials Science and Engineering, College of Engineering, Key Laboratory of Polymer Chemistry and Physics of Ministry of Education, Peking University, Beijing 100871, China
**Corresponding author*: yuhaifeng@pku.edu.cn

1.1 Introduction

Photochemical process often brings about great changes in materials properties, including refractive index, density, solubility, modulus of elasticity and dielectric parameters. These light-triggered performances are of particular importance in fabricating various kinds of intelligent light-driven devices. Photoresponsive groups are usually divided into three classes according to the photoreaction mechanism: photoisomerization, photo-crosslinking and photo-degradation. As a typical photoactive molecule which can photo-isomerize, along with its rigid rod shape of the thermally stable *trans* configuration, an azobenzene moiety is often introduced in photoresponsive liquid crystalline polymers (LCPs) as the photoactive mesogens, endowing the obtained LCPs with photocontrollable features.

As is known, an azobenzene group often transfers from its *trans* configuration to *cis* one under photoirradiation of actinic light with a certain wavelength and thermally returns back to the stable *trans*-isomer (Figure 1.1a). The lifetime of the *trans* isomer of an azobenzene strongly depends on the particular substitution. Furthermore, linearly polarized light (LPL) can be applied to control the orientation of the azobenzene moiety due to the Weigert effect (Figure 1.1b). The alignment direction is usually perpendicular to the polarization direction of LPL.

Photoresponsive properties are significantly correlated with the substituent groups on the aromatic rings of azobenzenes, which has been summarized by Natansohn and Kumar according to the lifetime

Liquid Crystalline Polymers, edited by Vikas Mittal

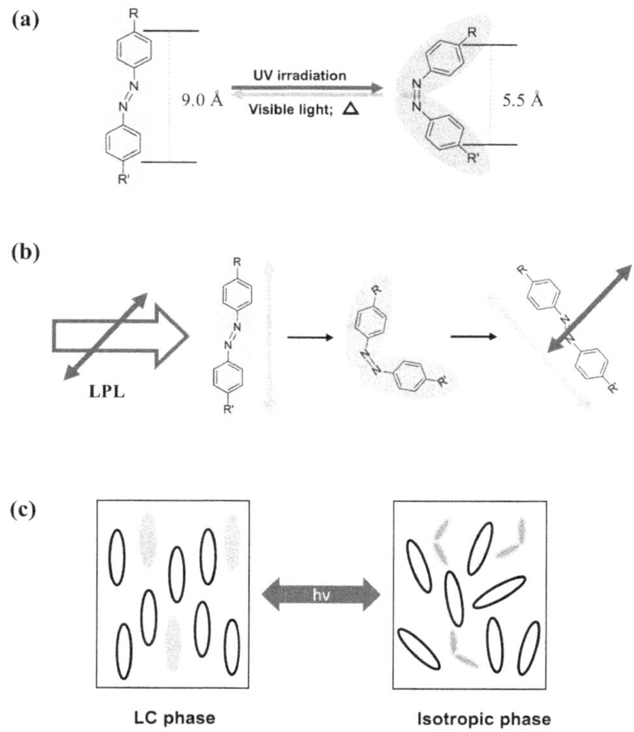

Figure 1.1 Interesting properties of azobenzene-containing materials. (a) Typical photoisomerization; (b) photoalignment under LPL irradiation; (c) photoinduced phase transition from a LC phase containing azobenzene chromophores to isotropic phase.

of *cis*-isomers [1-3]. Three kinds of azobenzenes are categorized as following: (I) ordinary azobenzenes without any electronic cloud migration, which show poor overlap of their n-π* and π-π* absorption peaks in their UV-vis absorption spectra with a relatively long lifetime of the *cis*-isomers and exhibit a relatively lower photoresponsive rate; (II) amino-azobenzenes with an electron-donating group at one end of the azobenzene, which show significant overlap of their n-π* and π-π* absorption peaks in the UV-vis absorption spectra with a shorter lifetime of the *cis*-isomers and exhibit a faster photoresponsive rate; (III) pseudo-stilbenes with an electron-donating group at one end and an electron-withdrawing group at the other end, which show high-degree overlap of their n-π* and π-π* absorption with a very short lifetime of the *cis*-isomers, and exhibit a quite rapid photoresponsive rate.

In some rod-like *trans*-azobenzene molecules with substituents of a soft spacers (like a long alkyl chain), liquid crystal (LC) phases are often observed in a range of temperature above the melting point, while such LC behaviors are seldom obtained in their *cis*-isomers because of the bent shape, which destabilizes the LC ordering. Therefore, photoinduced order-to-disorder transition often takes place as the *trans*-isomers exhibiting LC phases change into their *cis*-isomers in isotropic phase upon irradiation of actinic light. Such a photoinduced phase transition usually leads to great changes in physical and chemical properties of materials containing azobenzene mesogens. In addition, azobenzene chromophores can be used to photo-modulate the light-inert mesogens via the effect of molecular cooperative motion, although the latter is photo-inactive. Phase transition or alignment changes of photoresponsive LCs can be induced even if there is only a small molar proportion of azobenzene chromophores responding to external stimuli, as shown in Figure 1.1c. This has been applied to magnify the photo-triggered effect from a weak input signal.

Due to the fascinating photoresponsive properties of azobenzene chromophores, azobenzene-containing LCPs have been intensively studied as promising functional materials for various advanced applications. Here, the azobenzene acts as both a photoresponsive moiety and a mesogenic group. Thus, a large modulation of mechanical, optical and thermodynamic properties can be photocontrolled in such polymer films by photochemical phase transition, photoinduced molecular alignment and photo-triggered molecular cooperative motion, which greatly broadens their applications.

1.2 Homopolymers Containing Azobenzenes

Homopolymer often contains only one kind of photoresponsive mesogen in its side chain or main chain, incorporating the self-assembled LC properties with high-performance polymer materials possessing a film-forming nature, high processability, easy-fabrication characteristics and high corrosion-resistance. The combination of polymers with photoresponsive LCs is fascinating, because the independently advantageous properties of polymers and LCs are effectively incorporated together, as shown in Figure 1.2. Thus, it is convenient to manipulate the microscopic structures and even the macroscopic properties of materials with the photoresponsive LC moieties in the polymer processing. On one hand, the polymer provides the photoresponsive LCPs with good processability and stability. On the other hand, the photo-adjustable order-to-order or order-to-disorder transition of the anisotropic LCs may bring about controllable optical and mechanical properties, along with the unique features like photochemical phase transition, photo-triggered molecular cooperative motion and photoinduced alignment.

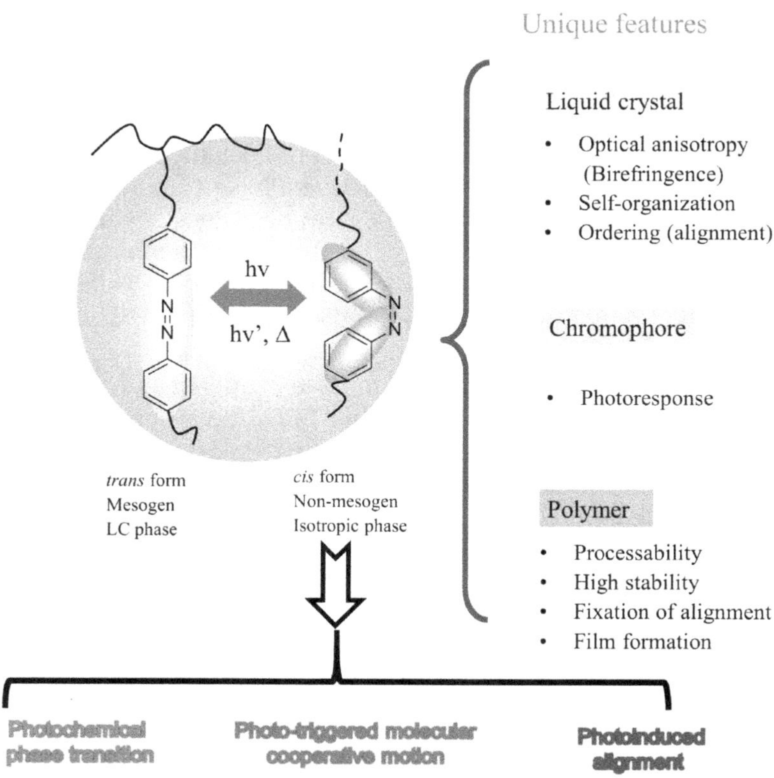

Figure 1.2 The unique features of LCPs by combination of the advantages of photoresponsive LCs and polymers.

Photoinduced phase transition is often observed in thin film of the azobenzene-containing homopolymer, as shown in Figure 1.3. As one typical example of photoresponsive LCPs, the homopolymer PM6ABOC2 presents a Schlieren texture of nematic LCs observed under polarizing optical microscopy (POM) at the temperature above the glass-transition temperature (T_g), but below the clearing point. The highly birefringent image can be rapidly transformed into a dark field upon UV irradiation for a short time (in a timescale of seconds) at room temperature, indicating the occurrence of order-to-disorder transition [4]. Interestingly, volume changes often occur in the LCP film when this photoinduced phase transition is induced, which has been widely used in the photomechanical motion.

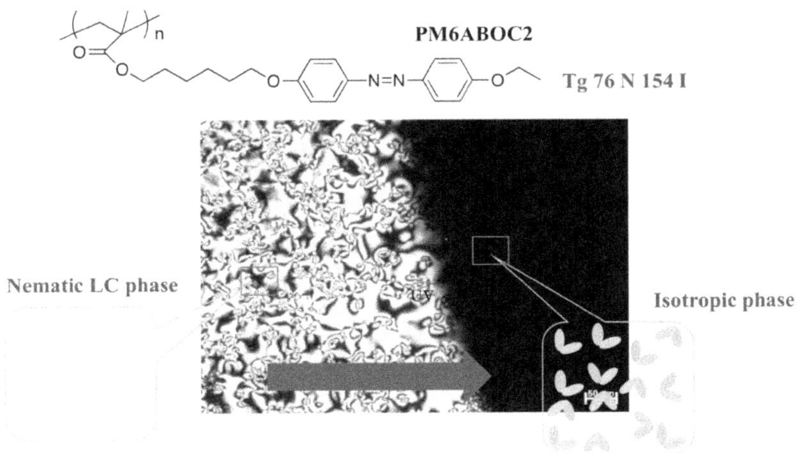

Figure 1.3 Photoinduced phase transition of the homopolymer (PM6ABOC2) at room temperature. Reproduced from Reference 4 with permission from Royal Society of Chemistry.

By utilizing these photoresponsive properties, optical storage of two-dimensional (2D) images have been successfully recorded in PA6ABOC2 film (Figure 1.4a), which was first reported by Ikeda and Tsutsumi [5]. As shown in Figure 1.4b, the photoresponsive LCPs initially underwent a rapid order-to-disorder transition upon UV irradiation, while they never exhibited an isotropic-to-nematic LC phase transition below their T_g. Such a phase transition in the UV-irradiated area is due to the formation of isotropic *cis*-azobenzene moieties. By this way, an optical image can be stored in the PLCP films by using a photomask. Then the thermally-induced *cis*-to-*trans* isomerization of azobenzene mesogens in the polymer film occurred in 24 h at room temperature. Although the *trans*-isomer recovered completely, the isotropic glassy domains induced in the irradiated area still remained unchangeable at room temperature even after several years. These results suggest that information can be recorded by the photoinduced phase transition and preserved below the polymer T_g even though the occurrence of the thermal *cis*-to-*trans* isomerization. The recorded images can be deleted by heating the sample film above its T_g, which can be easily used for rewritable storage of new images. Therefore, the photoresponsive LCP can be used as optical switching as well as 2D optical image storage materials.

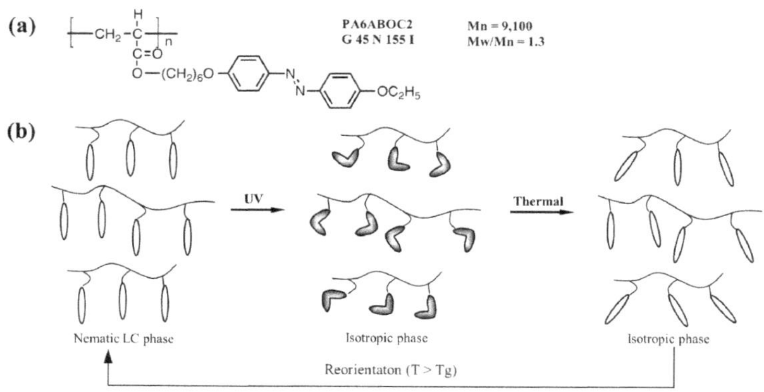

Figure 1.4 Molecular structure and properties of the photoresponsive LCP (PA6ABOC2) (a) and the mechanism of optical storage application (b).

To realize a complete three-dimensional (3D) storage, holographic gratings have been developed correspondingly. This unique technique enables concomitant recording of both phases and amplitudes of light waves. Thus, both refractive-index grating and surface-relief grating can be simultaneously obtained in photoresponsive LCPs [6,7], as shown in Figure 1.5. Being one of the most promising holographic recording materials, azobenzene-containing LCPs can be easily modulated into diverse patterns by adjusting the input light with wavelength, intensity, polarization, phase, interference and so on. Usually, holograms can be recorded by the photoinduced changes in orientation of mesogenic molecules in a periodic pattern obtained from interference of two coherent laser beams. Since a large refractive-index modulation can be obtained in the LCPs, the recorded hologram is phase-type and often shows a higher diffraction efficiency than the amplitude one [8]. Various 3D objects have been recorded in the film of azobenzene-containing LCPs, showing their promising applications for volume storage.

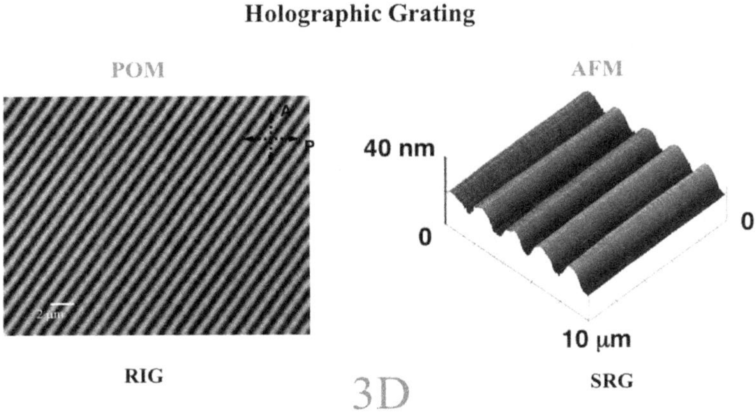

Figure 1.5 Holographic gratings of refractive-index grating (RIG) and surface-relief grating (SRG) recorded in photoresponsive LCPs containing azobenzene mesogens. A, analyzer; P, polarizer.

Actually, the above-mentioned holograms can be used as photonic devices and the incident light can be diffracted to reproduce the object light beams in the image reconstruction process. Following this optical feature, a rewritable thin-film optical diode based azobenzene-containing LCP (PM6ABOC2) has been reported by Zhang *et al.* [4]. As shown in Figure 1.6a, the thin-film lens recorded in the LCP film exhibited different colors within the exposed area when observed from different viewing angles (sample A-C). The photoresist grating (sample D) reveals the true spatial details of the surface-relief hologram by microscopic measurements (Figure 1.6b and Figure 1.6c). This kind of diode lens is essentially a diffractive device consisting of spatially chirped volume-phase gratings, which demonstrates unique physics of multifold photonic resonance modes excited by primary and secondary diffractions.

In addition to holographic recording, other methods to fabricate microscale grating structures have also been developed recently. As shown in Figure 1.7, hierarchically arranged stripes of photoresponsive LCPs (PM6ABOC2) were conveniently crafted via a flow-enabled self-assembly [9]. This method could produce photo-switchable LCP stripes and display two modes of deposition, namely, periodic primary stripes of large dimensions and regularly spaced secondary stripes of small dimensions situated between adjacent primary stripes. Different behaviors of photoinduced reversible phase transition were also observed in these hierarchical LCP stripes, due to the thickness difference between the primary and the secondary stripes.

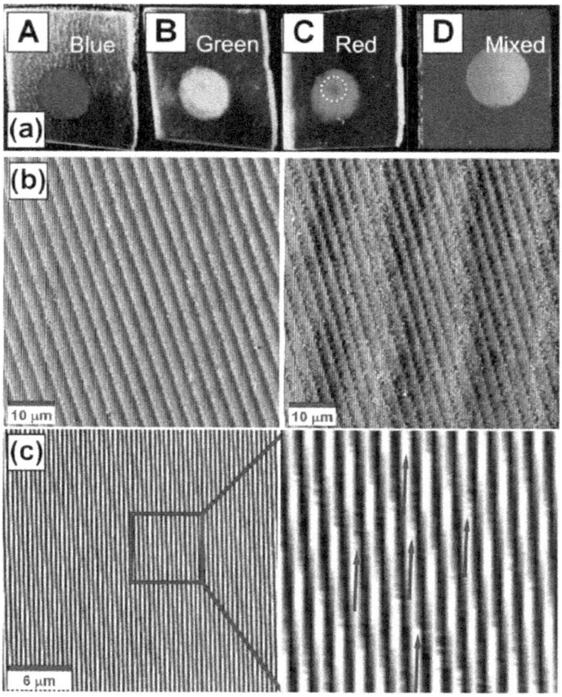

Figure 1.6 (a) Photographs of the LCP thin-film diode lens at different observation angles (A–C) under white-light illumination and that recorded as chirped grating in photoresist for comparison (D). (b) AFM images of the chirped holographic gratings on two different sites on the thin-film lens (sample D), showing interlaced configuration between small- and large-period gratings. (c) Enlarged AFM images showing the spiral features of each photoresist grating line. The blue arrows indicate the spiral surface facing the observation direction. Reproduced from Reference 4 with permission from Royal Society of Chemistry.

1.3 Supramolecular Liquid Crystalline Polymers Containing Azobenzenes

Self-assembly of supramolecules provides an efficient way to prepare functional materials by association of different properties of various

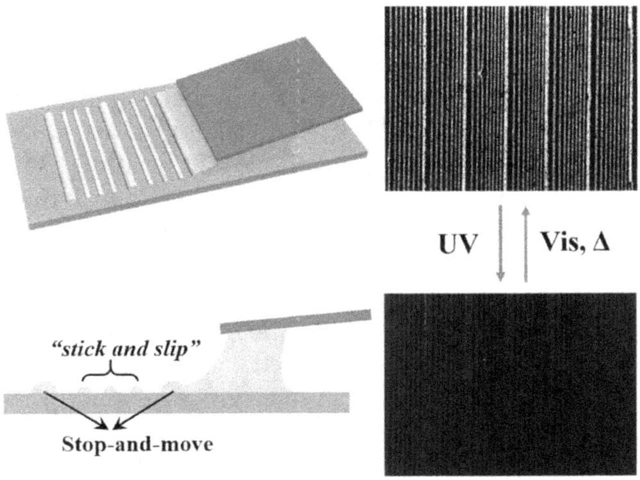

Figure 1.7 The method of flow-enabled self-assembly to prepare grating-like stipes and the photoresponsive properties of the micro-structures. Reproduced from Reference 9 with permission from American Chemical Society.

molecules and macromolecules [10]. Ever since Lehn [11] proposed and developed the concept of supramolecular chemistry, research studies on supramolecules have become a critical subject covering almost all kinds of fields in materials science, chemistry and biology [11]. Supramolecules are defined as the combination of molecules or macromolecules via non-covalent interactions, including hydrogen bonding, halogen bonding, π-π interaction, hydrophobic interaction, donor-acceptor interaction, metal-ion coordination and so on. Compared with traditional LCPs, SLCPs are fabricated via dynamic and reversible interactions, which renders their popularity on stimuli-responsive materials. Combining the photoresponsive properties of azobenzene and the self-assembled capability of pyridine group or carboxylic acid, the corresponding azopyridine or azocarboxylic acid has been widely used in preparation of photoresponsive supramolecules. Herein, we would mainly introduce supramolecular LCPs (SLCPs) formed by hydrogen bonding and halogen bonding.

The hydrogen-bonded supramolecular LCs system are often introduced by using the carboxylic acid or pyridyl group as the hydrogen-bond acceptor and donor [12]. Following these principles, various polymers containing pendent azopyridine have been synthesized as the hydrogen-bond acceptors. Thus, SLCPs can be easily fabricated by the hydrogen bonding interaction with certain carboxylic acids as the hydrogen-bond donor. For instance, Zhao et al. synthesized a series of polystyrene (PS) based block copolymers (BCs) containing azopyridine moieties (PS-b-PAzPy). A variety of carboxylic acids were added to transform the amorphous azopyridine (AzPy) block into one ordered LC block, which could further enhance the photoinduced orientation of the azopyridine moieties [13].

The supramolecules based on the hydrogen bonding between azopyridine and carboxyl groups could self-assemble into fibers, spherical particles, gels and so on. As shown in Figure 1.8a, SLCP microparticles were successfully fabricated with an azopyridyl polymer and sebacic acid by combining a simple self-organized precipitation

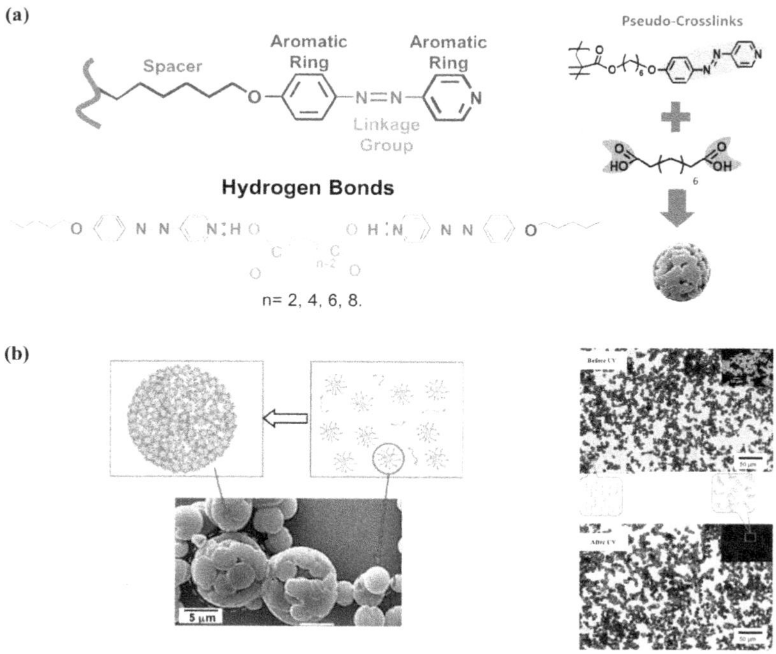

Figure 1.8 Microparticles fabricated with SLCPs showed controlled morphologies (a) and photoresponsive behaviors (b). Reproduced from Reference 16 with permission from Wiley and Reference 17 with permission from American Chemical Society.

method with hydrogen bonding interactions. [16,17] The surface morphologies of the SCLP microparticles and phase behaviors were studied by changing the molecular length of the diacids (Figure 1.8b). The fabricated SLCP microparticle possessed not only photoresponsive properties (due to azopyridyl groups), but also LC ordering, which might enable them to find advanced applications. In addition to microparticles, ultrathin free-standing Janus films were fabricated at air-water interfaces by using azopyridine derivatives and poly(acrylic acid) via multi-level self-assembly on molecular and microscopic scales (Figure 1.9), which showed distinct asymmetric water wetting abilities on different surfaces. [18]

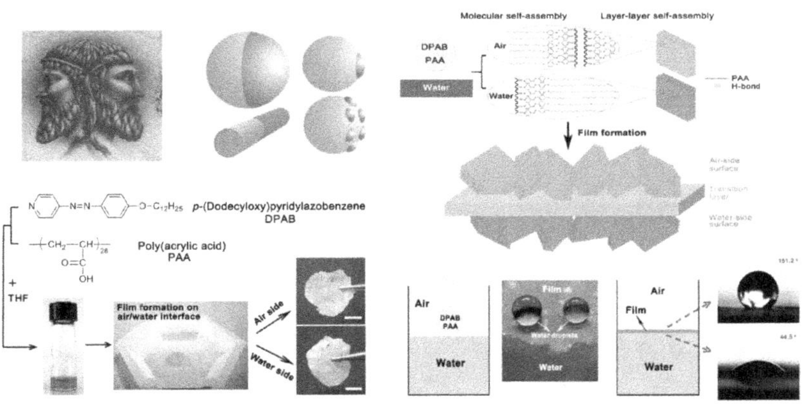

Figure 1.9 The "Jenus" film fabricated based on the SCLP complex of PAA and DPAB. Reproduced from Reference 18 with permission from Royal Society of Chemistry.

Due to the existence of pyridine group, the azopyridine can behave as a weak base in acidic environment. Recently, Yu and coworkers have carried out a series of studies on the self-assembled nanofibers with different morphologies based on amphiphilic low-molecular weight azopyridine derivatives [14,15]. The nanofibers were successfully obtained with different kind of acids and salts, such as aniline hydrochloride, alkylbenzenesulfate-based anionic surfactants. These fibers are expected to have applications in conductive nanomaterials and biomaterials.

Different from the azopyridine group being as the hydrogen-bond acceptor, the azocarboxylic acid moiety often serves as the hydrogen-bond donor and even forms hydrogen bonds themselves (dimer of carboxylic acids). Therefore, SLCPs containing pendent azocarboxylic acid show unique assembly properties. Chen *et al.* [19] fabricated multi-responsive reversible organogels with a carboxylic azo homopolymer (PM6AzCOOH) in organic solvent of dimethylsulfoxide (DMSO) by the interaction of H-aggregation and hydrogen bond [19]. The formed organogel could be destroyed by heating, addition of less polar solvent and irradiation of UV light. The reversible sol-gel process is presented in Figure 1.10.

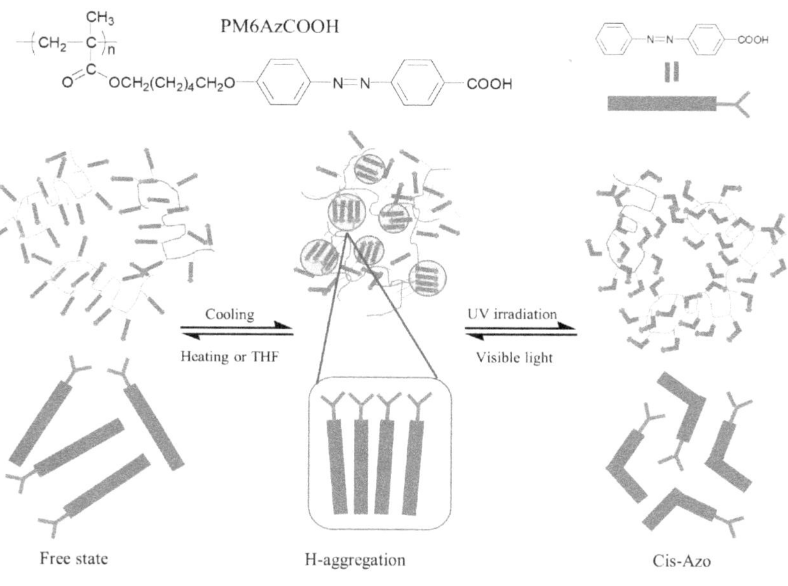

Figure 1.10 Multi-responsive organogel based on carboxylic azo homopolymer. Reproduced from Reference 19 with permission from Royal Society of Chemistry.

Interestingly, the special self-assembly properties of PM6AzCOOH also influence the fluorescence behaviors of azobenzene moieties in solution [20]. The polymer self-assembles into vesicles from a dimethyl formamide (DMF) solution when water was added dropwise and obvious fluorescence was observed. However, the fluorescence behavior would be diminished under UV irradiation due to the breaking of H-aggregation, while the aggregates transformed to a more stable micelle-like structure in alkaline solution and showed small changes under photoirradiation (Figure 1.11).

Another interaction which has been adopted in fabrication of supramolecular LCs is halogen bonding, which is the non-covalent type

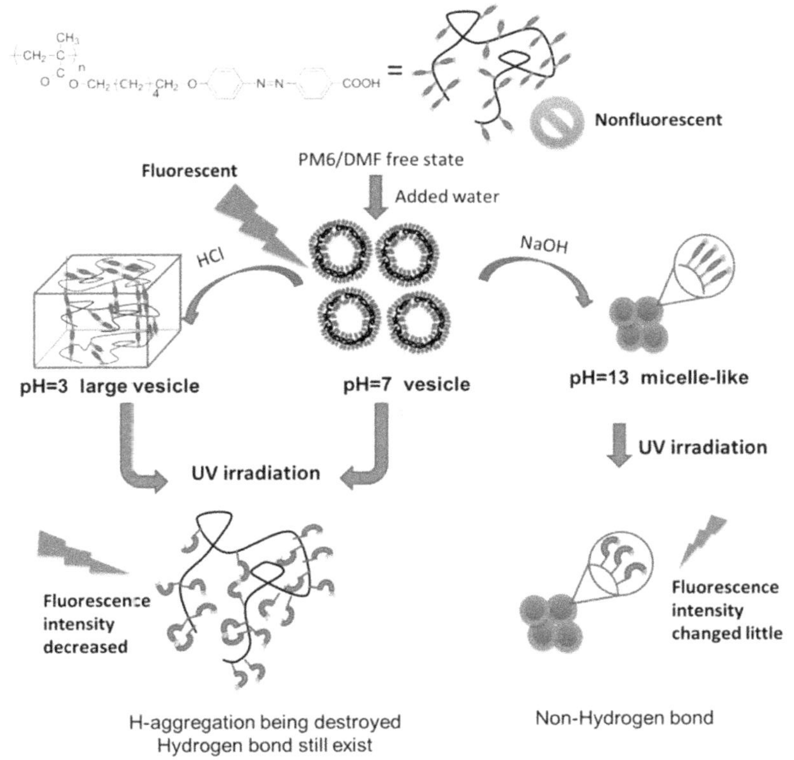

Figure 1.11 Self-assembly and fluorescence behaviors of the polymer containing azocarboxyl groups (PM6AZCOOH). Reproduced from Reference 20 with permission from Royal Society of Chemistry.

of interaction between halogen atoms and neutral or anionic Lewis bases [21]. Since halogen bonds have been introduced into LC system by Nguyen *et al.* [22], many researchers have made efforts to fabricate supramolecular LCs based on such interaction, as the halogen bonds are more directional with tunable strength compared with hydrogen bonds. Yu and coworkers first fabricated the halogen-bonded LCs by using azopyridine compounds and molecular bromide or molecular iodine. However, the reversible photoinduced phase transition was obtained only in iodine-bonded LCs upon UV irradiation, as shown in Figure 1.12 [23]. There was no obvious change in the bromide-bonded LCs even though UV light with a high intensity was irradiated for a long time (several hours).

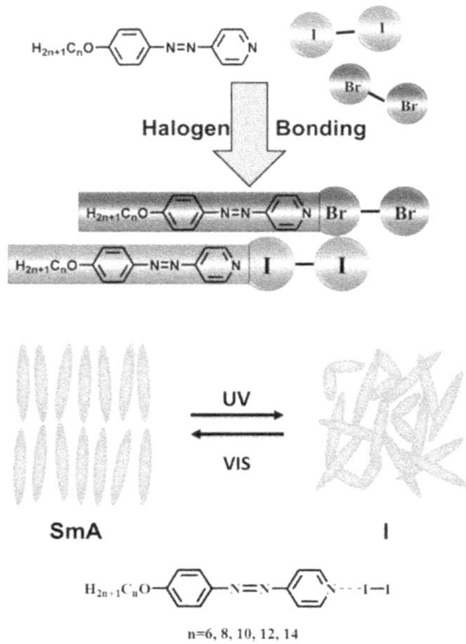

Figure 1.12 Fabrication of halogen-bonded LCs and their reversible photoinduced phase transition behaviors. Reproduced from Reference 23 with permission from Royal Society of Chemistry.

1.4 Liquid Crystalline Block Copolymers Containing Azobenzenes

Block copolymers (BCs) are usually made up of two or more polymer modules named "blocks" linked by chemical bonding [24]. Driven by competition between the incompatible but strongly-bonded blocks, microphase separation (MPS) often occurs in nanoscale. Thus, various nanostructures such as spheres, cylinders, double gyroid, lamella with a size range of 10~100 nm can be fabricated, as shown in Figure 1.13 [25]. It is believed that it is the thermodynamic equilibrium effect that determine the MPS structures in self-consistent field theory [24]. Besides, the volume content, thermal condition and thickness of the BC film affects the MPS structures, either. MPS helps broaden the applications of BCs in nanomaterials and nanoengineering [26-29].

Liquid crystalline block copolymers (LCBCs) can be obtained by introduction of LC ordering in at least one block of BCs. Therefore, LCBCs can combine the inherent MPS of BCs with the self-assembled

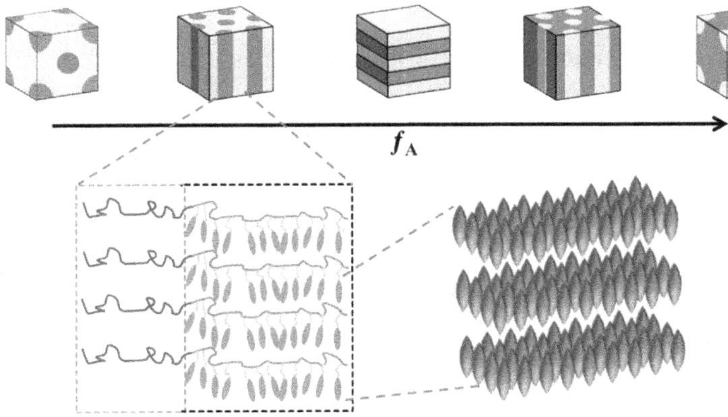

Figure 1.13 MPS structures of LCBCs with different volume content of the mesogenic block.

feature of mesogens, which brings about special novel interesting properties. More importantly, the trigger-responsive property of LCs can help us manipulate the ordered MPS nanostructures, which should broaden the potential applications of LCBCs. Therefore, photoresponsive LCBCs have attracted lots of research interest for several years. The photoresponsive LCBCs exhibit robust responsiveness upon photoirradiation, including photoinduced phase transition, photo-controlled surface and photo-triggered mass transfer, which is inherited from photoresponsive LCPs [3]. These enable photoresponsive LCBCs to find applications in photoelectronic materials [30], information storage [31-32], holographic grating [33-37], drug delivery materials [38-39], etc.

As shown in Figure 1.13, both the MPS of BCs and the photoresponsive property of azobenzene-based LCPs are combined together in one system of LCBCs, bringing about the following merits. Firstly, we can acquire different MPS structures by adjusting the volume ratio of the blocks. Secondly, the alignment of mesogens in the continuous phase significantly influences the MPS nanostructures in the separated phase due to the supramolecular cooperative motion (SMCM) effect. Thirdly, the photoresponsive feature of azobenzene mesogens make it possible to realize the photo-regulation of ordered nanostructures in non-contact way. In addition, the relative high value of Flory-Huggins–Staverman parameter (χ) in LCBCs, along with the self-assembly of the 1D or 2D ordered LCs, helps fabricate ordered nanostructures in large scale, which makes sure the reliability of their potential applications.

The hierarchical structures of LCBCs should be resulted from the interference between the self-assembly of mesogens and MPS of the polymer segments. The incompatible blocks in the minority phase disperse in the continuous phase of the polymer matrix, forming various nanostructures. In addition, thermal history and mechanical history significantly affect the MPS morphology as well, indicating that the MPS structures can be controlled by adjusting the processing technology. Consequently, the photoresponsive azobenzene mesogens make it possible for us to manipulate the MPS via all kinds of external stimulus such as electric field, magnetic field, force, light and etc., which are generally applied in mesogenic regulation.

Rubbing, as one of the popular methods to orient low-molecular weight LCs for LC display production has been introduced in LCBCs containing azobenzene mesogens in 2006. Yu *et al.* [40] prepared a series of PEO-based photoresponsive LCBCs, which forms nanocylinders upon phase segregation. With the assist of rubbing polyimide film, the orientation of the nanocylinder structures was elegantly realized in a large area (in a centimeter scale). As shown in Figure 1.14,

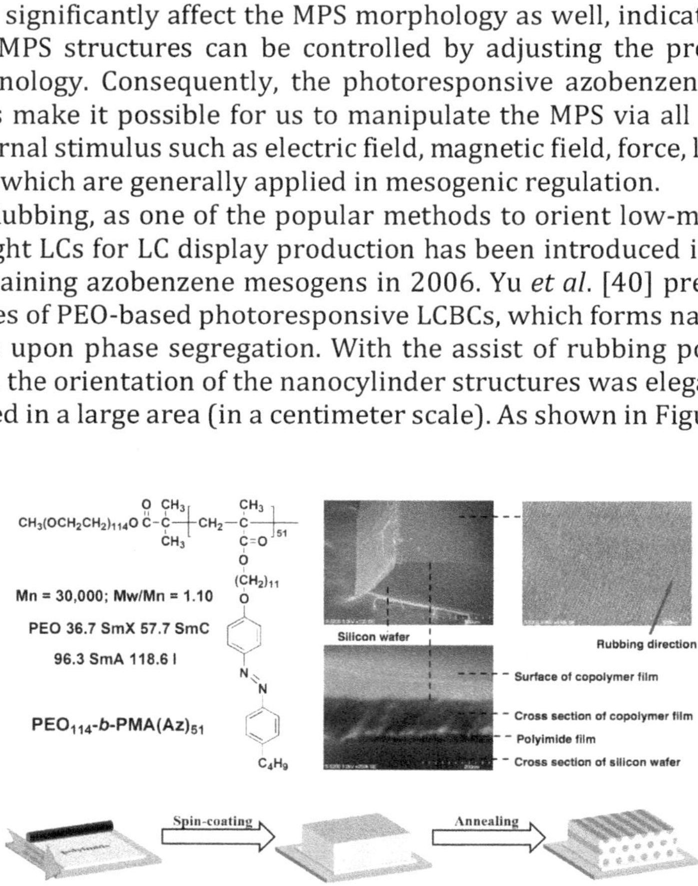

Figure 1.14 Rubbing induces the orientation of MPS nanocylinders in photoresponsive LCBCs by SMCM. Reproduced from Reference 40 with permission from Wiley.

an alignment layer was prepared by coating a polyvinyl alcohol (PVA) or a polyimide (PI) film on the glass slide or silicon wafer. After being mechanically rubbed, the surface of the alignment layer was coated with the LCBC film. Well-ordered cylinder nanostructures were successfully acquired after the film was gone through a thermal annealing process. The alignment direction is always parallel to the orientation direction of the azobenzene mesogens, along with the rubbing direction.

In addition, photoalignment of the azobenzene mesogens provides a convenient non-contact method for control of the LC alignment. The orientation of azobenzene mesogens can be adjusted by the polarization direction of the actinic light, which also has a significant impact on the orientation of MPS structures in LCBCs via SMCM, where the mesogenic block serves as the major phase. As shown in Figure 1.15, Morikawa *et al.* [41] studied the photoalignment of nanocylinders in photoresponsive LCBC films and revealed that whether the LCs along with the cylinders orient in-plane or out-of-plane depended on the thickness of LCBC films. Yu *et al.* [42] proposed an optical method to control a homogeneous in-plane alignment of PEO nanocylinders in PEO-based amphiphilic LCBCs. Firstly, PEO nanocylinder perpendicular to the direction of incident light was fabricated by thermal annealing; then this alignment was easily converted into in-plane nanocylinder orientation with the control of actinic polarized light. This method has advantages of clean, simple and convenient, providing a novel approach to get regular surface patterns even in curved surface, as shown Figure 1.16a.

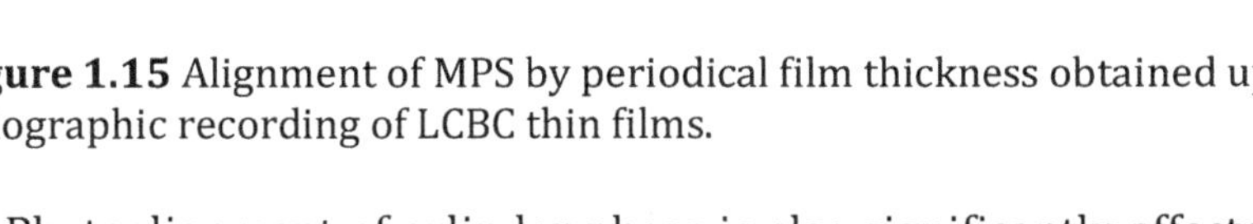

Figure 1.15 Alignment of MPS by periodical film thickness obtained upon holographic recording of LCBC thin films.

Photoalignment of cylinder phase is also significantly affected by the T_g and the crystallization temperature (T_c) of the minor phase in photoresponsive LCBCs. Seki's group has studied the photoresponsive properties of a PS-based LCBC containing smectic azobenzene mesogens as the major phase [43]. As shown in Figure 1.16b, out-of-plane nanocylinder orientation was observed after direct thermal annealing without any other treatment, which should be induced by the self-organization of the smectic azobenzene block. An in-plane orientation was induced by LPL irradiation along with the proper thermal treatment. Besides, the orientation direction of the nanocylinders would be changed corresponding to the polarization direction of LPL, keeping the direction perpendicular to the polarization direction. Highly ordered MPS nanostructures were also observed in a 1-dimension (1D) ordered nematic PEO-based LCBC, indicating that the thermal properties (mainly T_g and T_c) of the soft blocks which form the minor phase are the critical factor that contributes to the degree of ordering [44]. Thus, a series of poly-n-butyl methacrylate (PBMA)-based photoresponsive LCBCs were synthesized for fabricating ordered light-controllable MPS structures, because PBMA has a softer molecular chain and a much lower T_g than PEO. Whereas, PBMA segments aggregated at the air-polymer interface due to interface effect,

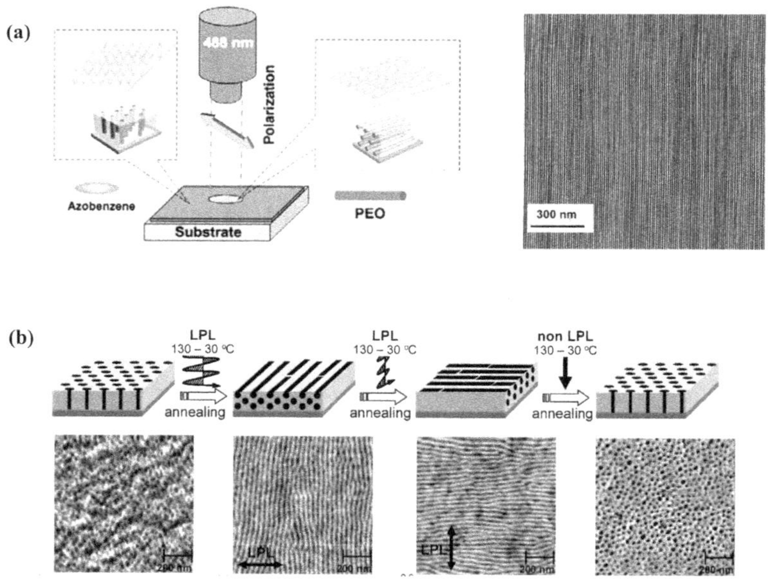

Figure 1.16 (a) Photocontrollable ordered MPS in large scale prepared by a PEO-based photoresponsive LCBC containing a smectic azobenzene mesogen. Reproduced from Reference 42 with permission from American Chemical Society. (b) Rewritable information storage by photocontrol of the orientation of nanocylinders in a PS-based photoresponsive LCBC. Reproduced from Reference 43 with permission from American Chemical Society.

inducing a disordered in-plane distribution after the thermal annealing [45]. However, as it is much easier for the soft PBMA segments to move and assemble induced by photo-triggered SMCM in LCBCs, a nearly real-time photoinduced re-orientation phenomenon was observed, and the ordered alignment was acquired in several minutes.

More recently, unpolarized light were also applied in manipulating the cylindrical MPS nanostructures. Vapaavuori *et al.* [46] fabricated a composite film of P4VP-b-PS/4-butyl-4′-hydroxyazobenzene (BHAB) by a dip-coating method [46]. Here, the spherical morphology of thin films dip-coated from a tetrahydrofuran (THF) solution at slow withdrawal rates in the dark convert to cylindrical morphology when dip-coated under illumination. This is attributed to volume expansion of the P4VP/BHAB phase due to *trans–cis* photoisomerization combined with a light-induced increase in BHAB uptake in the film (Figure 1.17a). Furthermore, planar orientation of the cylindrical

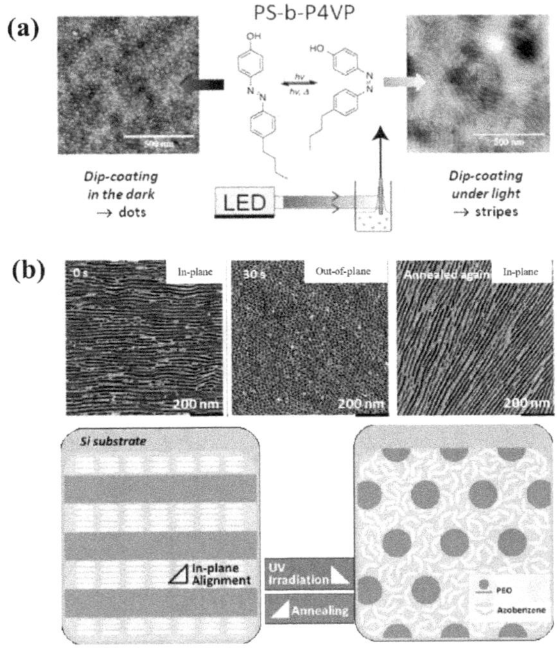

Figure 1.17 (a) In situ photocontrol of a P4VP-b-PS/BHAB film by dip-coating under irradiation of unpolarized light; (b) photoinduced orientation transition by unpolarized light and thermal recovery of a PEO-based PLCBC. Reproduced from Reference 46 and 47 with permission from American Chemical Society.

structures in PEO-based photoresponsive LCBCs containing azobenzene chromophores can be rapidly transformed to vertical alignment upon irradiation of unpolarized UV light [47]. It was considered that the *trans-cis* isomerization of azobenzene mesogens changed the steady state of the cylindrical alignment, and in the meanwhile provided sufficient free volume for the PEO segments to reorganize (Figure 1.17b). It used to be considered that it would be difficult to manipulate MPS structures if the photoresponsive azobenzene mesogenic block serves as the minor phase, before a series of photoresponsive hybrid oligo(dimethylsiloxane) LCPs that form periodic cylindrical nanostructures with periodicities between 3.8 and 5.1 nm was fabricated and studied by Nickmans *et al.* [48]. The azobenzene mesogens here can be oriented in-plane by exposure to linearly polarized light and out-of-plane by exposure to unpolarized light.

Internal phase nanostructures have also been fabricated in LCBCs by introduction of hydrogen bonds without any dopants. As shown in Figure 1.18, both out-of-plane and in-plane orientations of nanocylinders were successfully obtained in the PEO-based LCBCs (PEO-b-PM11UAZ) depending on the film thickness mentioned above. However, the newly-introduced urethane groups in the side chain of the hydrophobic block of LCBCs also interacted with the ether groups of hydrophilic PEO block, which should affect the crystallization process

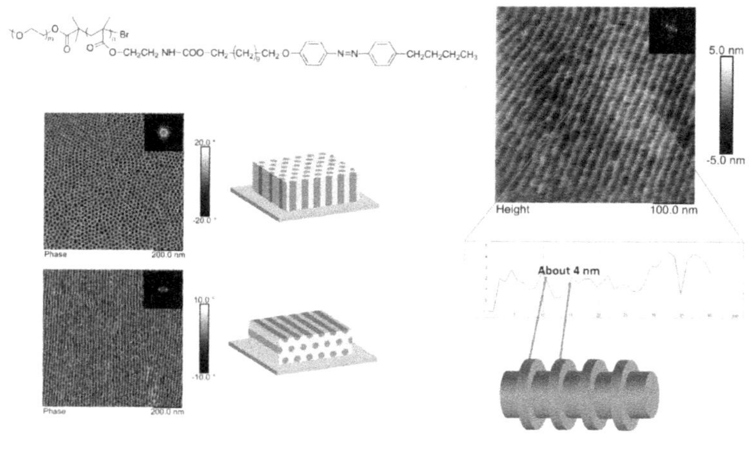

Figure 1.18 Morphologies of the out-of-plane alignment, in-plane alignment and the internal phase structure in the hydrogen-bond-containing LCBC. Reproduced from Reference 49 with permission from Wiley.

of the PEO blocks. Thus, a sub-10 nm inner-phase nanostructure was induced in the LCBC film [49].

There are scarce reports on photocontrollable lamellar phase in photoresponsive LCBCs containing azobenzenes up until now, except a study carried out by Beppu *et al.* [50]. The authors investigated the MPS and photoalignment behavior of a diblock copolymer consisting of two kinds of side chain LCPs (the mesogens of photoresponsive azobenzene and non-photoresponsive cyanobiphenyl, respectively). The nearly defect-free morphology of this LC-LC diblock copolymer after the photoalignment was a distinct feature when compared with those of coil (amorphous)-LC diblock copolymer films possessing the same azobenzene side chain, as mesogens exist in both minor and major phases and the effect of SMCM should be much stronger here.

Based on the photoresponsive properties and the controllable ordered MPS nanostructures, various applications of photoresponsive LCBCs have been explored in nanophotonics, nanotemplates and nanolithography. Holographic storage has been one of the most studied application of photoresponsive LCBCs for recent years. Phase transition or phase changes would take place if the azobenzene mesogens are irradiated by actinic light. Such photo-caused effect will be reflected and preserved on both surface and volume of the materials, and namely the optical information storage can be realized. Although homopolymers with azobenzene mesogens have also been applied in fabricating holographic gratings, photoresponsive LCBCs exhibit more advantages: (1) introduction of another block improves transmittance of film; (2) MPS in nanoscale prohibits light scattering; (3) macroscopic mass migration and surface undulation can be greatly restricted. For instance, Yu and coworkers have realized a holographic grating with a PMMA-based photoresponsive LCBC where the azobenzene mesogenic block play as the major phase (Figure 1.19) [51].

MPS nanostructures of photoresponsive LCBCs exhibit good potential to be applied in nanoengineering. The LCBC films with regularly ordered nanostructures can serve as nanotemplates in fabricating metallic and inorganic materials with special nanostructures [52-54]. It means that the feasibility of applying MPS in nanolithography and nanotemplates. Therefore, MPS and SMCM have been utilized to manipulate some special functions introduced by a certain block in designing novel photoresponsive LCBCs. In addition, self-assembly and applications of photoresponsive LCBCs in solution is another hot topic. Amphiphilic LCBCs usually form vesicles or micelles in solutions, before the sol-gel transition takes place induced by photoisomerization as the polarity of the mesogenic block changes [55-59]. These photoresponsive vesicles and micelles have the potential in medicine targeting delivery and controlled release [60].

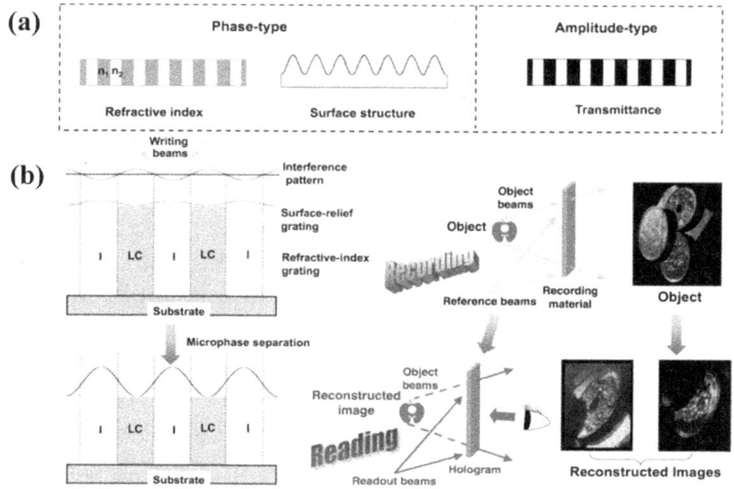

Figure 1.19 (a) Schematic illustration of the two categories of holographic gratings; (b) fabrication and enhancement of surface-relief grating based on the MPS of photoresponsive LCBC, and the reading-writing process of 3D images. Reproduced from Reference 51 with permission from American Chemical Society.

1.5 Photomechanical Liquid Crystalline Polymers Containing Azobenzenes

LCPs are known to be anisotropic and undergo great changes in order parameter during the phase transition, which has been expected to be used as artificial muscles ever since the prediction by de Gennes [61]. Thermomechanical responses were then widely achieved especially in nematic LCPs whose stress and responsive time were similar to natural muscles [62,63], indicating high potential of LCPs for applications such as motors, sensors, pumps and robots. However, compared to thermal response, deformation triggered by light has attracted more attention because of its characteristics of precise, remote and contactless control. It has been well-known for decades that azobenzene groups undergo *trans-cis* isomerization under irradiation of UV light. When incorporated into LC systems, the isomerization can reduce the order parameter of mesogens and even induce the occurrence of phase transition [64,65]. However, it is difficult to convert light energy into mechanical energy directly until the combination of azobenzene chromophore with LCPs.

1.5.1 Early Stage

The pioneering discovery of photomechanical effect was obtained in isotropic polymers containing azobenzene chromophores but with a low contraction rate (~0.25%) under UV irradiation [66]. Interestingly, Finkelmann *et al.* [67] demonstrated that the contraction rate could be enhanced to 20% in nematic LC system both experimentally and theoretically. The authors prepared LC elastomers according to their previously reported two-stage crosslinking method [68], in which the orientation was initially obtained by mechanical stretching and then fixed by further crosslinking. Li *et al.* [69] promoted a more efficient way by utilizing two glass slides with alignment layer known as LC cell. The authors indicated that the alignment layer could induce the mesogenic orientation and in-situ polymerization was initiated below the clearing point or by a longer wavelength photoirradiation to ensure the fixation of the oriented LCs by crosslinking. Basically, these two methods have been widely utilized in subsequent research studies, which greatly accelerated the development of photodeformable LCPs.

What is the mechanism of the photomechanical response in LCP systems? First, polymer backbone in the LC state is stretched with anisotropic radius of gyration which is spherical in isotropic phase [70-72]. Second, rod-shape *trans*-azobenzene can be aligned along with other photo-inactive mesogens while the *cis*-azobenzene molecules disrupt the ordered structure. Thus, upon UV irradiation, the *trans-cis* isomerization decreases the degree of orientation and polymer backbones tend to transform from ellipsoidal configuration to sphere, as shown in Figure 1.20. In consequence, cooperative movements of mesogenic molecules and polymer backbones result in contraction along the direction of orientation (director) and expansion perpendicular to it [62,69]. Actually, there are other changes such as the generation of free volume which may lead to opposite consequence in some situations and it will be discussed in following sections.

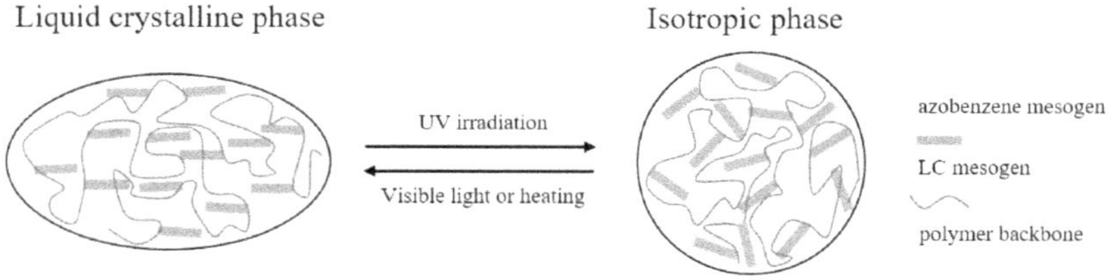

Figure 1.20 Change in configuration of polymer backbone during phase transition.

Gradually, it was realized that 2D contraction could not meet diversified application requirements, thus, more complicated 3D deformations were explored. Ikeda and coworkers first reported photoinduced bending motion of crosslinked LCP films [73-75]. The authors prepared crosslinked LCP with a high azobenzene content so that the transmission depth of the incident light was only about one micron. Under UV irradiation, the whole film bent towards the light source because the irradiated side of the film contracted along the nematic director with almost no changes in the remaining part. The change in the irradiated surface can be controlled either by the polarization direction of LPL [74] or initial alignment of mesogens [75]. Films without alignment consisted of many micro-sized domains in which azobenzene mesogens aligned in different directions. Only those mesogens parallel to direction of the incident LPL can absorb the photon energy and carry out the isomerization. As a result, the polarization direction of the incident light can control the surface contraction direction and eventually the bending direction [74]. On the other hand, benefiting from the development of alignment agent for mesogens, researchers can achieve homeotropic alignment [75] in LCP films in which the surface region tends to expand symmetrically in plane rather than contraction, so that the whole film bends away from the light source. Ikeda and coworkers have been successful at breaking the barriers toward 3D photoinduced deformation and inspiring other groups as pioneers.

Broer and coworkers prepared a series of crosslinked LCP films in which the direction of mesogens changed along the depth [76-78]. As shown in Figure 1.21, there are two typical modes, termed as twisted [76] and splayed [78] molecular alignment, respectively. In the twisted mode, mesogens near the two surfaces are oriented parallel to the surface but perpendicular to each other, while in splayed mode, mesogens at the two sides are arranged parallel and perpendicular to the surface. Inside the crosslinked LCP films, the orientation changes gently along the depth. Therefore, under UV irradiation, the films contract around one surface but expand at the other side along the long axis of the rectangular films which eventually results in a large 3D bending deformation. Concentration of azobenzene moieties in these experiments was about 5-20 mol% in the crosslinked LCPs and it was confirmed that photomechanical deformation could be triggered with only 2 mol% azobenzene concentration [79]. Thus, these research studies presented preparation, mechanism and strategies for photoinduced deformation control and laid the foundation for more complicated and creative attempts.

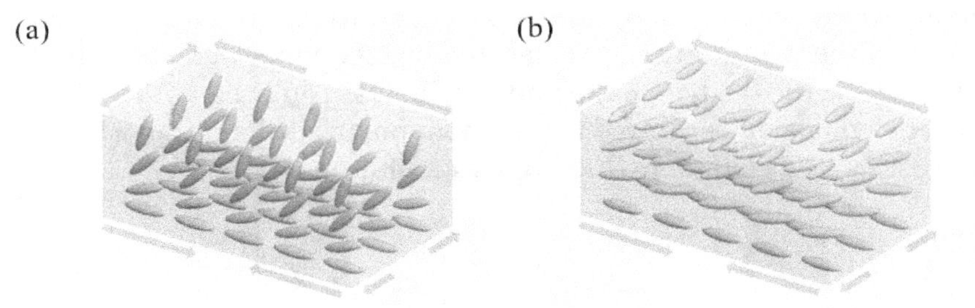

Figure 1.21 Molecule arrangement in (a) splayed and (b) twisted mode with arrows pointing out the stress direction at each surface.

1.5.2 Recent Progress

In recent years, the photomechanical LCP materials have witnessed a significant progress by introducing new techniques such as novel photoalignment, molecular modification and 3D printing. This enables researchers to explore more specific functions for practical applications.

Spatially Controlled Alignment of Mesogens

Orientation of LC molecules is vital to controlling the actuation behaviors of crosslinked LCPs or LC elastomers, but only recently researchers have paid attention to the significance of in-plane alignment. It has been predicted that LCPs containing defects generate complex strain around defects and undergo distinguishing deformation in response to external stimulus [80,81]. Then, there comes the problem that how to achieve complex alignment with defects in these crosslinked LCPs.

One approach is to utilize photoalignment materials to fabricate LC cells with complex orientation patterns. As shown in Figure 1.22, a simple but ingenious equipment was built [82]. LC cells with photoalignment agents which orient themselves parallel to the polarized direction of light were placed on a rotator. With the aid of a wedge-shape photomask, azimuthal, spiral, radial patterns were achieved with the polarized light set at an angle of 90°, 45°, 0° to the wedge. Upon external stimulus, films with azimuthal and radial alignments deformed into conical and saddle shape, respectively. Furthermore, if the linear polarizer was also placed on a rotator, more complex alignments could be realized by adjusting the rotation direction and speed ration of the two rotators, as exhibited by McConney *et al.* [83]. In the latter work, the wedge-shape photomask was replaced with cylindrical lens which concentrated the incident polarized light into a line for more accurate alignment and higher resolution.

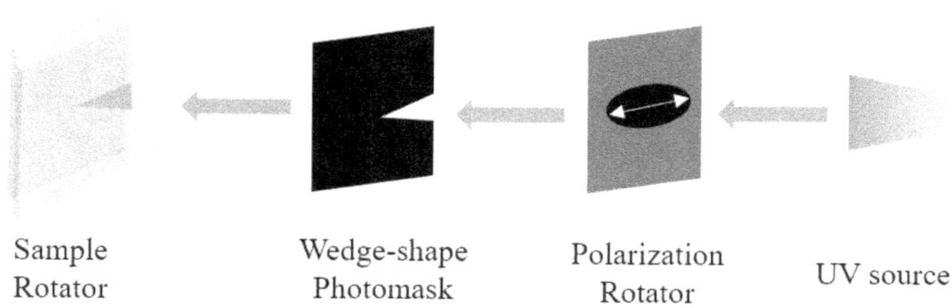

Figure 1.22 Illustration of device for preparing patterned alignment cells.

Although the above-mentioned device based on rotation is elaborate, it cannot implement arbitrary patterns and is limited in resolution. Thus, Ware *et al.* [84] developed an optical patterning system in which the polarized light was focused in an area as small as 0.01 mm^2. With the cooperation between the rotation of the polarization direction of light and the high-precision movement of stage on which the cell was fixed, the authors could achieve any pattern as they presented the grayscale image of the White Brothers and their first flight in history. Later, the authors utilized the system to prepare photodeformable LCP films [85] which was superior in integrating several defects or even different defects into one film. Accordingly, it allowed spatial control of deformation by localizing the UV irradiation.

Besides, the deformations here were sharper compared to the rotation method [82,83] and could afford heavier load [84] because optical patterning system showed a high resolution.

Another approach is to utilize photolithography to prepare channels to guide the orientation of mesogens. Xia *et al.* [86] fabricated desired patterns in photoresist with the aid of photomask and then replica molds twice to polydimethylsiloxane (PDMS) and then epoxy resin on glass slides for preparing LC cells [86]. It was demonstrated that mesogens in the cell tended to orientate along the channel to form pre-designed topological defects and achieve out-of-plane deformation.

In the studies mentioned above, the researches follow the basic principle to regulate the in-plane alignments into certain patterns with defects. Inspired by these results, there comes a question that can we achieve 3D control over alignment directly. Till now, such attempts, needing more ingenious structure design, haven't been seen.

Molecular Modification

In general, there exist two common limitations in most of the early research studies about azobenzene-containing crosslinked LCPs: large time consumption during *cis-trans* relaxation and response to UV light [65,68,73-76]. Several studies have been carried out employing chemical routes, aiming at breaking these shackles through molecular modification.

Broer and coworkers demonstrated novel LC networks showing fast relaxation of photoinduced deformation [87]. As shown in Figure 1.23, the authors designed two azobenzene-derivatives: the first one with a push-pull group and the second one forming tautomerizable azobydrazone structure to accelerate the *cis-trans* relaxation [88]. Strips with a splay-aligned configuration exhibited bending and relaxation immediately as the illumination was applied and removed at room temperature, while in traditional cases, strips tend to retain the bending shape when the light is turned off [73,75]. This phenomenon arose from both the accelerated thermal relaxation of *cis*-azobenzene and the photo-caused temperature increase. Furthermore, the authors created a buckled shape through fixing the two ends to a substrate but shorter than the length of the film. By applying a constant illumination parallel to the substrate, the device exhibited a continuous travelling wave with a feedback loop utilizing self-shadowing. Impressively, when the substrate was removed, the buckled shape film could walk in one direction continuously under constant illumination. It should be pointed out that the splay-aligned configuration contributed to not only larger deformation but also the control over walking direction by placing different side of the films on the top.

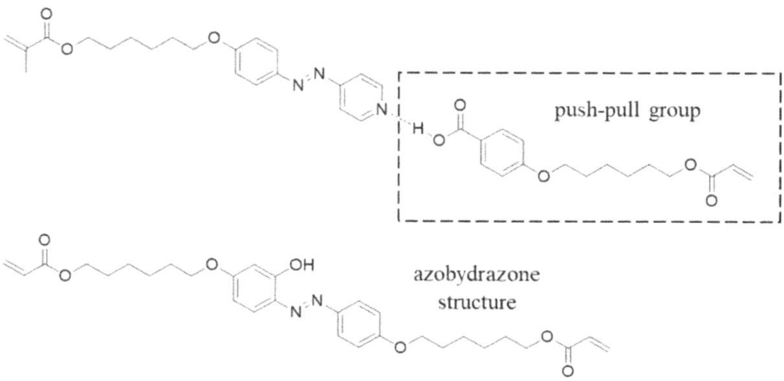

Figure 1.23 Chemical structures of azo-derivatives.

Generally, UV light is harmful to living organisms and less penetrating, thus, response to visible light is more preferred in practical applications. Besides, taking the low energy conversion rate into account, response to UV light is energy-inefficient. However, making use of visible light which occupies most of the spectrum of sunlight presents a way out. There are solutions that merge thermally-responsive LCPs and materials with photothermal effect together such as nano-gold particles, carbon nanotube or graphene. However, more intrinsically, researchers utilized azotolane to achieve visible light response [89-91]. It was demonstrated that LCPs containing azotolanes deformed under irradiation of visible light in coincidence with traditional UV responsible devices containing azobenzenes. Molecular modification requires a lot of repetitive work to design structures and synthesize compounds until obtaining satisfying results. However, it solves problems at source and plays irreplaceable roles in many unmentioned aspects such as lowering the glass transition temperature of polymer materials [92].

Novel Fabrication Techniques for 3D Structures

Novel techniques can bring vitality to various fields, so do in the field of photodeformable LCPs. Being a representative, 3D-printing is a process of fabricating 3D structures by layering materials according to the pre-designed digital program [93]. If the materials show response to external stimuli, the programmed action bring the fourth dimension of time, which is often called four-dimensional (4D) printing. These smart materials can be self-assembled materials, shape memory alloys/polymers and certainly LCPs [94]. Kotikian *et al.* [95] demonstrated that LC mesogens underwent shear-induced alignment as the ink flowed out the nozzle. Accordingly, the authors prepared uniaxially-aligned LC elastomers that contracted upon being heated into isotropic phase and could lift roughly 1000 times of their own weight. Furthermore, the authors also printed LC elastomers with complex spatially programed directors that deformed into cone and saddle forms. However, to our knowledge, 4D-printing with photoresponsive LCPs haven't been reported yet which may result from the photo-crosslinking step in current process to retain the orientation induced by shearing. It can be envisaged that there will be a great progress on this aspect in near future.

Besides, direct laser writing (DLW) is another breakthrough in fabricating 3D structures with a sub-micrometer resolution [96-99]. Zeng *et al.* [98] enhanced the resolution in DLW of LC elastomer structure by two-photon absorption, which refers to the photopolymerization initiated only in the focused area of two beams of light. The authors demonstrated that the orientation of mesogens was not destroyed during the writing process and the technique was suitable for preparing structures of LC elastomers from 1D lines to 2D rings and also 3D woodpiles shape. Combination of 3D printing or DLW with photoresponsive LCPs shows great potential, which may contribute to the question that how to adjust 3D alignment structure directly proposed in the first part of Section 1.5.2.

Application Design

Despite the extraordinary mechanical response of crosslinked LCPs or LC elastomers, concrete and practical applications are poorly developed except for light-driven motors designed by Yamada *et al.* [100]. Recently, there are several fascinating and creative devices utilizing photomechanical effect of azobenzene-containing LCPs, which demonstrates the great potential in practical applications.

Techniques to manipulate fluids in channels with dimensions of tens of micrometers is termed as microfluidics [101]. Tubular microactuators composed of a linear LCP were fabricated [102], which could direct the fluid motion under irradiation of gradient distributed light with the aid of capillary forces. Azobenzene mesogens tend to align along the propagation direction under irradiation of unpolarized light according to the Weigert effect. The realignment of the mesogens enlarged the cross-section area of the tube and the amplitude increased with the rise of light intensity. Along the long axis of tube, the light intensity was gradient distributed, thus, the tube formed a funnel-shaped structure and the fluids inside moved away from the radical center because of capillary forces. Furthermore, the microactuators were still effective when being made into "Y" or serpentine shape and the driving forces could overcome the gravity to elevate the liquid inside.

Priimagi and coworkers reported about biomimetic light-driven crosslinked LCPs in which the authors pursued self-regulation of the devices [79,103,104]. One of these works presented an automatic gripper which integrated an optical fiber at the center of rectangular LC elastomers with splay-aligned mesogens [79]. As a mirror was placed in front of the device, the reflected light induced the bending of the crosslinked LCP films like a gripper as illustrated in Figure 1.24a. Similarly, tiny objects that can reflect light would be caught when passed by the front of device and the whole process imitated the flytrap catching insects intelligently (Figure 1.24b). Another research paid attention to iris that closed in bright environment and opened in the dark [103]. The authors prepared crosslinked LCPs

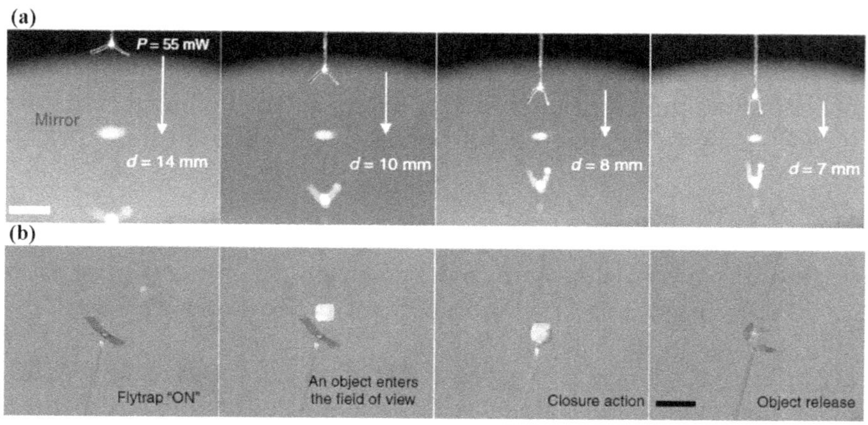

Figure 1.24 (a) Reflected light from mirror induced bending of the artificial flytrap; (b) Artificial flytrap catch objects "fly" onto it spontaneously. Reproduced from Reference 79 with the permission from Springer Nature.

with splayed radial molecular alignment and it was cut into 12 petal-shape segments from the center. Cooling down from the polymerization temperature, segments curled towards homeotropic layer similar to the open state of natural iris. Upon photoirradiation or placing in a brighter environment, the artificial iris bent reversely and re-formed a flat plane blocking out the irradiation. Through these works, the authors chased light-responsive devices that can interact with external environment and adjust themselves to the change of surroundings automatically, which the authors termed as the fifth generation of light robots in their review article [104].

There are many other excellent works not mentioned here such as pendulum generator based on oscillating bilayer cantilever [105] and photocontrollable gas permeation membranes [106]. All these applications step forward from merely contracting, bending, curling or twisting to more concrete concepts which involves so many aspects and shows boundless potential of crosslinked LCPs containing azobenzene moieties.

1.5.3 Liquid Crystal and Polymer Composition

Poor mechanical properties and complicated preparation methods of LCPs have been an obstacle to their practical applications. Composition with other materials may promote a way to solve these problems. Yu *et al.* [107] first incorporated photoresponsive LCPs with polymer-dispersed system and successfully fabricated composite films with any required size and good processability. As shown in Figure 1.25a, the photoactive LCPs microparticles were initially homodispersed in a water-soluble polymer (PVA) solution, which was then cast onto a flat glass substrate. During the volatilization of solvent, the LCP microparticles tend to aggregate asymmetrically due to difference in density between them. It was demonstrated that the mesogens were oriented along the stretching direction when the polymer matrix was mechanically stretched. Upon photoirradiation, phase transition occurred in the LCP microparticles, leading to contraction along the tensile axis. As a result of the asymmetric distribution, the composite film exhibited photoinduced bending towards the airside. Obviously, the fabricating method is applicable to various polymer matrix. Choosing shape-memory polymer such as polyurethane, the films presented triple shape-memory effect and could mimic blooming and fading of flowers [108]. Besides, by doping graphene oxide together with azobenzene LC mesogens into PVA matrix, the hybrid film showed full band NIR-Vis-UV response [109], as illustrated in Figure 1.25b.

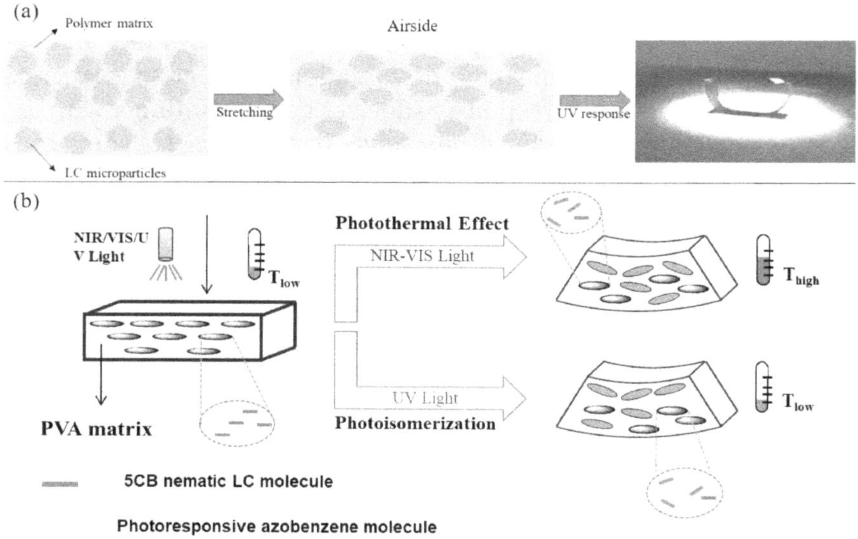

Figure 1.25 (a) Schematic illustration of the mechanical stretching and the UV response; (b) NIR-Vis-UV light response of the hybrid films though photothermal and photoisomerization effects. Reproduced from Reference 109 with the permission from American Chemical Society.

Another composition strategy is to directly deposit photoactive LC layer on general polymer substrate forming bilayer structures. As shown in Figure 1.26a, the LC layer here can be polymer, network, gel and even small molecules. With proper polishing treatment on the polymer substrate, the whole films are expected to exhibit deformation such as bending, expansion, contraction and twisting. Yu and coworkers demonstrated that azobenzene mesogens on the polished low-density polyethylene (LDPE) films can be aligned along the microgrooves whether as small molecules [110] or polymers [105]. Upon UV irradiation, these films bent towards the photoactive layer and reversed immediately when the light was removed. Furthermore, the authors designed an optical pendulum generator with copper coils in front of bilayer LCP/LDPE film [105], as shown in Figure 1.26b. Placed in a magnetic field, the device generated a voltage pulse upon photoirradiation and a damped oscillate voltage when the light was turned off (Figure 1.26c). The composite way provides a variety of possibilities for realizing more complex deformation and exploring fascinating applications in this promising field.

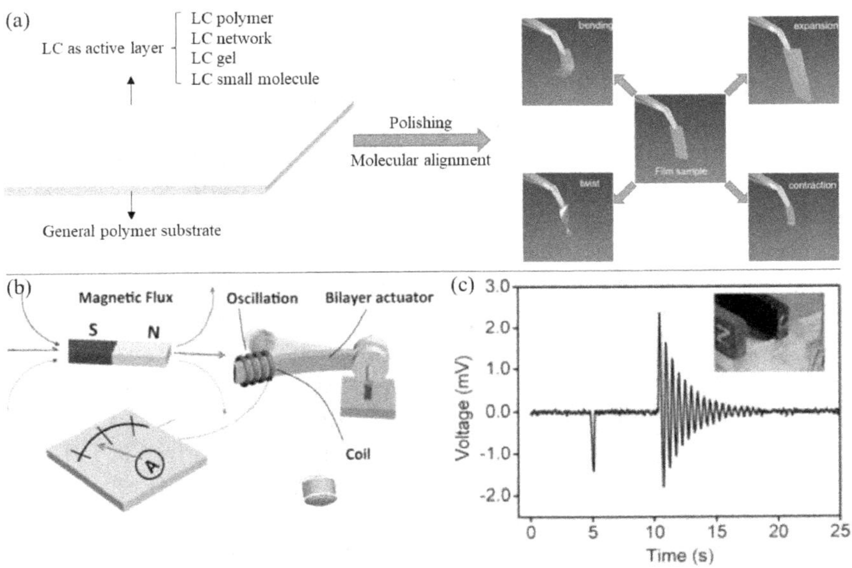

Figure 1.26 (a) Illustration of bilayer structure and its possible deformation; (b) schematic illustration of optical pendulum generator; (c) light-induced electric pulse signal and the successive damped oscillate voltage. Reproduced from Reference 105 with the permission from American Chemical Society.

Efforts have been made to utilize carbon nanotube (CNT) as alignment layer for its intrinsic 1D structure [111-113]. The addition of CNTs enhanced the mechanical properties of materials and endowed the LCP films with electrical conductivity. However, this method is still complicated during the preparation process and lacks control over film size compared to the polymer dispersed or the bilayer system. Moreover, there are contradictions in alignment direction of mesogens in two similar systems [111,112], which needs additional studies.

1.5.4 Dilemma

Along with the rapid development of photodeformable LCPs containing azobenzene moieties in recent years, there are also phenomena that cannot be explained by previous theories. As mentioned above, it is generally recognized that the photomechanical response of crosslinked LCPs or LC elastomers arises from reduction of order parameter upon *trans-cis* isomerization. However, there is a neglected phenomenon that the isomerization is accompanied by the molecular shape change which collectively appears as volume increase in macroscopic level [114]. Besides, the repeated *trans-cis-trans* isomerization requires a free volume to occur which also generate a macroscopic volume increase [115]. It is possible that these volume increases may be in conflict with the contraction induced by the decrease in order parameter so that the direction of photoinduced deformation is uncertain. For example, Priimagi *et al.* [116] reported that similar LCP films with only difference in the location of azobenzene moieties showed opposite bending direction upon UV irradiation. Specifically, films with azobenzene in the crosslinkers bent towards the light source while films with azobenzene as side chains bent away from light source. Similar phenomenon was also observed in side chain LCP fibers which bent away from light source [117], but there are also reports on opposite conclusion [115]. Further research studies are needed to explore factors such as azobenzene concentration and the crosslinking density in the crosslinked LCPs.

There is another contradiction concerning the large mismatch in time scale about different transitions in the photomechanical process. It has been demonstrated that *trans-cis* isomerization takes only seconds, but the reverse may take several hours in typical azobenzene compounds, which is corresponding to the time of macroscopic deformation and relaxation. However, when it comes to volume increase which is represented by the height change of thin azobenzene coatings or the performance in the salt brine (ups or downs due to the density change), it takes only less than 10 s for the relaxation of volume after removal of light [115]. Similar results have also been observed in the modulus of LCP networks that adapt to the change of irradiation condition immediately without delay [118], which may account for the fast relaxation of macroscopic deformation in systems with the bilayer structures [110]. Large efforts are needed to reveal the relationship between these variates and predict a unified theory to predict deformation behaviors.

1.6 Outlook

In this chapter, we introduced azobenzene-containing photoresponsive LCPs from aspects of homopolymers, supramolecules, block polymers and crosslinked polymer or elastomer as well as composite systems. Their applications covering from photonic materials, nanotemplates to photo-driven devices were also summarized. These materials have attracted significant attention in recent years due to two intrinsic advantages: large anisotropic response as LCs and utilizing light as a contactless, precise and clean stimulus. In addition to the applications mentioned above, the materials have shown great potential in various fields through microscopic to macroscopic uses such as biomedicine, haptic display and solar-energy harvesting [119-121]. It can be predicted that more interesting and inspiring devices will be designed in the coming years, However, in our view, more systematic research studies are also needed to understand quantitative relationships between stimulus and response which will prepare these materials for integration into more complicated systems.

References

1. Rau, H. (2002) *Photoisomerization of Azobenzenes*. Elsevier Science, The Netherlands.
2. Kumar, G. S., and Neckers, D. (1989) Photochemistry of azobenzene-containing polymers. *Chemical Reviews*, **89**(8), 1915-1925.
3. Natansohn, A., and Rochon, P. (2002) Photoinduced motions in azo-containing polymers. *Chemical Reviews*, **102**(11), 4139-4176.
4. Zhang, X. P., Zhang, J., Sun, Y. J., Yang, H., and Yu, H. F. (2014) Erasable thin-film optical diode based on a photoresponsive liquid crystal polymer. *Nanoscale*, **6**(7), 3854-3860.
5. Ikeda, T., and Tsutsumi, O. (1995) Optical switching and image storage by means of azobenzene liquid-crystalline films. *Science*, **268**(5219), 1873-1875.
6. Yu, H., Naka, Y., Shishido, A., Iyoda, T., and Ikeda, T. (2009) Effect of recording time on grating formation and enhancement in an amphiphilic diblock liquid-crystalline copolymer. *Molecular Crystals and Liquid Crystals*, **498**, 29-39.
7. Yu, H. F., Shishido, A., and Ikeda, T. (2008) Subwavelength modulation of surface relief and refractive index in preirradiated liquid-crystalline polymer films. *Applied Physics Letters*, **92**(10), doi: 10.1063/1.2894073.
8. Yu, H. F., Okano, K., Shishido, A., Ikeda, T., Kamata, K., Komura, M., and Iyoda, T. (2005) Enhancement of surface-relief gratings recorded on amphiphilic liquid-crystalline diblock copolymer by nanoscale phase separation. *Advanced Materials*, **17**(18), 2184-2188.

9. Li, X., Li, B., He, M., Wang, W., Wang, T. J., Wang, A., Yu, J., Wang, Z. L., Hong, S. W., Byun, M., Lin, S. L., Yu, H. F., and Lin, Z. Q. (2018) Convenient and robust route to photoswitchable hierarchical liquid crystal polymer stripes via flow-enabled self-assembly. *ACS Applied Materials & Interfaces*, **10**(5), 4961-4970.
10. Ikkala, O., and ten Brinke, G. (2002) Functional materials based on self-assembly of polymeric supramolecules. *Science*, **295**(5564), 2407-2409.
11. Lehn, J. M. (1990) Perspective in supramolecular chemistry - from molecular recognition towards molecular information-processing and self-organization. *Angewandte Chemie-International Edition*, **29**(11), 1304-1319.
12. Kato, T., and Frechet, J. M. J. (1989) New approach to mesophase stabilization through hydrogen-bonding molecular-interactions in binary-mixtures. *Journal of the American Chemical Society*, **111**(22), 8533-8534.
13. Cui, L., Dahmane, S., Tong, X., Zhu, L., and Zhao, Y. (2005) Using self-assembly to prepare multifunctional diblock copolymers containing azopyridine moiety. *Macromolecules*, **38**(6), 2076-2084.
14. Zhou, W. M., Kobayashi, T., Zhu, H. R., and Yu, H. F. (2011) Electrically conductive hybrid nanofibers constructed with two amphiphilic salt components. *Chemical Communications*, **47**(48), 12768-12770.
15. Zhou, W. M., and Yu, H. F. (2012) Conductive hybrid nanofibers self-assembled with three different amphiphilic salts. *ACS Applied Materials & Interfaces*, **4**(4), 2154-2159.
16. Liu, H., Kobayashi, T., and Yu, H. F. (2011) Easy fabrication and morphology control of supramolecular liquid-crystalline polymer microparticles. *Macromolecular Rapid Communications*, **32**(4), 378-383.
17. Yu, H. F., Liu, H., and Kobayashi, T. (2011) Fabrication and photoresponse of supramolecular liquid-crystalline microparticles. *ACS Applied Materials & Interfaces*, **3**(4), 1333-1340.
18. Zhang, H. B., Hao, R., Jackson, J. K., Chiao, M., and Yu, H. F. (2014) Janus ultrathin film from multi-level self-assembly at air-water interfaces. *Chemical Communications*, **50**(94), 14843-14846.
19. Chen, D., Liu, H., Kobayashi, T., and Yu, H. F. (2010) Multiresponsive reversible gels based on a carboxylic azo polymer. *Journal of Materials Chemistry*, **20**(18), 3610-3614.
20. Ren, H., Chen, D., Shi, Y., Yu, H. F., and Fu, Z. F. (2015) A carboxylic azo monomer and its homopolymer: synthesis, self-organization and fluorescence behaviour in solution. *Polymer Chemistry*, **6**(2), 270-277.
21. Metrangolo, P., Neukirch, H., Pilati, T., and Resnati, G. (2005) Halogen bonding based recognition processes: A world parallel to hydr-

ogen bonding. *Accounts of Chemical Research*, **38**(5), 386-395.

22. Nguyen, H. L., Horton, P. N., Hursthouse, M. B., Legon, A. C., and Bruce, D. W. (2004) Halogen bonding: A new interaction for liquid crystal formation. *Journal of the American Chemical Society*, **126**(1), 16-17.
23. Chen, Y. J., Yu, H. F., Zhang, L. Y., Yang, H. A., and Lu, Y. F. (2014) Photoresponsive liquid crystals based on halogen bonding of azopyridines. *Chemical Communications*, **50**(68), 9647-9649.
24. Fasolka, M. J., and Mayes, A. M. (2001) Block copolymer thin films: physics and applications. *Annual Review of Materials Research*, **31**(1), 323-355.
25. Bates, F. S., and Fredrickson, G. H. (1999) Block copolymers - Designer soft materials. *Physics Today*, **52**(2), 32-38.
26. Hamley, I. W. (2003) Nanostructure fabrication using block co-polymers. *Nanotechnology*, **14**(10), R39-R54.
27. Lazzari, M., and Lopez-Quintela, M. A. (2003) Block copolymers as a tool for nanomaterial fabrication. *Advanced Materials*, **15**(19), 1583-1594.
28. Park, C., Yoon, J., and Thomas, E. L. (2003) Enabling nanotechnology with self assembled block copolymer patterns. *Polymer*, **44**(22), 6725-6760.
29. Kim, H.-C., Park, S.-M., and Hinsberg, W. D. (2009) Block copolymer based nanostructures: materials, processes, and applications to electronics. *Chemical Reviews*, **110**(1), 146-177.
30. Yu, H. (2014) Recent advances in photoresponsive liquid-crystalline polymers containing azobenzene chromophores. *Journal of Material Chemistry C*, **2**(17), 3047-3054.
31. Ikeda, T., and Tsutsumi, O. (1995) Optical switching and image storage by means of azobenzene liquid-crystal films. *Science*, **268**(5219), 1873-1875.
32. Hirai, T., Hirano, T., and Komasawa, I. (2000) Preparation of Y_2O_3: Eu^{3+} phosphor fine particles using an emulsion liquid mem-brane system. *Journal of Material Chemistry*, **10**(10), 2306-2310.
33. Hasegawa, M., Yamamoto, T., Kanazawa, A., Shiono, T., and Ikeda, T. (1999) A dynamic grating using a photochemical phase transition of polymer liquid crystals containing azobenzene derivatives. *Advanced Materials*, **11**(11), 675-677.
34. Hasegawa, M., Yamamoto, T., Kanazawa, A., Takeshi Shiono, A., and Ikeda, T. (1999) Photochemically induced dynamic grating by means of side chain polymer liquid crystals. *Chemistry of Materials*, **11**(10), 2764-2769.
35. Hasegawa, M., Yamamoto, T., Kanazawa, A., Shiono, T., Ikeda, T., Yu, N., Akiyama, E., and Takamura, Y. (1999) Real-time holographic grating by means of photoresponsive polymer liquid crystals with a flexible siloxane spacer in the side chain. *Journal of Material Chem-*

istry, **9**(11), 2765-2769.

36. Yamamoto, T., Yoneyama, S., Tsutsumi, O., Kanazawa, A., Shiono, T., and Ikeda, T. (2000) Holographic gratings in the optically isotropic state of polymer azobenzene liquid-crystal films. *Journal of Applied Physics*, **88**(5), 2215-2220.
37. Yamamoto, T., Ohashi, A., Yoneyama, S., Hasegawa, M., Tsutsumi, O., Kanazawa, A., Takeshi Shiono, A., and Ikeda, T. (2001) Phase-type gratings formed by photochemical phase transition of polymer azobenzene liquid crystal. 2. Rapid switching of diffraction beams in thin films. *Journal of Physical Chemistry B*, **105**(12), 2308-2313.
38. Addison, T., Cayre, O. J., Biggs, S., Armes, S. P., and York, D. (2010) Polymeric microcapsules assembled from a cationic/zwitterionic pair of responsive block copolymer micelles. *Langmuir*, **26**(9), 6281-6286.
39. Chang, X., and Dong, C. M. (2013) Synthesis of hyperbranched polypeptide and PEO block copolymer by consecutive thiol-yne chemistry. *Biomacromolecules*, **14**(9), 3329-3337.
40. Yu, H., Li, J., Ikeda, T., and Iyoda, T. (2006) Macroscopic parallel nanocylinder array fabrication using a simple rubbing technique. *Advanced Materials*, **18**(17), 2213-2215.
41. Morikawa, Y., Nagano, S., Watanabe, K., Kamata, K., Iyoda, T., and Seki, T. (2006) Optical alignment and patterning of nanoscale micro-domains in a block copolymer thin film. *Advanced Materials*, **18**(7), 883-886.
42. Yu, H., Iyoda, T., and Ikeda, T. (2006) Photoinduced alignment of nanocylinders by supramolecular cooperative motions. *Journal of the American Chemical Society*, **128**(34), 11010-11011.
43. Morikawa, Y., Kondo, T., Shusaku Nagano, A., and Seki, T. (2007) Photoinduced 3D ordering and patterning of microphase-separated nanostructure in polystyrene-based block copolymer. *Chemistry of Materials*, **19**(7), 1540-1542.
44. Yu, H., Kobayashi, T., and Hu, G.-H. (2011) Photocontrolled microphase separation in a nematic liquid–crystalline diblock copolymer. *Polymer*, **52**(7), 1554-1561.
45. Nagano, S., Koizuka, Y., Murase, T., Sano, M., Shinohara, Y., Amemiya, Y., and Seki, T. (2012) Synergy effect on morphology switching: Real-time observation of photo-orientation of microphase separation in a block copolymer. *Angewandte Chemie-International Edition*, **51**(24), 5884-5888.
46. Vapaavuori, J., Grosrenaud, J., Pellerin, C., and Bazuin, C. G. (2015) In situ photocontrol of block copolymer morphology during dip-coating of thin films. *ACS Macro Letters*, **4**(10), 1158-1162.
47. Wang, T., Li, X., Dong, Z., Huang, S., and Yu, H. (2017) Vertical orientation of nanocylinders in liquid-crystalline block copolymers directed by light. *ACS Applied Materials & Interfaces*, **9**(29), 24864-

24872.

48. Nickmans, K., Bogels, G. M., Sanchez-Somolinos, C., Murphy, J. N., Leclere, P., Voets, I. K., and Schenning, A. P. H. J. (2017) 3D Orientational control in self-assembled thin films with sub-5 nm features by light. *Small*, **13**(33), 1701043.
49. Huang, S., Pang, L., Chen, Y., Zhou, L., Fang, S., Yu, H. (2018) Hydrogen bond induces hierarchical self-assembly in liquid-crystalline block copolymers. *Macromolecular Rapid Communications*, **39**(6), 1700783.
50. Beppu, K., Nagashima, Y., Hara, M., Nagano, S., and Seki, T. (2017) Photoalignment of vertically oriented microphase separated lamellae in LC-LC diblock copolymer thin film. *Macromolecular Rapid Communications*, **38**(13), doi: 10.1002/marc.201600659.
51. Yu, H., Naka, Y., Shishido, A., and Ikeda, T. (2008) Well-defined liquid-crystalline diblock copolymers with an azobenzene moiety: Synthesis, photoinduced alignment and their holographic properties. *Macromolecules*, **41**(21), 7959-7966.
52. Li, J., Blake, J., and Delaney, C. (2007) Template- and vacuum-ultraviolet-assisted fabrication of a Ag-nanoparticle array on flexible and rigid substrates. *Advanced Materials*, **19**(9), 1267-1271.
53. Watanabe, S., Fujiwara, R., Hada, M., Okazaki, Y., and Iyoda, T. (2007) Site-specific recognition of nanophase-separated surfaces of amphiphilic block copolymers by hydrophilic and hydrophobic gold nanoparticles. *Angewandte Chemie-International Edition*, **46**(7), 1120-1123.
54. Chen, A., Komura, M., Kamata, K., and Iyoda, T. (2008) Highly ordered arrays of mesoporous silica nanorods with tunable aspect ratios from block copolymer thin films. *Advanced Materials*, **20**(4), 763-767.
55. Wang, G., Tong, X., and Zhao, Y. (2004) Preparation of azobenzene-containing amphiphilic diblock copolymers for light-responsive micellar aggregates. *Macromolecules*, **37**(24), 8911-8917.
56. Tong, X., Wang, G., Soldera, A., and Zhao, Y. (2005) How can azobenzene block copolymer vesicles be dissociated and reformed by light? *Journal of Physical Chemistry B*, **109**(43), 20281-20287.
57. Su, W., Han, K., Luo, Y., Wang, Z., Li, Y., and Zhang, Q. (2010) Formation and photoresponsive properties of giant microvesicles assembled from azobenzene-containing amphiphilic diblock copolymers. *Macromolecular Chemistry and Physics*, **208**(9), 955-963.
58. Yang, J., Levy, D., Deng, W., Keller, P., and Li, M. H. (2005) Polymer vesicles formed by amphiphilic diblock copolymers containing a thermotropic liquid crystalline polymer block. *Chemical Communications*, **34**(34), 4345-4347.
59. Yang, J., Pinol, R., Gubellini, F., Levy, D., Albouy, P. A., Keller, P., and Li, M. H. (2006) Formation of polymer vesicles by liquid crystal am-

phiphilic block copolymers. *Langmuir*, **22**(18), 7907-7911.

60. Mabrouk, E., Cuvelier, D., Brochardwyart, F., Nassoy, P., and Li, M. H. (2009) Bursting of sensitive polymersomes induced by curling. *Proceedings of the National Academy of Sciences of the United States of America*, **106**(18), 7294-7298.
61. De Gennes P.G. (1974) *The Physics of Liquid Crystals*, Clarendon Press, UK.
62. Wermter, H. and H. Finkelmann. (2001) Liquid crystalline elastomers as artificial muscles. *e-Polymers*, **1**(1), 111-123.
63. Thomsen, D. L., Keller, P., Naciri, J., Pink, R., Jeon, H., Shenoy, D., and Ratna, B. R. (2001) Liquid crystal elastomers with mechanical proper-ties of a muscle. *Macromolecules*, **34**(17), 5868-5875.
64. Ikeda, T. (2003) Photomodulation of liquid crystal orientations for photonic applications. *Journal of Materials Chemistry*, **13**(9), 2037-2057.
65. Hogan, P. M., Tajbakhsh, A. R., and Terentjev, E. M. (2002) Uv manipulation of order and macroscopic shape in nematic elastomers. *Physical Review E*, **65**(4), 041720.
66. van der Veen, G., and Prins, W. (1971) Photomechanical energy conversion in a polymer membrane. *Nature Physical Science*, **230**, 70-72.
67. Finkelmann, H., Nishikawa, E., Pereira, G. G., and Warner, M. (2001) A new opto-mechanical effect in solids. *Physical Review Letters*, **87**(1), 015501.
68. Kupfer, J., and Finkelmann, H. (1991) Nematic liquid single crystal elastomers. *Die Makromolekulare Chemie, Rapid Communications*, 12(12), 717-726.
69. Li, M. H., Keller, P., Li, B., Wang, X., and Brunet, M. (2003) Light-driven side-on nematic elastomer actuators. *Advanced Materials*, **15**(7-8), 569-572.
70. Kirste, R. G., and Ohm, H. G. (1985) The conformation of liquid-crystalline polymers as revealed by neutron scattering. *Macromolecular Rapid Communications*, **6**(3), 179-185.
71. D'Allest, J. F., Maissa, P., Ten Bosch, A., Sixou, P., Blumstein, A., Blumstein, R., Teixeira, J., and Noirez, L. (1988) Experimental evidence of chain extension at the transition temperature of a nematic polymer. *Physical Review Letters*, **61**(22), 2562-2565.
72. Leroux, N., Keller, P., Achard, M. F., Noirez, L., and Hardouin, F. (1993) Small-angle neutron-scattering experiments on side-on fixed liquid-crystal polyacrylates. *Journal De Physique*, **3**(8), 1289-1296.
73. Ikeda, T., Nakano, M., Yu, Y., Tsutsumi, O., and Kanazawa, A. (2003) Anisotropic bending and unbending behavior of azobenzene liquid-crystalline gels by light exposure. *Advanced Materials*, **15**(3), 201-205.

74. Yu, Y., Nakano, M., and Ikeda, T. (2003) Directed bending of a polymer film by light. *Nature*, **425**, 145.
75. Kondo, M., Yu, Y., and Ikeda, T. (2006) How does the initial alignment of mesogens affect the photoinduced bending behavior of liquid-crystalline elastomers? *Angewandte Chemie-International Edition*, **45**(9), 1378-1382.
76. Harris, K. D., Cuypers, R., Scheibe, P., van Oosten, C. L., Bastiaansen, C. W. M., Lub, J., and Broer, D. J. (2005) Large amplitude light-induced motion in high elastic modulus polymer actuators. *Journal of Materials Chemistry*, **15**(47), 5043-5048.
77. Harris, K. D., Bastiaansen, C. W. M., and Broer, D. J. (2007) Physical properties of anisotropically swelling hydrogen-bonded liquid crystal polymer actuators. *Journal of Microelectromechanical Systems*, **16**(2), 480-488.
78. van Oosten, C. L., Bastiaansen, C. W. M., and Broer, D. J. (2009) Printed artificial cilia from liquid-crystal network actuators modularly driven by light. *Nature Materials*, **8**(8), 677-682.
79. Wani, O. M., Zeng, H., and Priimagi, A. (2017) A light-driven artificial flytrap. *Nature Communications*, **8**, 15546.
80. Modes, C. D., Bhattacharya, K., and Warner, M. (2010) Disclination-mediated thermo-optical response in nematic glass sheets. *Physical Review E*, **81**(6), 060701.
81. Modes, C. D., and Warner, M. (2011) Blueprinting nematic glass: Systematically constructing and combining active points of curvature for emergent morphology. *Physical Review E*, **84**(2), 021711.
82. de Haan, L. T., Sanchez-Somolinos, C., Bastiaansen, C. M. W., Schenning, A. P. H. J., and Broer, D. J. (2012) Engineering of complex order and the macroscopic deformation of liquid crystal polymer networks. *Angewandte Chemie-International Edition*, **51**(50), 12469-12472.
83. McConney, M. E., Martinez, A., Tondiglia, V. P., Lee, K. M., Langley, D., Smalyukh, I. I., and White, T. J. (2013) Topography from topology: Photoinduced surface features generated in liquid crystal polymer net-works. *Advanced Materials*, **25**(41), 5880-5885.
84. Ware, T. H., McConney, M. E., Wie, J. J., Tondiglia, V. P., and White, T. J. (2015) Voxelated liquid crystal elastomers. *Science*, **347**(6225), 982-984.
85. Ahn, S.-k., Ware, T. H., Lee, K. M., Tondiglia, V. P., and White, T. J. (2016) Photoinduced topographical feature development in blueprinted azobenzene-functionalized liquid crystalline elastomers. *Advanced Functional Materials*, **26**(32), 5819-5826.
86. Xia, Y., Cedillo-Servin, G., Kamien, R. D., and Yang, S. (2016) Guided folding of nematic liquid crystal elastomer sheets into 3d via patterned 1D microchannels. *Advanced Materials*, **28**(43), 9637-9643.
87. Gelebart, A. H., Jan Mulder, D., Varga, M., Konya, A., Vantomme, G.,

Meijer, E. W., Selinger, R. L. B., and Broer, D. J. (2017) Making waves in a photoactive polymer film. *Nature*, **546**(7660), 632-636.

88. Brode, W. R., Gould, J. H., and Wyman, G. M. (1952) The relation between the absorption spectra and the chemical constitution of dyes .25. Phototropism and cis-trans isomerism in aromatic azo compounds. *Journal of the American Chemical Society*, **74**(18), 4641-4646.
89. Yin, R., Xu, W., Kondo, M., Yen, C.-C., Mamiya, J.-i., Ikeda, T., and Yu, Y. (2009) Can sunlight drive the photoinduced bending of polymer films? *Journal of Materials Chemistry*, **19**(20), 3141-3143.
90. Cheng, F., Zhang, Y., Yin, R., and Yu, Y. (2010) Visible light induced bending and unbending behavior of crosslinked liquid-crystalline polymer films containing azotolane moieties. *Journal of Materials Chemistry*, **20**(23), 4888-4896.
91. Liu, Y., Wu, W., Wei, J., and Yu, Y. (2016) Visible light responsive liquid crystal polymers containing reactive moieties with good processability. *ACS Applied Materials & Interfaces*, **9**(1), 782-789.
92. Zhou, H., Xue, C., Weis, P., Suzuki, Y., Huang, S., Koynov, K., Auernhammer, G. K., Berger, R., Butt, H.-J., and Wu, S. (2016) Photoswitching of glass transition temperatures of azobenzene-containing polymers induces reversible solid-to-liquid transitions. *Nature Chemistry*, **9**, 145-151.
93. Jin, C., O-Chang, K., Wonjin, J., Ju, L. H., and Myoung-Woon, M. (2015) 4D printing technology: A review. *3D Printing and Additive Manufacturing*, **2**(4), 159-167.
94. Leist, S. K., and Zhou, J. (2016) Current status of 4D printing technology and the potential of light-reactive smart materials as 4D printable materials. *Virtual and Physical Prototyping*, **11**(4), 249-262.
95. Kotikian, A., Truby, R. L., Boley, J. W., White, T. J., and Lewis, J. A. (2018) 3D printing of liquid crystal elastomeric actuators with spatially programed nematic order. *Advanced Materials*, **30**(10), 1706164.
96. Maruo, S., and Fourkas, J. T. (2008) Recent progress in multiphoton microfabrication. *Laser & Photonics Reviews*, **2**(1-2), 100-111.
97. von Freymann, G., Ledermann, A., Thiel, M., Staude, I., Essig, S., Busch, K., and Wegener, M. (2010) Three-dimensional nanostructures for photonics. *Advanced Functional Materials*, **20**(7), 1038-1052.
98. Zeng, H., Martella, D., Wasylczyk, P., Cerretti, G., Lavocat, J.-C. G., Ho, C.-H., Parmeggiani, C., and Wiersma, D. S. (2014) High-resolution 3D direct laser writing for liquid-crystalline elastomer microstructures. *Advanced Materials*, **26**(15), 2319-2322.

99. Schwarzle, D., Hou, X., Prucker, O., and Ruhe, J. (2017) Polymer microstructures through two-photon crosslinking. *Advanced Materials*, **29**(39), 1703469.
100. Yamada, M., Kondo, M., Mamiya, J.-i., Yu, Y., Kinoshita, M., Barrett, C. J., and Ikeda, T. (2008) Photomobile polymer materials: Towards light-driven plastic motors. *Angewandte Chemie-International Edition*, **47**(27), 4986-4988.
101. Whitesides, G. M. (2006) The origins and the future of microfluidics. *Nature*, **442**(7101), 368-373.
102. Lv, J.-a., Liu, Y., Wei, J., Chen, E., Qin, L., and Yu, Y. (2016) Photocontrol of fluid slugs in liquid crystal polymer microactuators. *Nature*, **537**(7619), 179-184.
103. Zeng, H., Wani, O. M., Wasylczyk, P., Kaczmarek, R., and Priimagi, A. (2017) Self-regulating iris based on light-actuated liquid crystal elas-tomer. *Advanced Materials*, **29**(30), 1701814.
104. Zeng, H., Wasylczyk, P., Wiersma, D. S., and Priimagi, A. (2017) Light robots: Bridging the gap between microrobotics and photo-mechanics in soft materials. *Advanced Materials*, **30**(24), 1703554.
105. Tang, R., Liu, Z., Xu, D., Liu, J., Yu, L., and Yu, H. (2015) Optical pendulum generator based on photomechanical liquid-crystalline actuators. *ACS Applied Materials & Interfaces*, **7**(16), 8393-8397.
106. Glowacki, E., Horovitz, K., Tang, C. W., and Marshall, K. L. (2010) Photoswitchable gas permeation membranes based on liquid crystals. *Advanced Functional Materials*, **20**(17), 2778-2785.
107. Yu, H., Dong, C., Zhou, W., Kobayashi, T., and Yang, H. (2011) Wrinkled liquid-crystalline microparticle-enhanced photoresponse of pdlc-like films by coupling with mechanical stretching. *Small*, **7**(21), 3039-3045.
108. Zhou, L., Liu, Q., Lv, X., Gao, L., Fang, S., and Yu, H. (2016) Photoinduced triple shape memory polyurethane enabled by doping with azobenzene and go. *Journal of Materials Chemistry C*, **4**(42), 9993-9997.
109. Cheng, Z., Wang, T., Li, X., Zhang, Y., and Yu, H. (2015) Nir–vis–uv light-responsive actuator films of polymer-dispersed liquid crystal/graphene oxide nanocomposites. *ACS Applied Materials & Interfaces*, **7**(49), 27494-27501.
110. Liu, Z., Tang, R., Xu, D., Liu, J., and Yu, H. (2015) Precise actuation of bilayer photomechanical films coated with molecular azobenzene chromophores. *Macromolecular Rapid Communications*, **36**(12), 1171-1176.
111. Wang, W., Sun, X., Wu, W., Peng, H., and Yu, Y. (2012) Photoinduced deformation of crosslinked liquid-crystalline polymer film oriented by a highly aligned carbon nanotube sheet. *Angewandte Chemie-International Edition*, **51**(19), 4644-4647.
112. Sun, X., Wang, W., Qiu, L., Guo, W., Yu, Y., and Peng, H. (2012) Unus-

ual reversible photomechanical actuation in polymer/nanotube compo-sites. *Angewandte Chemie-International Edition*, **51**(34), 8520-8524.

113. Wang, M., Sayed, S. M., Guo, L.-X., Lin, B.-P., Zhang, X.-Q., Sun, Y., and Yang, H. (2016) Multi-stimuli responsive carbon nanotube incorporated polysiloxane azobenzene liquid crystalline elastomer composites. *Macromolecules*, **49**(2), 663-671.
114. Liu, D. Q., Bastiaansen, C. W. M., den Toonder, J. M. J., and Broer, D. J. (2012) Light-induced formation of dynamic and permanent surface topologies in chiral-nematic polymer networks. *Macromolecules*, **45**(19), 8005-8012.
115. Liu, D., and Broer, D. J. (2015) New insights into photoactivated volume generation boost surface morphing in liquid crystal coatings. *Nature Communications*, **6**, 8334.
116. Priimagi, A., Shimamura, A., Kondo, M., Hiraoka, T., Kubo, S., Mamiya, J.-I., Kinoshita, M., Ikeda, T., and Shishido, A. (2012) Location of the azobenzene moieties within the cross-linked liquid-crystalline polymers can dictate the direction of photoinduced bending. *ACS Macro Letters*, **1**(1), 96-99.
117. Cheng, Z., Ma, S., Zhang, Y., Huang, S., Chen, Y., and Yu, H. (2017) Photomechanical motion of liquid-crystalline fibers bending away from a light source. *Macromolecules*, **50**(21), 8317-8324.
118. Kumar, K., Schenning, A. P. H. J., Broer, D. J., and Liu, D. (2016) Regulating the modulus of a chiral liquid crystal polymer network by light. *Soft Matter*, **12**(13), 3196-3201.
119. Hiscock, T., Warner, M., and Palffy-Muhoray, P. (2011) Solar to electrical conversion via liquid crystal elastomers. *Journal of Applied Physics*, **109**(10), 104506.
120. Camargo, C. J., Campanella, H., Marshall, J. E., Torras, N., Zinoviev, K., Terentjev, E. M., and Esteve, J. (2012) Batch fabrication of optical actuators using nanotube-elastomer composites towards refreshable braille displays. *Journal of Micromechanics and Microengineering*, **22**(7), 075009.
121. Kocer, G., ter Schiphorst, J., Hendrikx, M., Kassa, H. G., Leclere, P., Schenning, A. P. H. J., and Jonkheijm, P. (2017) Light-responsive hierarchically structured liquid crystal polymer networks for harnessing cell adhesion and migration. *Advanced Materials*, **29**(27), 1606407.

Chapter 2

Developments in the Use of Liquid Crystalline Polymers for Generating Nanohybrids

J.-D. Marty

Laboratoire des Interactions Moléculaires et Réactivités Chimiques et Photochimiques, Université de Toulouse, CNRS UMR 5623, 31062 Toulouse Cedex 09, France
marty@chimie.ups-tlse.fr

2.1 Introduction

Mixtures of liquid crystals (LCs) and nanoparticles (NPs) have aroused great interest in the scientific community [1-3] and can give rise to applications in various technologies such as highly-sensitive LC-based sensors[4]. First studies have involved the doping of LC materials with preformed nanoparticles. As trivial as it may sound, a homogeneous material involving NPs and LCs is not, in any way, straightforward to synthesize. One of the key parameters is the interaction between the particles and the LC matrix. A weak interaction may lead to LC phase disruption or segregation of NPs in the LC phase defects, whereas a high level of interaction enables the achievement of perfect dispersions of NPs in LC matrices (Figure 2.1A). Eventually, if the interaction between the NPs and the LC is strong enough to obtain a functional hybrid that behaves as a liquid crystalline functional unit, complex mesomorphic hybrids can be obtained [5].

In-situ (direct growth) approach is also possible. In this case, dispersion of a NP precursor should be achieved first. Once again, if precursor-interactions are weak, the dispersion may occur within the defects of the LC structure (Figure 2.1B) and homogeneous dispersions are expected for higher interaction strengths (Figure 2.1B). Irrespective of the strength of the interactions, the *in-situ* approach needs an additional step to obtain NPs from the precursor by using various synthetic pathways (reaction of the liquid-crystal compound, additional reactants, light or other means, etc.), as shown in

Liquid Crystalline Polymers, edited by Vikas Mittal

Figure 2.1C.

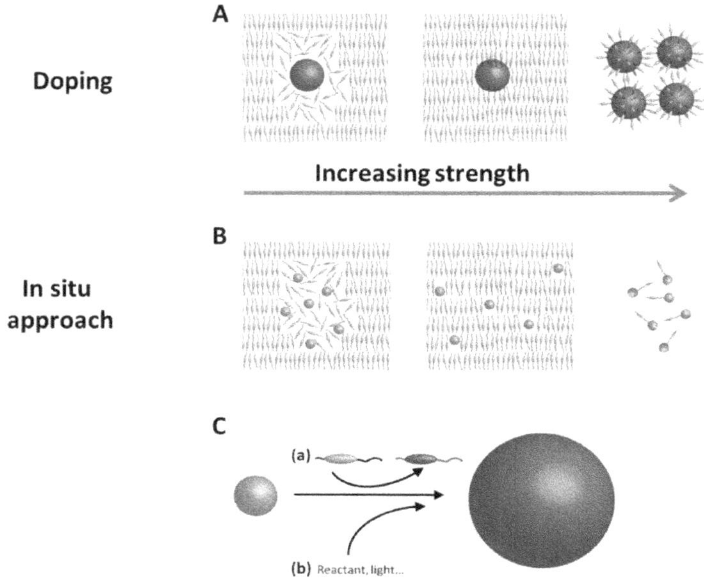

Figure 2.1 A) Schematic organization of NPs (in red) and LC (in blue) depending on their interaction strength. B) Schematic organization of a precursor of NPs (in green) and LC (in blue) depending on their interaction strength. C) Schematic reaction from precursor (green) to NPs (in red) either by reaction with the LC molecules (in blue before reaction, in purple after) (a) or by addition of an external reactant or trigger factor (b).

In this context, polymers forming liquid crystal phases are materials of peculiar interest for the stabilization or the *in-situ* synthesis of nanoparticles. Their specific structure, which combines properties intrinsic to LCs and those of polymers, is mainly responsible for the results discussed in the next section. Briefly, the most common structures forming such LC materials are given in Figure 2.2. The monomer units consist of low molar mass calamitic or discotic mesogens which either form part of the main chain in the polymer backbone or attached as side groups. The nature of the mesophase is largely dependent on various factors, the most important being the type of mesogen, the spacers and the backbone itself. In the case of repeating calamitic units, nematic-, smectic- or cholesteric-like mesophases are generally observed [6-7]. For polymers containing discotic liquid crystal units, different kinds of mesophases can be formed including the hexagonal columnar, sanidic nematic and columnar nematic [8-10].

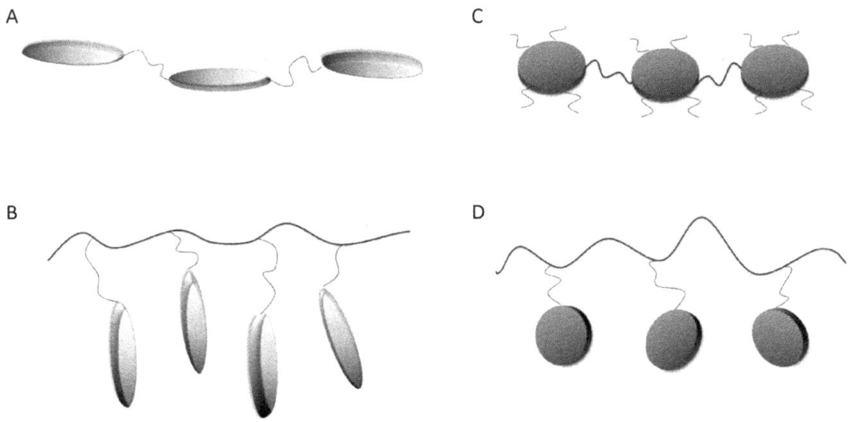

Figure 2.2 Structures for the most common liquid crystal polymers. The mesogens can be of a rod-like (calamitic) or disc-like nature. These units can be attached to the polymer backbone (A and C) or as side groups (B and D).

In addition, liquid crystal main-chain and side-chain dendritic structures (branched or hyperbranched) have also been synthesized (Figure 2.3) [11-14]. The attachment of mesogens can occur *via* covalent or non-covalent interactions [15-16].

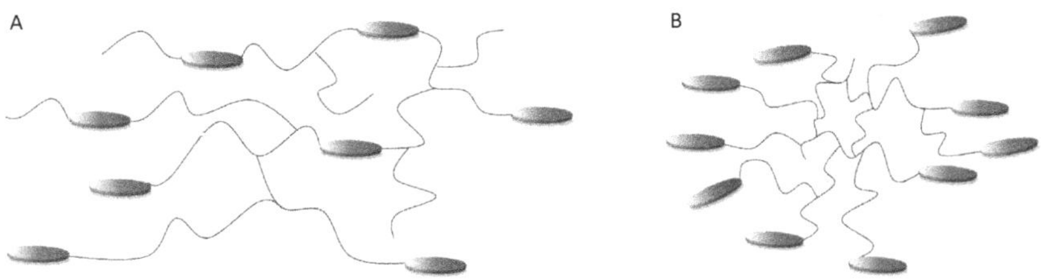

Figure 2.3 Schematic representation of the two general classes of liquid crystal hyperbranched polymers; main-chain (A) and side-chain (B).

Therefore, this chapter will focus on the use of liquid crystalline polymers (LCPs) or elastomers (LCEs) in order to obtain intimate structures of NPs and LCs. We will first describe the doping of NPs in LC media. Secondly, the direct growth within a LC matrix will be described. This overview will, therefore, highlight the various strategies to incorporate different types of NPs, e.g. metallic or metal oxides, in different types of LCs.

2.2 Stabilization of Preformed Nanoparticles

Some practical applications of nanoparticles, i.e. solar cells, optical sensing devices or actuation can take advantage from the collective behavior of ordered NPs. In this context, LCP are organized materials of significant interest to achieve such a goal. Nevertheless, doping LCP or LCE with preformed NPs to obtain hybrid LCPs-NPs composites can be complicated by several challenges. First, addition of nanoparticles with controlled size and shape as well as narrow dispersity is necessary. Another crucial factor of this approach is the homogeneity of the NPs dispersion in the LC matrix as NP-NP interactions should be at least partially counter-balanced by NP-polymer interactions. This can be achieved by using mesogens on the polymer chain with functional groups able to interact with NPs or surrounding the NPs with liquid crystalline ligands. This is true for all types of NPs. The NP choice usually depends on the desired application, the LCP is then accordingly tailored.

2.2.1 Critical Role of LC/NP Interactions

CdSe spherical quantum dots (QDs) [17] and CdS nanorods [18] were successfully redispersed inside a side-chain liquid crystal polymer by using poly[4-(n-acryloyloxyalkoxy)]benzoic acid with a smectic C-type phase. Alkoxybenzoic acid functionalities ensure in both cases the interaction with the inorganic fillers. This interaction is clearly shown for CdSe by infrared analysis: upon increasing the concentration of CdSe, the intensities of bands related to the carbonyl group ν(CO) in the H-bonded dimers decreased, while new bands relative to carboxylate anions appeared. A change in the interlayer spacing due to the presence of CdSe NPs was also evidenced by X-ray scattering measurement. From these results, a composite nanostructure where polymer layers alternated with QDs highly charged layers was suggested (Figure 2.4a, b). The study of the optical properties of these materials also confirmed the good dispersion of the NPs inside the LC matrix [19]. Thermal behavior of the LC composite was also modified by the presence of CdSe NPs: for up to 20 wt% QDs content, the LC-Isotropic transition and associated transition enthalpy slightly decreased. At higher content (i.e. 40 wt%), the LC transition disappeared. The NPs content in a LC matrix is a limiting factor, so it must be carefully chosen in order to preserve the liquid crystalline properties of the composite materials.

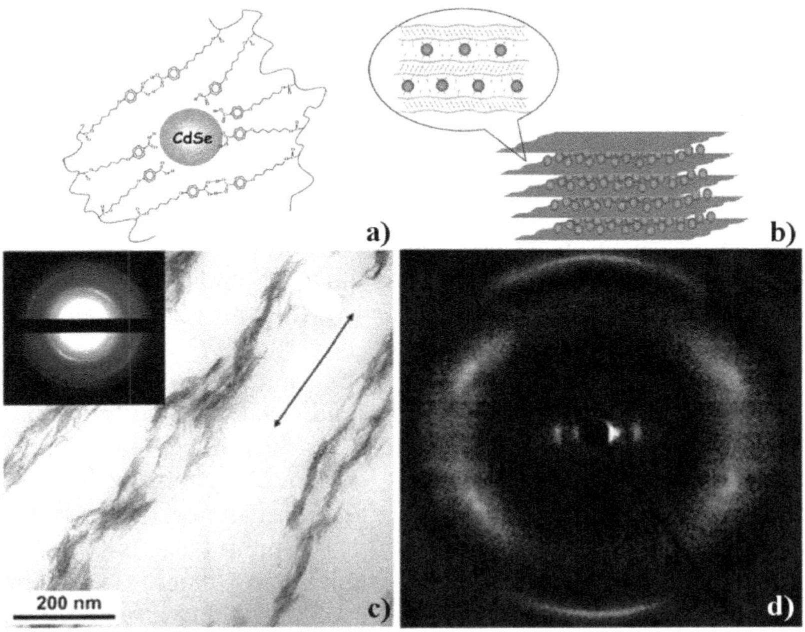

Figure 2.4 Schematic representation of a) a side-chain LC polymer stabilizing CdSe NPs via alkoxy benzoic functionalities and b) the localization of CdSe between polymer nanolayers. Reproduced from Reference 17 with permission from American Chemical Society. c) TEM images and d) X-ray pattern (right) of uniaxially oriented films of the liquid crystal polymer with CdS nanorods. Reproduced from Reference 18 with permission from American Chemical Society.

The case of the CdS nanorods described by the same authors is less straightforward [18]. The IR spectra of the composite showed no sign of the presence of the ligands used for the nanorods synthesis and that the number of H-bonded dimers in the composite decreased: functional groups of the polymer replaced ligands on the surface of the NPs. X-ray analysis showed a change of 0.5 nm in the interlayer spacing of the material once nanorods were incorporated which is less than expected considering the size of inserted nanorods. Moreover, the LC thermal behavior was only slightly influenced by the presence of the nanorods. This suggested weak interactions between the nanorods and the LC matrix. This was first confirmed by TEM images showing 100-200 nm low density domains of nanorods evenly distributed in the matrix. Interestingly, the uniaxial stretching of the composite films caused the formation of wire-like structures oriented along the deformation (Figure 2.4c), thus, confirming an interaction between the nanorods and the mesogens. These interactions were confirmed by replacing the LC polymer by an amorphous material made of norbornene-methyl methacrylate copolymer without functional groups. Indeed, a significant phase separation with larger and much denser domains was observed in the latter case. Once again, successful dispersion of the nanorods in the LC matrix was proved by the optical properties of the composites: the CdS nanorods emission peak in the LC matrix and the peak of a homogeneous solution were nearly the same, whereas it was red-shifted in the amorphous matrix.

In previous examples, efficient dispersion was obtained due to specific NP-LC interactions and the liquid crystal character helps to organize the inorganic filler. Another critical point is, for a chosen application, to carefully adjust the NP concentration, location and order, as these directly influence the performance of the device. Hence, for hybrid photovoltaic devices, the main issue is to achieve large interfacial areas and highly ordered nanostructures in which the semiconductor NPs are ideally located. Liquid crystalline block-copolymers are well suited to favour oriented ordered domain formation. Besides, a high NP content is necessary, to guarantee a sufficiently high current density. Li *et al.* [20] have described the dispersion of ZnO QDs in a liquid crystal donor-acceptor copolymer poly[3-(6-(cyanobiphenyloxy)thiophene)-alt-4,7-(benzothiadiazole)], P3HbpT-BTD. The composite films were prepared with a high dopant content, (1:3 wt% P3HbpT-BTD: QDs), by spin-coating and annealing of highly concentrated dichlorobenzene solutions. The authors observed a relation between the annealing conditions, composite film microstructure and device performance. In particular, films annealed while being in the mesomorphous state achieved power conversion efficiencies comparable to the best values in the literature (ca. 2%). X-ray diffraction and TEM images showed highly ordered ZnO domains and well developed interpenetrating networks if annealing was conducted in the liquid crystal state; on the contrary, films annealed while in the crystalline or isotropic state were characterized by extensive ZnO aggregation. Therefore, the enhanced charge separation and transport efficiency could be attributed to the well-ordered percolated network. To improve the performance of such device, a more intimate mixture of the liquid crystalline block-copolymer and inorganic NPs was obtained by using NPs surfaces modified by liquid crystalline ligands, 4′-hydroxy-[1,1′-biphenyl]-4-carbonitrile (cbp) [21]. The role of the liquid crystalline ligands was to guarantee a better miscibility with the polymer owing to the intermolecular interactions arising from their similar structure. The authors compared the power conversion efficiencies of hybrid solar cells formed by conjugated rod–rod diblock copolymers containing cyanobiphenyl mesogenic pendant groups, polythiophene-b-poly{3-[10-(4′-cyanobiphenyloxy)-decyl]thiophene} (P3HT-b-PTcbp), as electron donors, and by ZnO or CdS QDs as electron acceptors. All devices containing QDs with liquid crystalline ligands possessed better power energy conversion. The authors attributed the improved performances to the improved order of the active layer due to the presence of liquid crystalline moieties on the conducting polymer and to the better nanoparticle dispersion, evidenced by TEM images, when liquid crystalline cbp are present on the QDs surface.

2.2.2 Stimuli Responsive Materials

New applications involving composite materials aim at controlling different properties, such as electric polarization, magnetization and shape, either directly or indirectly by various stimuli (electric, magnetic fields or stress). In this context, LCP that exhibit an anisotropic response to a magnetic field are of peculiar interest: the mesogens assume specific orientation depending on their magnetic susceptibility. Brochard and de Gennes [22] first suggested to use magnetic particles in order to enhance this magneto-orientational response of LCs. For this purpose, liquid crystalline silicones doped with Co nanorods were recently synthesized [23-24]. Interestingly in this work, the use of rod shaped NPs interacting with LCPs bearing acidic function enhanced the anchoring efficiency of LCs on the surface of NPs and thus favoured the dispersion of Co nanorods within the polymer matrix. The microscopic structure of these composites was investigated by X-ray scattering. Even if homogeneous at a macroscopic level, inhomogeneities still existed at a microscopic level. Bundles of nanorods with a hexagonal order were observed (Figure 2.5). The strong ferromagnetic interactions between nanorods were not completely overcome even in the presence LCP bearing carboxylic acids interacting groups. Nevertheless, the lateral distances between the nanorods were larger than their diameter indicating that polymer chains surrounded them [24].

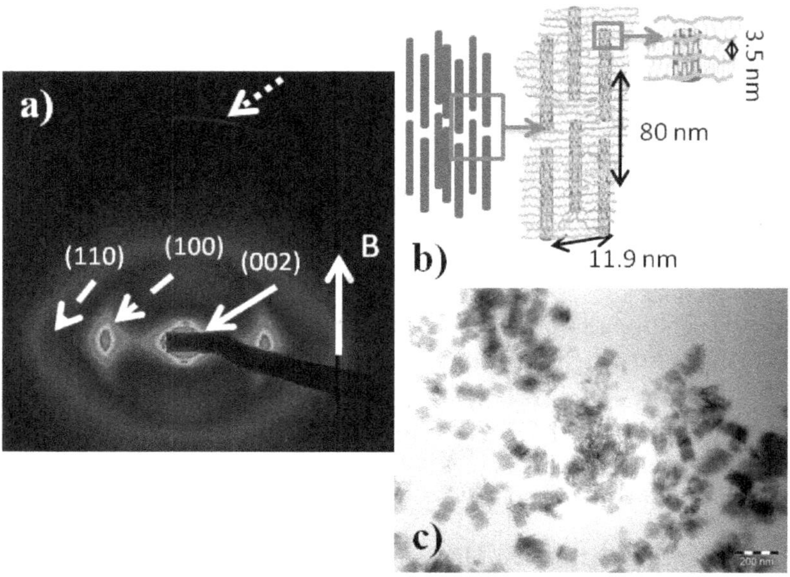

Figure 2.5 a) SAXS pattern of the polymer composite containing carboxylic interacting groups doped with 0.8 wt% Co nanorods. The solid white arrow points to the (002) reflection due to the organization of the Co nanorods in rows within the bundles whereas the dashed arrows point to the (100) and (110) reflections of the hexagonal 2-dimensional lattice. The dotted white arrow points to the smectic reflection from the mesogens; b) schematic representation of the nanorod bundles in the composite; c) typical TEM picture of a magnetic liquid crystal composite. Reproduced from Reference 24 with permission from American Chemical Society.

For these materials, the use of strong magnetic Co nanoparticles instead of weak magnetic iron oxide NPs may guarantee higher torque with smaller NPs. These materials possessed both liquid crystalline and magnetic properties, the samples being ferromagnetic in the smectic state. The almost square hysteresis curves of those materials proved that cobalt nanorods could be efficiently aligned in a magnetic field (3T) during cooling down from the isotropic state to the smectic phase. An effect due to the anisotropic matrix, which was also sensitive to a magnetic field, was observed: a quite low magnetic field (50 mT) was sufficient to orient the nanorods when these were dispersed in the LCP compared to an amorphous matrix. This behavior was interpreted as a possible signature of cooperative effects between the LCP and the magnetic nanorods [23]. The orientational coupling between liquid crystalline groups and nanorods was further investigated. For this purpose, different composites were compared, with and without groups interacting with the nanorods by measuring the nematic order parameter of the mesomorphous groups under a magnetic field. The polymer containing carboxylic groups was more easily oriented in the presence of magnetic nanorods at low magnetic field, while the effect was low, almost inexistent, when no ligand was present on the liquid crystalline matrix [24]. This underlines, as in the previous section, the critical role of NP/LC interaction in order to obtain materials with optimal properties.

Aimed at obtaining a magneto-responsive materials, liquid crystalline elastomers (LCEs) have been doped with magnetic NPs. In this material, the inorganic and component sensitive to magnetic field should act on the elastic and deformable matrix producing phenomena such as shape change or tuneable and anisotropic elastic modulus. Vice versa, the elastic force acting on the NPs could orient them and change the magnetic properties of the composite material with respect to the free NPs. These different approaches were described by various research studies [25-31]. For instance, Song *et al.* [25] synthesized a liquid crystalline polysiloxane elastomer with covalently bonded dopamine anchored ferrite NPs with an average diameter of 6 nm. LC transparent films were obtained by spin-coating. Annealing the film in presence of a 0.35 T magnetic field led to the formation of a LCEs with a higher transparency with a nematic order parameter measured at 0.42. The LC films showed super-paramagnetic behavior at 300 K. This example proves that the magnetic NPs do not hamper the elastomer from retaining its liquid crystalline properties and the mesogens orientation inside the matrix. The composites contained 7 wt% of ferrite NPs. Based on the same idea where the magnetic NPs are directly linked to the elastomer matrix and play the role of crosslinking agents, a main chain LCE doped with iron oxide NPs was obtained by Garcia-Marquez *et al.* [27]. Main chain LCE was chosen as these can undergo more important shape changes than their comb-like homologues and was synthesized *via* a one-step reaction, by mixing the di-olefinic mesogen with the hexamethyl trisiloxane spacer and 3.3 nm ferrite

NPs covered with 4-(undec-10-enyloxy) benzoic acid as cross-linkers. The synthesized composites were mesomorphous and presented a magnetic behavior. Similarly, Riou *et al.* [26] described the synthesis (*in-situ* and *ex-situ*) and magnetic characterization of a LCE embedding Co nanorods NPs. [26] Despite their overall interest, the properties of the previous systems were hampered by the fact that samples were obtained as polydomain or poorly macroscopically oriented samples. In order to prove that dispersing magnetic particles within LCE is an efficient way to induce a macroscopic magnetic reorganization and the associated shape and magnetic memory features, the synthesis of monodomain samples is a prerequisite.

In order to achieve a shape change by the wireless action of a magnetic field, a LCE possessing a reversible magnetic memory was synthetized by Haberl *et al.* [28-29]. The authors prepared a nanocomposite with mutually crosslinked ellipsoidal silica coated iron oxide nanoparticles and LCE to form a covalent network. Different composites with NPs content ranging from 0.5 to 10 wt% (named E0.5-E10) were synthesized and the structural, mechanical and magnetic properties were analyzed. After uniaxial deformation, anisotropy was found for the magnetic susceptibility, magnetic coercivity and remnant magnetization. From X-ray diffraction experiments, it appeared that these properties were correlated with the particle shape and their orientation distribution. In tensile tests, the effect of particles on the modulus of the material was evaluated and three different regions were found. The authors concluded that the most effective modulus increase occurred in the softening region of the nanocomposite [28].

For these examples, the samples are strained in the isotropic phase at 80 °C and subsequently cooled down into the liquid crystalline phase in order to achieve an anisotropic soft magnet. Heating the sample brings it back to its initial disordered state. Nevertheless, these examples require crossing the liquid crystalline to isotropic transition, that is, the use of thermal energy to exploit the magneto-elastic coupling. Therefore, these do not take advantage of the full potential of the liquid crystalline state.

Another approach for inducing wireless shape change is to exploit the local heating induced by an alternating magnetic field in the vicinity of magnetic nanoparticles. Kaiser *et al.* [30] prepared magnetic composites based on the monodomain LCE matrices and magnetite spherical nanoparticles. Upon exposure to the alternating magnetic field, the inductive local heating induce a nematic/isotropic transition and led to a 27% reversible shape change [30]. In this example, hydrophobic N-oleoylsarcosine stabilized NPs, compatible with the organic medium, were used and no special care to guarantee interactions with the liquid crystalline groups was necessary. The monodomain elastomer (order parameter of 0.63 as deduced from X-ray experiments) was obtained through a two-step crosslinking procedure. The sample reversible contraction after application of an alternating magnetic field increased with increasing nanoparticle concentration up to a maximum of 74% in the presence of 1.64% NPs (volume fraction). Furthermore, the actuation depended on the magnetic field intensity, which can be subsequently modified in order to adjust the final dimensions.

Riou *et al.* [31] proposed an alternative novel strategy in which the system remains in the liquid crystalline state at room temperature, exploiting its ability of orienting/disorienting under mechanical stress to switch the orientation of magnetic nanorods and, therefore, the magnetic properties of the composite. Dispersion of preformed monodisperse cobalt nanorods into a liquid crystalline polymer was followed by alignment and crosslinking in the presence of an external magnetic field (Figure 2.6). The magnetic liquid crystalline elastomers exhibited a nematic phase with clearing and glass transition temperatures around 90 °C and 0 °C, respectively. The influence of mechanical deformation, applied at room temperature, on the structural and magnetic behavior of both non-oriented and oriented elastomers was then studied and remarkable magneto-elastic coupling was observed on these highly anisotropic magnetic materials: the nanorod alignment can be switched at will at room temperature by weak mechanical stress, leading to a change of more than 50% of the remnant magnetization ratio and of the coercive field.

Beyond the synthesis of magnetic composite, doping monodomain LCE with other inorganic NPs enable the formation of hybrid materials with improved properties. For instance, Montazani *et al.* [32] incorporated 3 nm gold NPs in a monodomain LCE in concentrations up to 0.1 mol% [32]. Doping stiffened the elastomer, but the actuator strain rate was significantly improved at higher nanoparticle concentrations, due to the enhanced thermal conductivity and by consequence to the more efficient heat transfer to the matrix. Domenici *et al.* [33-34] successfully created a conductive layer of $PbTiO_3$ or molybdenum based nanowires in monodomain LCEs. The elastomers retained the typical shape memory property of monodomain LCEs, even though their maximum elongation was slightly lower than in the absence of nanoparticles. Despite the fact that the preparation method was the same, spherical NPs and nanowires were differently aligned inside the elastomer. The possibility of electromechanical actuation of these materials was suggested and in the case of ferroelectric nanoparticles another mechanism for actuation, i.e. by applying external voltage perpendicularly to the thin layer of nanoparticles, was proposed.

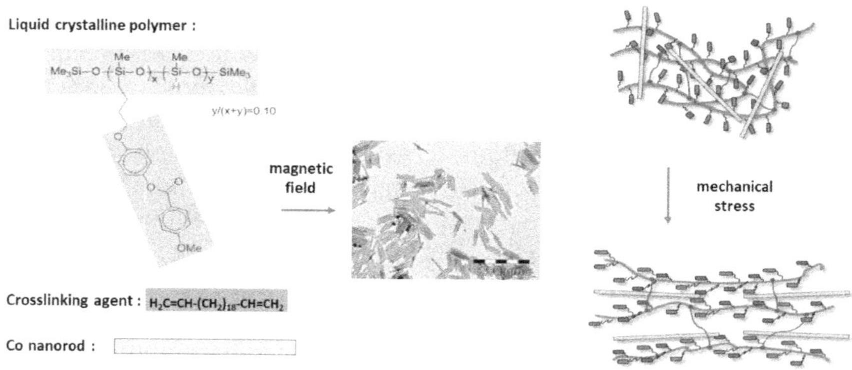

Figure 2.6 Chemical formula of the liquid crystalline polymer and the crosslinker, and TEM image of the Co nanorods used for the synthesis of the hybrid material. Right: Schematic representation of the resulting material with strain-induced tunable orientation of the magnetic nanorods (adapted from reference [31]).

2.3 In-situ Synthesis of Liquid Crystal/Nanoparticle Hybrid Materials

The potential of inorganic nanoparticles with liquid crystalline organic coronas has generated significant research efforts focusing on controlling both the NP core composition, size and shape and the nature of the LC organic corona. The possibility of growing NPs directly in presence of LCP instead of using preformed particles present several advantages, the first one being to avoid the presence of additional chemical reagents issued from the pristine NPs and, thus, avoiding tedious purification process. The synthesis can be performed in presence of a suitable solvent, subsequently generating the LC composite after complete removal of the solvent (Section 2.3.1). Of greater interest is the synthesis of NPs in the absence of solvent. This second strategy enables to take benefit of their organized structure and to use them as nanoreactors for synthesizing nanoparticles with controlled size and shape. Though lyotropic phases have been extensively used to obtain such a template effect, [35,36], thermotropic phases are seldomly described in literature (Sections 2.3.2 and 2.3.3).

2.3.1 Formation of Hybrid through Solvent Mediated Methods

Numerous synthetic methods for the formation of NPs in solution with controlled size and functionalization are described in literature. Nevertheless, these often present a lack of flexibility and control on NP parameter like core size. The NP shape is often obtained through limitative growth conditions that necessitate the use of specific solvent, reducing and/or functionalizing agent. Therefore, using a chosen LC compound as a stabilizing agent often hampered the synthesis of NPs with controlled size and shape. In this context, few examples have described synthetic process in solution using directly LCP as stabilizing agent. Zadoina *et al.* [37] adapted the synthesis of Co NPs starting from bis(bis(trimethylsilyl)amido)cobalt(II), $Co[N(SiMe_3)_2]_2$, in the presence of a siloxane based LCP bearing a carboxylic acid side-chain groups and hexadecyl amine using different ratio between the different reactants. The precursor decomposition was carried under 3 bar H_2 in toluene solution. Spherical and/or rod-like Co NPs were obtained. For a low content of acid ligand, anisotropic NPs started to form and a majority of nanorods formed when hexadecyl amine ligand was added in the solution at Co:acid ligand:amine ligand molar ratio 1:0.16:1. After solvent removal, all NPs were well dispersed inside the matrix which retained its mesomorphous character. The material demonstrated expected ferromagnetic properties. By using a similar strategy, superparamagnetic γ-Fe_2O_3 NPs have been prepared by oxidation in THF solution of a second generation polypropylene imine LC dendrimer – iron (II) complex. NPs with an average diameter of about 2.5 nm were obtained. After removal of the solvent, a liquid crystalline columnar phase was identified by X-ray diffractometry measurements. The electron magnetic resonance spectra of the NPs within the LC phase were compared to the one of bulk γ-Fe_2O_3. It was demonstrated that LC NPs possessed enhanced effective magnetic and uniaxial anisotropy induced by the particle surface and shape effect compared to bulk NPs.

2.3.2 In-situ Formation within a Thermotropic Phase

As stated above, LCP exhibit an organized structure that can be of prime interest for synthesizing nanoparticles with controlled size and shape. However, as discussed in Section 2.1, the direct synthesis of NPs in a thermotropic LC medium is hampered by different challenges. First, the addition of the NP precursor upon the liquid crystalline phase can disrupt the LC phase. Moreover, the *in-situ* process to generate NPs within a LC medium in a homogeneous way is not an easy task. Most studies involve the *in-situ* reduction of metal precursors *via* oxidation of the LC medium in order to obtain the desired NPs. Sputtering or electrodeposition techniques were also used. Alternatively the diffusion of small molecules in a gas phase (water, dihydrogen) have been also reported. Indeed, the use of gas phase enables to minimize the disruption of the liquid crystalline phase. Lastly, if the interactions between the formed NPs and the LC phase are poor, segregation of the formed NPs may occur. Once again the specificity of LCP towards small LCs molecules (high viscosity medium and viscoelastic properties lead to higher interaction with NPs surface, etc.) made them material of choice to obtain controlled growth process, as illustrated in the following examples.

Lee *et al.* [38] were the first to report the growth of metallic NPs inside a thermotropic LC based on LCP. A polyester matrix polymer consisting of hydrophilic PEG branches attached to a hydrophobic aromatic polyester backbone was used for this purpose (Figure 2.7). The structure tended to be organized in smectic-like domains with the PEG branches occupying the inter-layer domains. Subsequently, an aqueous solution of $HAuCl_4$ was allowed to diffuse into the hydrophilic domains. Then, the gold ions were reduced to gold NPs upon addition of a hydrazine solution, as observed in the TEM micrographs with a diameter in the 8-14 nm range. The mesomorphous properties of the LC/NP hybrid were not discussed. In this first example, the formation of gold necessitates the addition of a limited quantity of solvent (here water) to the liquid crystal materials in order to introduce the NP precursor and the reducing agent. Therefore, a partial disruption of the mesomorphic order at localized sites during the nanocomposite preparation should occur. To solve this problem, Barmatov *et al.* [39] used a different strategy:

metal precursor was directly reduced by the LC matrix. LC polymer with cyano- or methoxy-biphenylmesogenic side-fragments was used in this study (Figure 2.7). A silver-olefin complex was first mixed with a solution of polymer in THF and the solvent was then evaporated. After heating, Ag^{+} ions were reduced by olefin ligands, inducing the formation of isotropic Ag NPs with sizes in the 5-30 nm

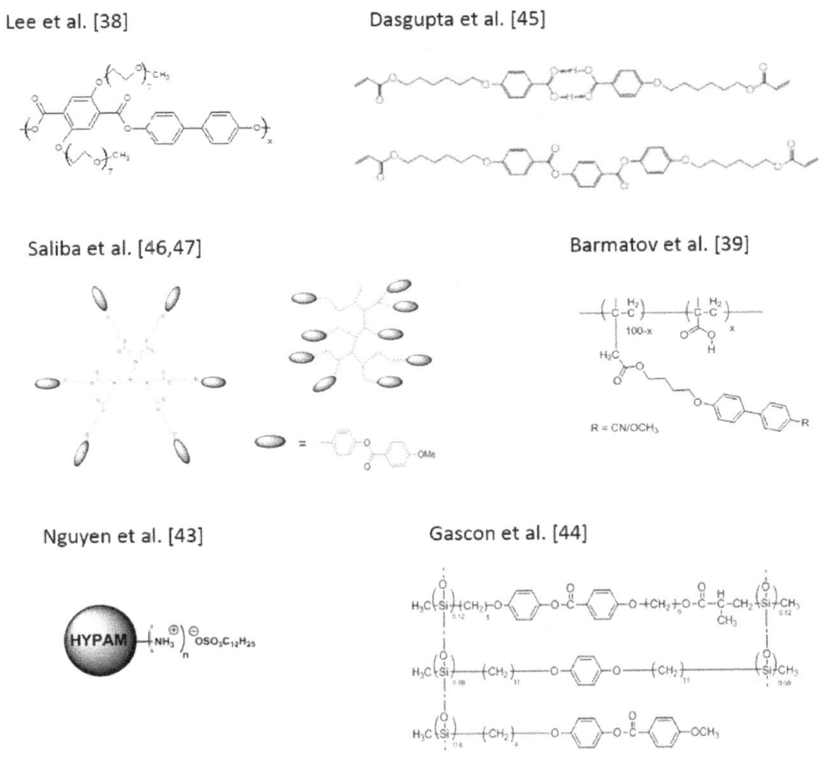

Figure 2.7 Examples of liquid crystalline polymer structures used to perform the *in-situ* synthesis of NPs (adapted from References [38-39] and [43-47]).

range. Since NPs presented large size distribution, the control over the NP morphology was somehow limited. Additionally, the mesomorphous behavior was strongly disturbed by the presence of NPs, whereas the materials based on methoxybiphenyl mesogens remained relatively stable up to a NP content of 15 wt%, the ones based on cyanobiphenyl mesogens lost the mesomorphic properties above a NP content higher than 2 wt%. This should be related to the strong interactions between terminal cyano groups and silver surfaces, which significantly disturbs the packing of the mesogenic groups and results in the rapid loss of LC behavior.

2.3.3 Towards Anisotropic Growth of Nanoparticles

The first steps towards the control of NP morphology have involved the use of mixtures of ionic liquids and NPs precursors. These mixtures possess mesomorphic properties and lead after reduction, to gold, copper or silver nanoparticles [40-41]. While nanospheres were obtained under isotropic conditions, anisotropic structures (copper platelets, leaf-like Au nanostructures) were obtained in a LC state. Thus, even if particle size and thickness can be tuned by varying the reaction temperature, no direct relation between the LC structure and the morphology of the synthesized NPs was clearly demonstrated. A clear correlation between the LC phase and the NP size and shape can be found in a study by Mallia *et al.* [42]. The system involved amphiphilic low-molecular-weight mesogens, which were capable of reducing the metallic salt without the need of an external reducing agent. The gold ions interacted with the ammonium groups in the liquid crystal molecule. The size and shape control was achieved by either varying the concentration of the gold chloride precursor or by changing the organization of LC groups.

In this context, liquid crystalline polymers/elastomers were demonstrated to be the materials of choice in order to gain control over the morphology of the *in-situ* generated NPs. Nguyen *et al.* [43] used branched thermotropic liquid crystals obtained from the ionic interactions between hyperbranched polyamidoamine of different molar masses and a surfactant sodium dodecyl sulfate (Figure 2.7). The ionic nature of these mesophases allowed enhancing the interaction of the salt precursor with the liquid crystalline phases to obtain homogeneous materials. These ionic complexes presented columnar and lamellar phases and these phases were used to generate gold NPs *in-situ*. To obtain gold NPs, a flux of dihydrogen was then introduced to ensure the reduction of the $HAuCl_4$ salt. The templating effect of the mesophase resulted in the formation of NPs with a spherical shape. As illustrated in Figure 2.8, the size was dictated by the local organization of the mesophase (as already observed by Mallia *et al.* [42]), but also by the average molar mass of the hyperbranched complex.

Gascon *et al.* [44] demonstrated that not only the liquid crystalline nature can have a strong influence on the nanoparticle size, but also the local organization of a given mesophase. For this, liquid crystalline elastomers, based on side chain polysiloxane (Figure 2.7) with the same chemical structure but different macroscopic organizations (polydomain and monodomain) were synthesized. In the smectic phase of a monodomain polymer, Au NPs of 3 nm diameter were obtained, whereas larger Au NPs (50 nm) were observed when

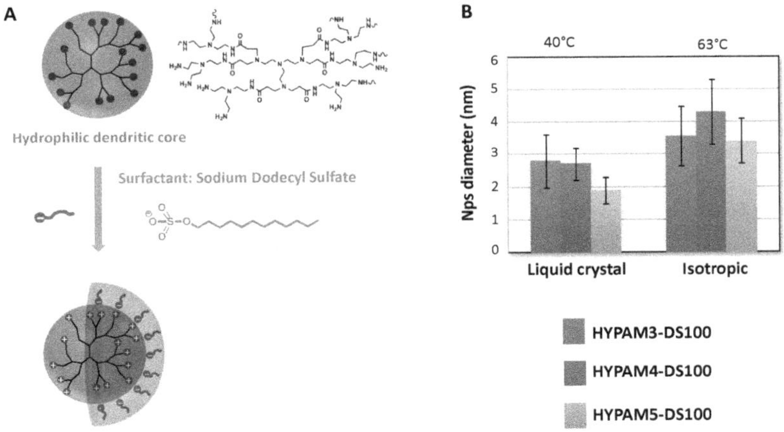

Figure 2.8 Controlled growth of gold nanoparticle within in liquid crystalline medium. (A). Formation of hyperbranched polyamidoamine-dodecyl sulfate complexes (HYPAM-DS). For their formation, three hyperbranched polymers with increasing molar masses (HYPAM 3, HYPAM4 and HYPAM 5) were complexed with dodecyl sulfate (with a 100% of amine function complexed) and named HYPAM3-DS100, HYPAM4-DS100 and HYPAM5-DS100 (B). Average diameter of gold nanoparticles synthesized within the HYPAM-DS polymers as a function a) of the polymer organization (lamellar liquid crystal at 40°C or isotropic at 63°C) and b) of the molar mass of the hyperbranched complex (adapted from Reference [43]).

the same experiment was performed in the isotropic state. The observed differences may be related to the formation of NPs within the structural defects of the LC polydomain samples due to their intrinsic nature, presenting more defects than monodomain materials. More recently Dasgupta *et al.* [45] synthesized nanoporous smectic liquid crystalline polymer networks (Figure 2.7) and used them as a template for the photochemical synthesis of silver nanoparticles. The authors showed that the diameter of the NPs was controlled by the length of the crosslinking agent used during synthesis. Indeed, the length allowed controlling the periodicity of the layered morphology and therefore of the *in-situ* generated silver nanoparticles. These materials have been used to fabricate patternable organic/inorganic hybrid materials.

The control of NP growth within the LC mesophase has been presented by Saliba *et al.* [46,47]. This approach involved the *in-situ* hydrolysis of dicyclohexylzinc, [Zn(Cy_2)] leading to ZnO NPs with volatile cyclohexane as the unique side product. To favor interactions between the LC and [Zn(Cy_2)] or ZnO NPs, LC compounds having an oligomeric or a polymeric structure containing various amine and amide chemical functions were used (Figure 2.7 and Figure 2.9). These interactions avoided the disruption of the LC organization and, therefore, increased the templating effect of the LC phase. Two different types of LCs which differed by their backbone type were

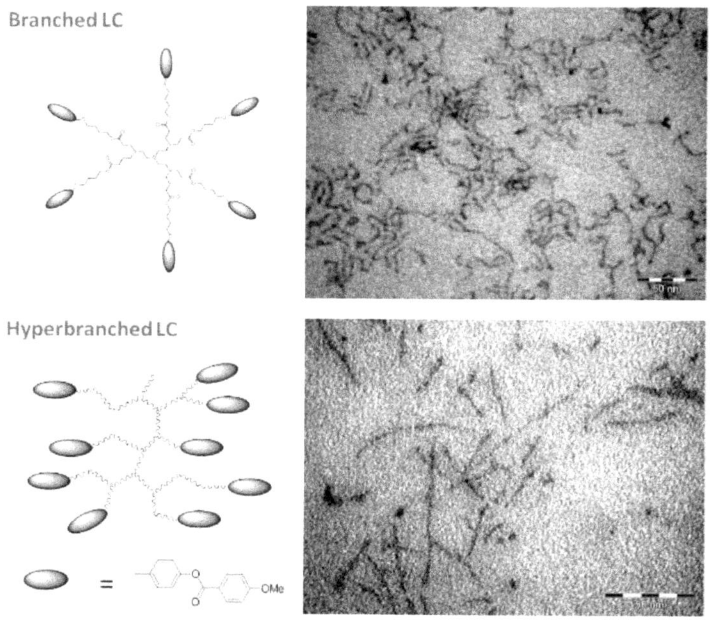

Figure 2.9 Anisotropic growth of ZnO nanoparticles within a liquid crystalline medium. TEM micrographs of ZnO NPs synthesized in a liquid crystalline state in presence of a branched LC (up, scale bar = 50 nm) or in hyperbranched LC (down, scale bar = 50 nm) (adapted from References [46-47]).

investigated (Figure 2.9). Both targeted LCs exhibited a nematic phase close to- or at room temperatures. Under isotropic conditions, hydrolysis of the $[Zn(Cy)_2]$ precursor led to isotropic NPs with an average size of the NPs being around 5.4 ± 0.7 nm and 2.7 ± 0.3 nm, respectively.

As shown in Figure 2.9, when the experiments were performed in the nematic phase state (*i.e.* at lower temperatures), anisotropic ZnO nano-worm-like or nano-wire structures were obtained (respectively in branched LC and hyperbranched LC). These exhibited an average width of 2.5 ± 0.2 nm and 2.7 ± 0.4 nm, respectively. The lengths varied from a few nanometers up to around 100 nm, when the branched LC was used, while lengths from ca. 10 to 200 nm were obtained, when the hyperbranched LC was used. Thus, a direct correlation between the morphology of the nanostructures and the structural characteristics of the LC was demonstrated. From this, in order to tailor the NP morphology using a LC phase, different criteria should be fulfilled. Primarily, the chemical reaction leading to the NPs should not disrupt the LC organization. Thus, the LC molecules should not play the role of reactants and side products should be avoided as much as possible throughout the NP formation. Next, interactions between the LC molecules, NP precursor and possibly synthesized NPs should favour the templating effect of the LC phase. Finally, the use of relatively high viscosity LCs should prevent a fast disruption of the organization during the NP formation.

2.4 Conclusions and Future Perspectives

Due to their specific properties (organized medium, viscoelastic properties, etc.), liquid crystalline polymers or elastomers are materials of specific interest for the stabilization or *in-situ* synthesis of nanoparticles. Nevertheless having access to homogeneous hybrid LCP/NPs materials requires designing of LCP bearing specific moieties which enable the interaction with NPs and favor the stability of NPs within a LC medium. Moreover liquid crystalline polymer can present different mesomorphic organizations (nematic, lamellar, columnar, etc.) which make them perfect candidates for the organization of the NPs within the LC structure in a precise and controlled fashion. Therefore, numerous examples have described the use of LCP to stabilize preformed NPs of perfectly controlled size and shape. Incorporation of NPs (for instance with magnetic properties) within the LC structure enables to obtain materials presenting, for instance, magneto-elastic coupling. Alternatively, *in-situ* strategies have also been used with LCP. In these cases, the difficulty to perform chemistry in a LC medium without disrupting the LC order during the synthesis process poses a significant challenge. Different strategies have been used to circumvent this problem (use of gas reactant, direct reduction of LC, etc.). Interestingly, the *in-situ* strategy enables to achieve in some cases a control over NPs growth (size or/and shape) and first examples of anisotropic growth have, thus, been demonstrated. In the latter case, the viscoelastic properties of the LCP are of prime importance to obtain such a control which is not obtained by using small LC molecules. Further studies will take advantage of the different kinds of organization that LCP can generate (columnar, smectic, etc.), thus, leading to the generation of commercially relevant materials.

References

1. Bisoyi, H. K., and Kumar, S. (2011) Liquid-crystal nanoscience: an emerging avenue of soft self-assembly. *Chemical Society Reviews*, **40**, 306-319.
2. Hegmann, T., Qi, H., and Marx, V. M. (2007) Nanoparticles in liquid crystals: synthesis, self-assembly, defect formation and potential applications. *Journal of Inorganic and Organometallic Polymers and Materials*, **17**, 483-508.
3. Nealon, G. L., Greget, R., Dominguez, C., Nagy, Z. T., Guillon, D., Gallani, J. L., and Donnio, B. (2012) Liquid-crystalline nanoparticles: Hybrid design and mesophase structures. *Beilstein Journal of Organic Chemistry*, **8**, 349-370.
4. Guzman, O., Abbott, N. L., and de Pablo, J. J. (2005) Quenched disorder in a liquid-crystal biosensor: adsorbed nanoparticles at confining walls. *Journal of Chemical Physics*, **122**(18):184711.
5. Cseh, L., and Mehl, G. H. (2006) The design and investigation of room temperature thermotropic nematic gold nanoparticles. *Journal of American Chemical Society*, **128**, 13376-13377.
6. Blumstein, A. (1978) *Liquid Crystalline Order in Polymers*, Academic Press, USA.
7. Ciferri, A. (1982) *Polymer Liquid Crystals*, Academic Press, USA.
8. Ebert, M., Herrmannschonherr, O., Wendorff, J. H., Ringsdorf, H., and Tschirner, P. (1988) Evidence for a biaxial nematic phase in sanidic aromatic polyamides with 1,4,7-trioxaoctyl side chains. *Makromolekulare Chemie-Rapid Communications*, **9**, 445-451.
9. Herrmannschonherr, O., Wendorff, J. H., Kreuder, W., and Ringsdorf, H. (1986) Structure of the mesophase of a discotic main

chain polymer. *Makromolekulare Chemie-Rapid Communications*, **7**(2), 97-101.

10. Ringsdorf, H., Wustefeld, R., Zerta, E., Ebert, M., and Wendorff, J. H. (1989) Induction of liquid crystalline phases: Formation of discotic systems by doping amorphous polymers with electron acceptors. *Angewandte Chemie-International Edition*, **28**, 914-918.
11. Percec, V., Chu, P. W., and Kawasumi, M. (1994) Toward "willow-like" thermotropic dendrimers. *Macromolecules*, **27**(16), 4441-4453.
12. Percec, V., and Kawasumi, M. (1992) Synthesis and characterization of a thermotropic nematic liquid crystalline dendrimeric polymer. *Macromolecules*, **25**(5), 3843-3850.
13. Bauer, S., Fischer, H., and Ringsdorf, H. (1993) Highly branched liquid crystalline polymers with chiral terminal groups. *Angewandte Chemie-International Edition,* **32***(11)*, 1589-1592.
14. Sunder, A., Quincy, M. F., Mulhaupt, R., and Frey, H. (1999) Hyperbranched polyether polyols with liquid crystalline properties. *Angewandte Chemie-International Edition,* **38**(19), 2928-2930.
15. Marcos, M., Alcala, R., Barbera, J., Romero, P., Sanchez, C., and Serrano, J. L. (2008) Photosensitive ionic nematic liquid crystalline complexes based on dendrimers and hyperbranched polymers and a cyanoazobenzene carboxylic acid. *Chemistry of Materials*, **20**, 5209-5217.
16. Felekis, T., Tziveleka, L., Tsiourvas, D., and Paleos, C. M. (2005) Liquid crystals derived from hydrogen-bonded supramolecular complexes of pyridinylated hyperbranched polyglycerol and cholesterol-based carboxylic acids. *Macromolecules*, **38**(5), 1705-1710.
17. Shandryuk, G. A., Matukhina, E. V., Vasil'ev, R. B., Rebrov, A., Bondarenko, G. N., Merekalov, A. S., Gas'kov, A. M., and Talroze, R. V. (2008) Effect of H-bonded liquid crystal polymers on CdSe quantum dot alignment within nanocomposite. *Macromolecules*, **41**, 2178-2185.
18. Ezhov, A. A., Shandryuk, G. A., Bondarenko, G. N., Merekalov, A. S., Abramchuk, S. S., Shatalova, A. M., Manna, P., Zubarev, E., and Talroze, R. V. (2011) Liquid-crystalline polymer composites with CdS nanorods: Structure and optical properties. *Langmuir*, **27**, 13353-13360.
19. Vasilets, V. N., Savenkov, G. N., Merekalov, A. S., Shandryuk, G. A., Shatalova, A. M., and Tal'roze, R. V. (2011) Immobilization of quantum dots of cadmium selenide on the matrix of a graft liquid-crystalline polymer. *Polymer Science Series A,* **53**, 521-526.
20. Li, F., Chen, W., and Chen, Y. W. (2012) Mesogen induced self-assembly for hybrid bulk heterojunction solar cells based on a liquid crystal D-A copolymer and ZnO nanocrystals. *Journal of Materi-*

als Chemistry, **22**, 6259-6266.

21. Shi, Y. Q., Li, F., Tan, L. C., and Chen, Y. W. (2013) Hybrid bulk heterojunction solar cells based on the cooperative interaction of liquid crystals within quantum dots and diblock copolymers. *ACS Applied Materials & Interfaces*, **5**, 11692-11702.
22. Brochard, F., and de Gennes, P.G. (1970) Theory of magnetic suspensions in liquid crystals. *Jornal de Physique*, **31**, 691-708.
23. Zadoina, L., Lonetti, B., Soulantica, K., Mingotaud, A. F., Respaud, M., Chaudret, B., and Mauzac, M. (2009) Liquid crystalline magnetic materials. *Journal of Materials Chemistry*, **19**, 8075-8078.
24. Riou, O., Lonetti, B., Davidson, P., Tan, R. P., Cormary, B., Mingotaud, A. F., Di Cola, E., Respaud, M., Chaudret, B., Soulantica, K., and Mauzac M. (2014) Liquid crystalline polymer-Co nanorod hybrids: Structural analysis and response to a magnetic field. *Journal of Physical Chemistry B*, **118**, 3218-3225.
25. Song, H. M., Kim, J. C., Hong, J. H., Lee, Y. B., Choi, J., Lee, J. I., Kim, W. S., Kim, J. H., and Hur, N. H. (2007) Magnetic and transparent composites by linking liquid crystals to ferrite nanoparticles through covalent networks. *Advanced Functional Materials*, **17**, 2070-2076.
26. Riou, O., Zadoina, L., Lonetti, B., Soulantica, K., Mingotaud, A. F., Respaud, M., Chaudret, B., and Mauzac, M. (2012) In situ and ex situ syntheses of magnetic liquid crystalline materials: A comparison. *Polymers*, **4**, 448-462.
27. Garcia-Marquez, A., Demortiere, A., Heinrich, B., Guillon, D., Begin-Colin, S., and Donnio, B. (2011) Iron oxide nanoparticle-containing main-chain liquid crystalline elastomer: towards soft magnetoactive networks. *Journal of Materials Chemistry*, **21**, 8994-8996.
28. Haberl, J. M., Sanchez-Ferrer, A., Mihut, A. M., Dietsch, H., Hirt, A. M., and Mezzenga, R. (2013) Strain-induced macroscopic magnetic anisotropy from smectic liquid-crystalline elastomer-maghemite nanoparticle hybrid nanocomposites. *Nanoscale*, **5**, 5539-5548.
29. Haberl, J. M., Sanchez-Ferrer, A., Mihut, A. M., Dietsch, H., Hirt, A. M., and Mezzenga. R. (2013) Liquid-crystalline elastomer-nanoparticle hybrids with reversible switch of magnetic memory. *Advanced Materials*, **25**, 1787-1791.
30. Kaiser, A., Winkler, M., Krause, S., Finkelmann, H., and Schmidt, A. M. (2009) Magnetoactive liquid crystal elastomer nanocomposites. *Journal of Materials Chemistry*, **19**, 538-543.
31. Riou, O., Lonetti, B., Tan, R. P., Harmel, J., Soulantica, K., Davidson, P., Mingotaud, A. F., Respaud, M., Chaudret, B., and Mauzac, M. (2015) Room-temperature, strain-tunable orientation of magnetization in a hybrid ferromagnetic Co nanorod–liquid crystalline elastomer nanocomposite. *Angewandte Chemie-International Edition*, **54**, 10811-10815.
32. Montazami, R., Spillmann, C. M., Naciri, J., and Ratna, B. R. (2012)

Enhanced thermomechanical properties of a nematic liquid crystal elastomer doped with gold nanoparticles. *Sensors and Actuators A: Physical*, **178**, 175-178.

33. Domenici, V., Zupancic, B., Laguta, V. V., Belous, A. G., V'Yunov, O. I., Remskar, M., and Zalar, B. (2010) $PbTiO_3$ nanoparticles embedded in a liquid crystalline elastomer matrix: Structural and ordering properties. *Journal of Physical Chemistry C*, **114**, 10782-10789.
34. Domenici, V., Conradi, M., Remskar, M., Virsek, M., Zupancic, B., Mrzel, A., Chambers, M., and Zalar, B. (2011) New composite films based on MoO_3-x nanowires aligned in a liquid single crystal elastomer matrix. *Journal of Materials Science*, **46**, 3639-3645.
35. Saliba, S., Mingotaud, C., Kahn, M. L., and Marty, J. D. (2013) Liquid crystalline thermotropic and lyotropic nanohybrids. *Nanoscale*, **5**, 6641-6661.
36. Hegman, T., Qi, H., and Marx, V. M.(2007) Nanoparticles and liquid crystals: Synthesis, self-assembly, defect formation and potential applications. *Journal of Inorganic and Organometallic Polymers and Materials*, **17**, 483-508.
37. Zadoina, L., Soulantica, K., Ferrere, S., Lonetti, B., Respau, M., Mingotaud, A. F., Falqui, A., Genovese, A, Chaudret, B., and Mauzac, M. (2011) In situ synthesis of cobalt nanoparticles in functionalized liquid crystalline polymers. *Journal of Materials Chemistry*, **21**, 6988-6994.
38. Lee, J. W., and Jin, J. (2003) Formation of gold nanoparticles within a liquid crystalline polymeric matrix. *Journal of Nanoscience and Nanotechnology*, **3**, 219-221.
39. Barmatov, E. B., Pebalk, D. A., and Barmatova, M. V. (2004) Influence of silver nanoparticles on the phase behavior of side-chain liquid crystalline polymers. *Langmuir*, **20**, 10868-10871.
40. Dobbs, W., Suisse, J. M., Douce, L., and Welter R. (2006) Electrodeposition of silver particles and gold nanoparticles from ionic liquid-crystal precursors. *Angewandte Chemie-International Edition*, **45**, 4179-4182.
41. Taubert, A. (2004) CuCl nanoplatelets from an ionic liquid-crystal precursor. *Angewandte Chemie-International Edition*, **43**, 5380-5382.
42. Mallia, V. A., Vemula, P. K., John, G., Kumar, A., and Ajayan, P. M. (2007) In situ synthesis and assembly of gold nanoparticles embedded in glass-forming liquid crystals. *Angewandte Chemie-International Edition*, **46**, 3269-3274.
43. Nguyen, H. H., Valverde Serrano, C., Lavedan, P., Goudouneche, D., Mingotaud, A. F., Lauth-de Viguerie, N., and Marty, J. D. (2014) Mesomorphic ionic hyperbranched polymers: Effect of structural parameters on liquid-crystalline properties and on the formation of gold nanohybrids. *Nanoscale*, **6**, 3599-3610.

44. Gascon, I., Marty, J. D., Gharsa, T., and Mingotaud, C. (2005) Formation of gold nanoparticles in a side-chain liquid crystalline network: Influence of the structure and macroscopic order of the material. *Chemistry of Materials*, **17**, 5228-5230.
45. Dasgupta, D., Shishmanova, I. K., Ruiz-Carretero, A., Lu, K., Verhoeven, M., van Kurigen, H. P. C., Portale, G., Leclere, P., Basttiaansen, C. W. M., Broer, D. J., and Schenning, A. P. H. J. (2013) Patterned silver nanoparticles embedded in a nanoporous smectic liquid crystalline polymer network. *Journal of American Chemical Society*, **135**, 10922-10925.
46. Saliba, S., Coppel, Y., Achard, M. F., Mingotaud, C., Marty, J. D., and Kahn, M. L. (2011) Thermotropic liquid crystals as templates for anisotropic growth of nanoparticles. *Angewandte Chemie-International Edition*, **50**, 12032-12035.
47. Saliba, S., Coppel, Y., Mingotaud, C., Marty, J. D., and Kahn, M. L. (2012) ZnO/liquid crystalline nanohybrids: From solution properties to the control of the anisotropic growth of nanoparticles. *Chemistry: A European Journal*, **18**, 8084-8091.

Chapter 3

Main Chain LCPs and Their Applications in Electronics

Suhana Mohd. Said,* Muhammad Akmal Kamarudin, Nik Muhd. Jazli Nik Ibrahim and Mohamad Syafie Mahmood

Department of Electrical Engineering, Faculty of Engineering, University of Malaya, Malaysia
**Corresponding author*: smsaid@um.edu.my

3.1 Introduction

Liquid crystalline polymers (LCPs) provide interesting physical characteristics, owing to both polymeric structure and liquid crystalline mesophase, giving rise to targeted applications which exploit these special characteristics. In this chapter, applications of LCPs in the electronics sector will be discussed within the context of their fundamental physical origins and chemical properties.

Liquid crystal (LC) materials were first discovered in 1888 by Friedrich Reinitzer, while working with cholesteryl benzoate [1]. This discovery led to the birth of a new field that explores the physics and chemistry of LC materials together with the development of many applications of LCs especially in displays technology [2]. However, the field of LC polymers (LCPs) is relatively new compared to small molecule LCs. Initially, in 1940, the first type of LCPs developed were based on biological products such as DNA and polypeptides [3]. Since then, scientists have succeeded in synthesizing various structures of LCPs either by preparing main chain or side chain LCPs, of which the main chain LCPs were first to be synthesized in the form of Kevlar [4]. So far, only the main chain LCPs have found real life applications. In this chapter, we will largely focus on the main chain LCPs and their applications in electronics.

It has been known that the LC phase of the LCPs is similar to low molecular liquid crystal, in exhibiting liquid crystalline mesophases such as the nematic, smectic, cholesteric and discotic phases, as shown in Figure 3.1. The difference between low molecular weight LCs and LCPs is viscosity, formability and long-term retention of LCs.

Liquid Crystalline Polymers, edited by Vikas Mittal

LCPs are more viscous than low molecular weight LCs (viscosity is drastically lowered in LC state unlike common plastics) and they have excellent shapeability. There are two major types of polymer liquid crystals, lyotropic LC and thermotropic LC. The lyotropic LC is a multicomponent LC which is formed by dissolving in a solvent, thus, it is a substance which becomes a LC phase depending on concentration change. Research aiming at high molecular weight and high elastic modulus materials has been initiated as a result of commercialization of high strength/high modulus fiber (Kevlar) obtained by spinning lyotropic LC. Many of the tissues formed in the ecosystem also have a lyotropic LC structure and are now applied for the functionalization of cosmetics and the like.

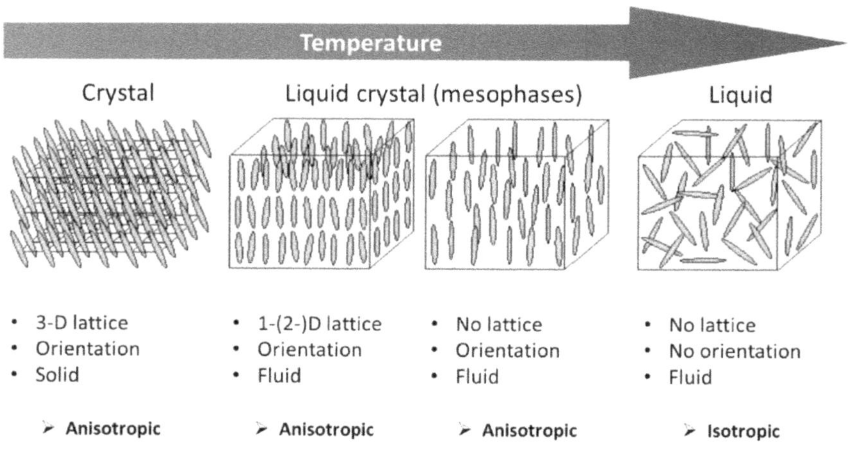

Figure 3.1 Schematic diagram of thermotropic LCs as a function of temperature.

On the other hand, a thermotropic LC is a substance which becomes a LC phase depending on a temperature change. As the temperature is raised from the crystal phase, it becomes a LC phase through a glass transition and it becomes a liquid phase when the temperature is further raised. Thermotropic LCPs are structurally composed of a main chain type liquid crystal in which aromatic atomic groups such as p-phenylene are connected in a linear form and an atomic group similar to a low molecular liquid crystal acts as a side chain group.

In order to develop various LC phases, the main chain LCPs are typically composed of a rod-shaped rigid chain and the side chain

LCPs composed of low molecular weight mesogen group showing liquid crystal phase covalently bonded to the main chain. Main chain LCPs are particularly known as high-performance LCPs with high strength and high elasticity [5-7]. The main chain type LCP realizes high strength and high elasticity by incorporating rigid mesogen molecules into the main chain and arranging the molecules by using a LC field. Typical examples are polyester showing thermotropic liquid crystallinity and aromatic polyamide showing lyotropic liquid crystallinity. The high mechanical stress is also determined by the fabrication process, for example the molding process in the molten state allows assembly of the main chain by the shear stress induced by flow. On the other hand, side chain type polymer liquid crystals are more responsive to external stimulus than main chain type high molecular liquid crystals, therefore, their application as light, electronic/electric functional materials are preferred. Side chain mesogenic groups are formed from the main chain whilst giving a characteristic property peculiar to the macromolecules such as mechanical strength and an increase in glass transition temperature. It is aimed to provide mobility similar to low molecular liquid crystals via a spacer such as a methylene chain. As mentioned earlier, the LC molecules include a rigid group (such as an aromatic ring) having a rod shape or a plate shape called a mesogenic group. In this respect, Figure 3.2 shows a comparison between the structures of main and side chain LCPs.

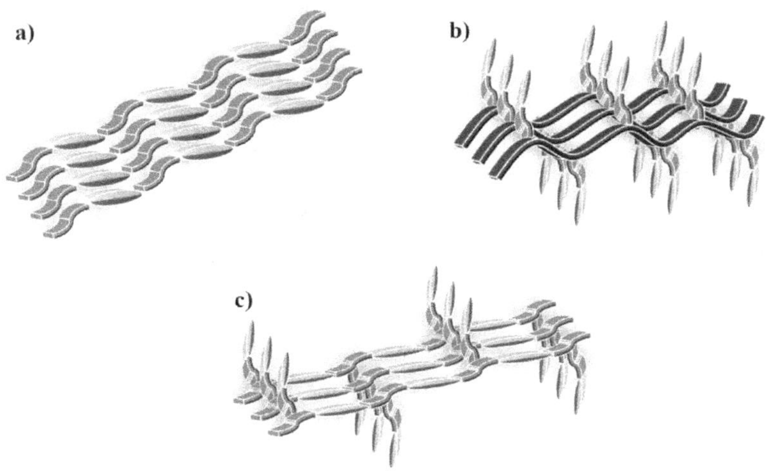

Figure 3.2 Comparison between a) main chain LCPs, b) side chain LCPs and c) combined main chain/side chain LCPs.

In the field of LCPs, molecular design allows fine tuning of the structural parameters such as main chain structure, molecular weight, spacer length, mesogen group structure, degree of substitution, copolymer composition in accordance to the desired application. The main chain structure predominantly determines the properties of the LCP as a macromolecule, whereas the side chain moiety including the spacer is largely involved in the properties of LCs. In general, if the length of the spacer is too short, it becomes difficult to form a LC phase under the influence of the main chain, and when it has a certain length, it shows a nematic phase. In addition, it tends to form a smectic LC phase with a higher order when it becomes longer. Mesogenic groups have the role of LC inducing ability due to their rigidity and are responsive to external field. For example, to increase the electric field responsiveness, a structure with large dielectric anisotropy or a structure with a large permanent dipole can be introduced in the form of photochromic group in order to impart photoresponsiveness. For the discussions relevant in this chapter, we will adopt the formalism for rod-like LCs, where the longitudinal shape of the molecule gives rise to a geometric anisotropy, which correspondingly gives rise to anisotropy of its physical parameters such as dielectric, optical and magnetic anisotropy, with respect to the properties in a direction parallel and perpendicular to the director. For example, the smectic A phase, which forms positional ordering through an orthogonal layered structure is uniaxially dielectric, with a dielectric permittivity tensor that can be resolved into a direction parallel to the LC director.

3.2 Applications of LCPs

LCPs have found widespread applications in industry due to their properties of outstanding strength, temperature resistance, dimensional stability and gas barrier. As these are fundamentally different than other polymers, new processing technology is needed in order to use LCPs in cost effective products. Composite thermoplastics containing LCPs have produced cost effective and high mechanical strength blends which can be used as polymeric films, sheet, tubes and molded articles in high barrier containers, electronic substrates and lightweight structures.

Main chain LCPs, generally composed of rigid mesogenic moieties, are especially useful as high modulus fibers, reinforcement of optical fibers or specific molded parts. On the other hand, side chain LCPs are of particular use as electroactive materials in optics and optoelectronics. A preferred director orientation can be induced by an applied electric or magnetic field in the nematic phase and then frozen into a glassy state, which is useful for nonlinear optical phenomena, such as lasing. Meanwhile, LC polyacrylates copolymer are able to produce chiral smectic C* LC polymers which are pyroelectric and potentially piezoelectric, and are particularly attractive as information storage media.

3.2.1 Electronic Applications

This chapter will focus on electronic applications of LCPs, as their electroactivity is particularly useful for their applications as functional materials in electron, ion, molecular transporting, sensory, catalytic, optical as well as bioactive materials. Their capability for self-assembly in a range of LC mesophases allow both order and mobility, at molecular, supramolecular and macroscopic levels. In addition, their ability to respond to external stimuli such as temperature, pressure and electromagnetic fields results in a macroscopic change (such as light polarization or magnetic spin), thus, making them useful as electro-optical sensors or data multiplexing/storage elements. In more recent times, LC compounds have found growing interest in biochemical systems as their self-assembly properties are able to closely support mechanisms in living cells, such as controlled drug delivery and protein binding.

Although there are similarities between small molecule LCs and LCPs, there are also vast differences in various properties which make LCPs more desirable for specific applications. Due to the polymeric structure, the elasticity and viscosity increase significantly with increase in the molecular weight, thus, the time required to realign the molecules become longer [8]. From the displays technology point of view, this is a disadvantage, but it can be exploited to other applications which require reliable self-assembly, as will be explained in this chapter. Furthermore, it has been reported that an increase in the viscosity could help to reduce structural defects in polymers through the LCP self-healing capabilities. By using an electric field, shear stress, surface orientation treatment or the like, as a means of assisting the self-assembly mechanism of LCPs, a homogeneous transparent stable LC monodomain can be easily obtained. High mechanical properties unique to macromolecules will give sufficient strength and impact resistance to the resulting LC material. Furthermore, there is an added advantage in LCPs with respect to their thermal properties. Generally, crystallization is suppressed as the molecular weight increases and the LC-isotropic transition temperature tends to rise, thus, resulting in an expanded temperature range of the LC phase through polymerization. Furthermore, the polymer has a relatively high glass transition temperature and is able to freeze a LC structure, thus, retaining its fluidity as a LC glass state. This characteristic is particularly advantageous with respect to the use of LCPS as data recording elements or lasers.

3.2.2 Charge Transfer in LCPs

Conduction in LCPs

LC materials are usually known as insulating dielectrics. However, due to impurities in the material during synthesis, these materials show ionic conduction with conductivities in the range of 10^{-8} - 10^{-12} Ω^{-1} cm^{-1} [9]. Charge injection from the cell electrodes might cause ions to migrate from the substrates into the material and render it conductive. It is also possible to tune the electronic property of LCs by incorporating ionic groups to induce ionic transport mechanism. In addition to ionic conduction, there are also LCs showing electronic conduction and anisotropic conductivity [10]. In rod-like LCs, it has been shown that the conductivity parallel to the director, $\sigma_{\parallel}$, is larger than the conductivity perpendicular to the director, $\sigma_{\perp}$. This can be attributed to the ionic mobility which is greater along the director, $\mu_{\parallel}$, rather than perpendicular to it, $\mu_{\perp}$. Owing to the molecular order and close-packed structure of LCs, the charge carrier mobility (from 10^{-3} cm^2 V^{-1} s^{-1} up to 0.1 cm^2 V^{-1} s^{-1}) is several magnitudes higher than that of amorphous material (10^{-3} cm^2 V^{-1} s^{-1} or less) [11]. However, when the LC materials exist in crystal phase, the presence of grain boundaries increases the charge trapping sites and, as a result, the charge carrier mobility is greatly reduced. One way to overcome this issue is through polymerization of the LC mesogens into LCPs. Interestingly, Yazaki *et al.* [12] reported that nanostructured LCs could show both ionic and electronic conduction properties by synthesizing smectic phase LC containing π-conjugated terthiophene group and imidazolium ionic group.

In order to understand the capability of LCPs in electronics applications, it is important to consider the impact of their capability of self-assembly on electronic charge transfer. Generally, in an organic semiconductor, holes and electrons are transported *via* the highest occupied molecular orbital (HOMO) and the lowest unoccupied molecular orbital (LUMO). This difference between the levels will introduce an electronic bandgap, which is a function of not only the individual molecular structure, but also their packing in the liquid crystalline phase. The process parameters, for example, in producing thin film devices from these LCPs are critical in defining their self-assembly characteristics and to ensure minimal defects in the morphology. An example is the work by McCulloch *et al.* [13], who reported that for the semiconducting liquid crystalline thieno[3,2-*b*]thiophene polymer, the enhancement in charge carrier mobility was achieved through highly organized morphology from processing in the mesophase. Transistor devices were successfully fabricated, which exhibited robust performance when exposed to the effects of ambient and low-humidity. The produced thin films exhibited a morphology of relatively large crystalline domain sizes (~200 nm), thus, indicating the prospect for single-crystal polymer transistors. Charge carrier field-effect mobilities of 0.2–0.6 $cm^2\ V^{-1}\ s^{-1}$ were achieved under nitrogen.

Thus, self-organization by the formation of liquid crystalline phases is a key strategy to control the order and packing of organic semiconductors and to suppress defect formation. Self-assembly in LCPs may be used in particular to enhance the performance of electronic devices such OLEDs, OPVs and OFETs. Strategies such as inducing a desired molecular orientation in a mesophase and locking in this orientation through crystallization, glass formation or crosslinking are useful. A LCP may also operate in the mesophase at room temperature, where the inherent self-assembly in the mesophase may be exploited, for example, chiral nematics for lasing. A key advantage of liquid crystallinity is the ability to form monodomain samples without defects caused by grain barriers between neighboring crystalline domains.

Charge Transfer Mechanism

Columnar, smectic and even nematic liquid crystals are recognized as potential organic semiconductors capable of charge transport, as these are able to self-assemble into highly ordered domains when processed into thin films. Generally, a thermally-activated hopping model is used to describe charge transport in LCs and LCPs (Figure 3.3). Charge hopping is enabled via traps, single molecules or repeat units of polymer chains which allow via π–π interactions. Disorder-based empirical models, as derived by Monte Carlo simulations, quantify mobility to be dependent on temperature and electric field, thus, correlating between structural and electronic disorder. A higher temperature will provide a higher kinetic energy which will be able to overcome the barriers for hopping between sites. In turn, the structural disorder is coupled to electronic conformation of the molecules. This disorder can often be described by a Gaussian distribution, where the disorder is dependent on variations on structural defects and conformation and polarizability of molecule.

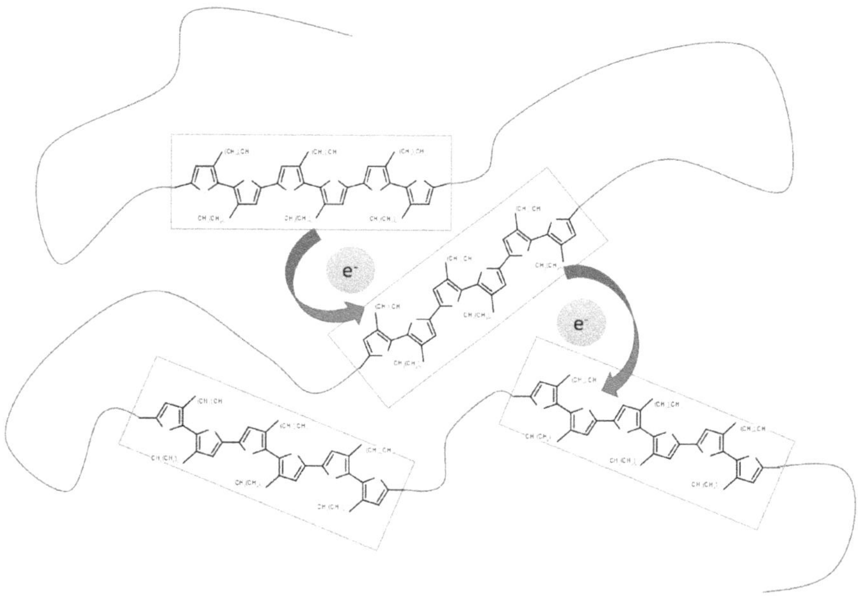

Figure 3.3 Example of charge hopping mechanism of conductive polymer. The molecular structure in boxes are representations of conductive polymers or LCPs.

Charge Transfer LCP Materials

Combining electronic functions and LC properties is an undeveloped, but attractive field. The ability to manipulate electron transfer in LCPs due to their self-assembly, can be exploited to realize new types of devices. Pure LCs possessing high conductivity are difficult to synthesize, however, some attempts to synthesize such materials have been reported. A SCLCP having a conjugated backbone similar to a polyacetylene structure has been reported [14]. The polymer exhibited a smectic liquid crystallinity at 92-105 °C. The film doped with iodine showed conductivity in the order of 10^{-2} S/cm. Nevertheless, the report did not mention the effect of molecular orientation on the conductivity. It will become a more interesting material if it is possible to control the conductivity by the orientation of the mesogen. Also, side chain type high molecular weight LCs having carbazole molecules, known to show electricity in mesogens, have been synthesized [15].

A typical semiconducting polymer is the regioregular poly(3-hexylthiophene) (P3HT), which shows hole mobility values of up to 0.3 $cm^2\ V^{-1}\ s^{-1}$, as shown in Figure 3.4. It comprises of a backbone of closely spaced layers of highly conjugated aromatic groups, separated by lamellar side chains which are solubilizing and insulating [16]. Apart from the fundamental molecular structure, its charge mobility is highly dependent on the connectivity between its microcrystalline domains. LCPs' ability to self-assemble assists in improving the field effect mobilities up to 1 $cm^2\ V^{-1}\ s^{-1}$. Such polymers such as poly(2,5-bis(3-alkylthiophen-2-yl)thieno [3,2-b]thiophene) (pBTTT) exhibit a mesophase above room temperature due to a lower side chain density compared to P3HT, which allows interdigitation leading to a highly ordered smectic phase. In the smectic mesophase, the polymer backbone is able to organize into a highly ordered and defect-free packing, which can be annealed and frozen when cooled to room temperature. Thus, such an annealing strategy enabled through the self-assembly capability of LCPs can be used to improve the OFET performance, which the P3HT polymer lacks.

Stacked Triphenylene based Discotic LCPs

An interesting variation in LCPs is based on discotic liquid crystals (DLCs), which have been long been identified as promising 1-D charge transfer materials, due to self-assembled columnar phase which can act as nanowires. Specifically, triphenylene is a common choice as the core for DLCs for the following reasons: (i) it possess C3 symmetry, (ii) it is relatively easy to synthesize (iii) it demonstrates high chemical and thermal stability and (iv) it demonstrates a rich variety of mesophases. The electron-rich nature of triphenylene-based discotics allows doping strategies with electron acceptors which lead to the formation of p-type semiconductors. Triphenylene-based DLCs have been extensively studied for their various physical properties including one-dimensional charge transfer [17], one-dimensional energy transfer [18], electroluminescence [19], ferroelectric switching [20] and surface mediated alignment and self-assembly for templating purposes.

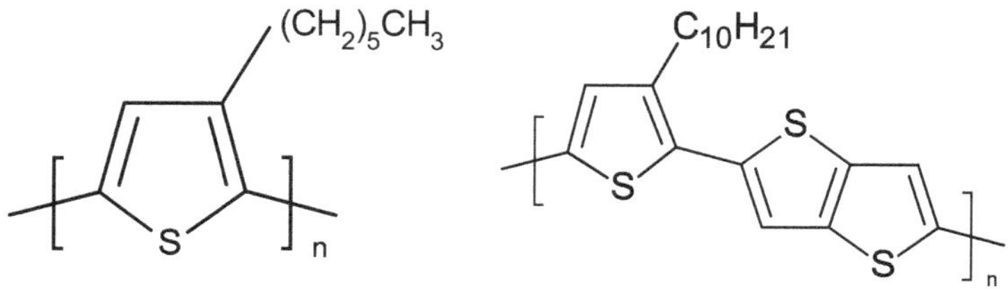

Figure 3.4 Molecular structure of regioregular (left) P3HT and (right) PBTTT.

Discotic polymers can also be sub-classified into side chain polymers, main chain polymers, elastomers and networks. For this chapter, the focus will be on main chain DLC polymers, as listed in Table 3.1. Difunctional triphenylene derivatives are the forerunner molecules for the preparation of the main chain polymers. The functional groups are often hydroxyl (-OH) moieties which may be converted easily to ethers (R-O-R') and esters (R-COO-R'). Several synthetic methods including polymer homologous reactions, free radical polymerizations, optical polymerizations, ring-opening metathesis polymerizations and condensation polymerizations have been used to prepare such polymers.

Table 3.1 Transition temperatures (°C) of 2,7- and 2,11- (also referred to as 3,6-) linked triphenylene based main chain polymers

Compound	Z	n	M_w	Phase transition
D2		11	24000	Cr 98 Col 118 I
D3	CH2	9	9000	Cr 93 Col 120 I
D4	CO	8	39797	g 56 Col_h 211 I
D5	CO	10	9656	g 40 Col_h 198 I
D6	CO	12	8818	g 56 Col_h 211 I

Some examples of discotic LCPs are listed in Figure 3.5 [21]. All of these polymers melt at about 93 °C and clear between 100 and 200 °C. **D1** shows a discotic nematic phase between the glass transition at

210 °C and isotropic transition at 31 °C. The nematic phase transforms into a columnar phase on doping with the electron-acceptor TNF. The complex exhibits the thermal behavior, Col_h 130 I.

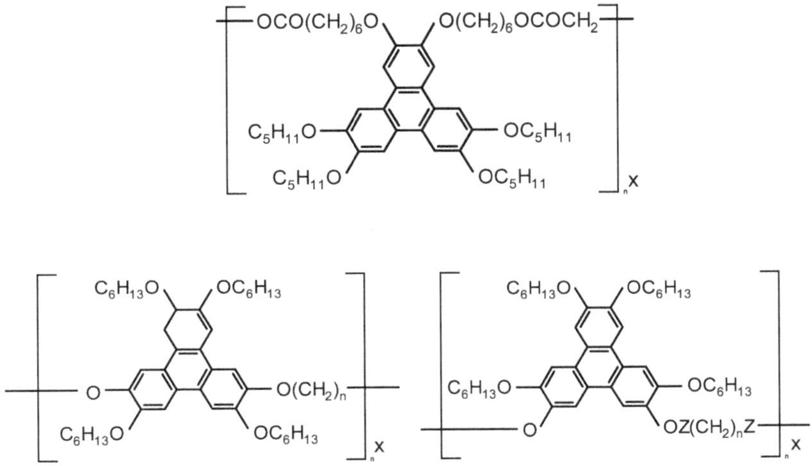

Figure 3.5 Molecular structure of LCP based on discotic liquid crystals. These polymers (**DLC2** to **DLC6**) form an ordered columnar mesophase between the glass transition and clearing temperatures.

Chromatographic separation of the products furnished the diblock copolymer **C1** and the triblock copolymer **C2**. Both polymers were found to be liquid crystalline at room temperature. The columnar phase of copolymer **C1** clears at 86 °C, and that of **C2** at 66 °C. The triblock copolymers **C3–C5**, having polyethyleneoxide (PEO2000) blocks at both ends of the triphenylene block linked via ester linkages were also prepared [22,23] (Figure 3.6). The thermal behavior of these main chain block polymers is summarized in Table 3.2.

One possible application of the triphenylene based discotic LCPs is as compensation films. For example, the Fuji Photo Film Company of Japan developed these films to improve the viewing angle of LCDs [24], where a negative birefringence compensation film is useful to suppress the narrow viewing angle of twisted nematic LCDs. The films are prepared by aligning the reactive monomer homeotropically in the nematic phase followed by photopolymerization. Triphenylene benzoate derivatives with one to six acrylate end groups or epoxide groups have been used to prepare anisotropic networks.

The crosslinked polymers exhibited no thermal transition up to 200 °C.

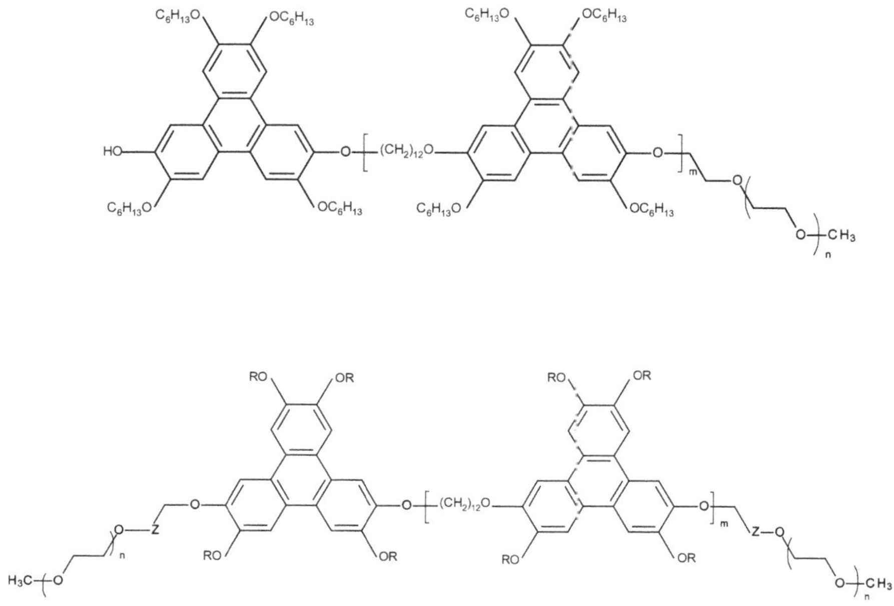

Figure 3.6 Molecular structure of diblock copolymer and triblock copolymer.

Table 3.2 Transition temperatures (°C) of triphenylene containing main chain block copolymers

Compound	n	M	Z	$\overline{M_w}$/g mol^{-1}	Phase transition
C1	16	10-12		~10000	<20 LC 86 I
C2	16	10-12	CH2	~10000	<20 LC 66 I
C3	~4-13	~12	CO		Cr 43.9 Col_h 72.1 I
C4	~4-13	~33	CO		Cr 38-48 Col_h 85.7 I
C5a	~12	~45	CO	18700	Cr 39.2 Col_h 81.4 I
C5b	~12	~45	CO	18500	Cr 38.4 Col_h 85.7 I

Electroluminescence in LCPs

Fluorene-type polymers are an important class of conducting polymers, owing to their efficient emission, high stability and relatively high charge mobility. Their electroluminescent properties can be tuned through blending with other emissive materials, in order to provide full colour emission through energy transfer to longer wavelength emissions. The LCP poly[(9,9-dioctylfluorenyl-2,7-diyl)-*co*-bithiophene] (F8T2), as shown in Figure 3.7, has shown promise for both transistor and optoelectronic applications, as its liquid crystal phase is stable across a wide temperature range and demonstrates good hole transport characteristics. In particular, the optical absorption spectra of the F8T2 solution can be tuned in order for it to act as an effective optical sensor, thus, paving the way for solution processable thin film optical sensors. Furthermore, the optical band gap of the F8T2 solutions can be significantly decreased with increasing molarity. The direct allowed band gap (E_{gd}) values of the F8T2 polymer are lower than the indirect allowed band gap (E_{gid}) values for all molarities. F8T2 liquid crystalline polymer can be used in the fabrication of the organic field effect transistors (OFETs), organic electronics, organic metal-semiconductor (MS) diodes, printed electronics and organic light emitting diodes (OLEDs) [25].

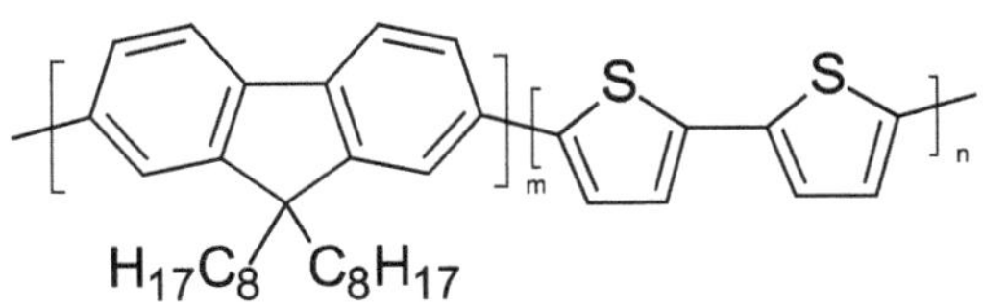

Figure 3.7 Molecular structure of F8T2.

3.2.3 Effect of Processing Parameters for LCP Performance

Chemical Oxidative Polymerization

As elaborated above, the capability for LCPs to conduct charges is highly dependent on the alignment of its director and, hence, the fabrication process employed which can enable the assembly of directors. For example, N,N,-dialkylimidazolium ionic liquids (ILs) have been shown to self-assemble into well-ordered liquid crystalline mesophases, producing a range of structures including 1D lamellae, 2D columnar and 3D cubic phases. Chemical oxidative polymerization of 1-decyl-3- methylimidazolium, in which a thiophene moiety is appended to the terminus of the alkyl chain, has been shown to produce a random coil and water-soluble polymer. The structural ordering of this LCP was further improved by designing a bifunctional monomer which retains the thiophene moiety positioned at the alkyl chain terminus and contains a vinyl moiety on the imidazolium headgroup. The thiophene moiety was expected to form pi-pi stacking, through the van der Waals interactions of the decyl chains and the intermolecular hydrogen bonding between the imidazolium anion and cation, thus, forming a supramolecular organization of the LCP, which can be assembled directly on the electrode surface. The result was a nematic mesophase, which possessed fibril structure when examined under atomic force micrograph. In the neutral state, the electrical conductivity of the LCP was an order of magnitude higher than that of PEDOT - poly(3,4-ethylenedioxythiophene). After treating with iodine, the conductivity value was further increased to 2.36 S cm^{-1}. In addition, recent studies have shown that composites which were formed from polythiophenes and insulating polymers, such as polystyrene, showed improved electrical conductivity [26]. An example is also depicted in Figure 3.8. Thus, these findings demonstrate the use of chemical oxidative polymerization process to induce specific morphologies of LCPs which are able to self-assemble on substrates and are useful in improving electrical conductivity.

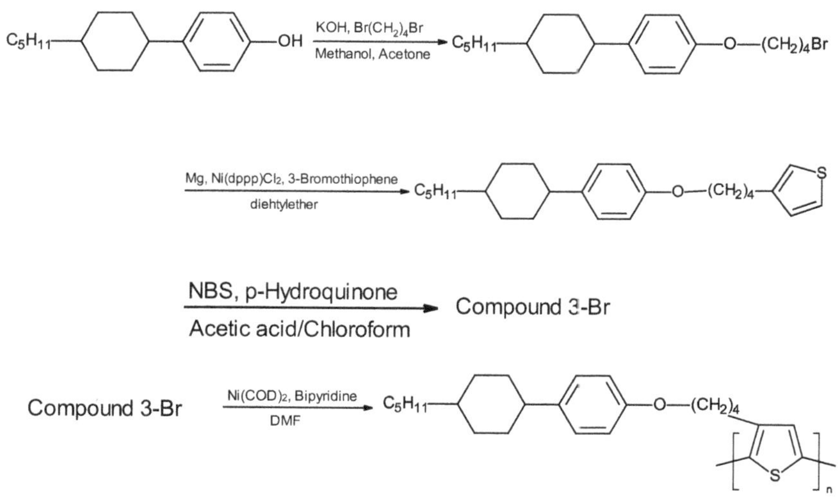

Figure 3.8 An example of polymerization of thiophene liquid crystalline conductive polythiophene derivative through dehalogenative polycondensation [27].

Photoalignment of LCPs

Typically, alignment of nematic and smectic LCs is enabled by rubbing the polymer alignment layers, such as doped polyimide, partially converted polyphenylenevinylene and poly(3,4-ethylenedioxythiophene) poly(styrenesulfonate), which are transparent and water soluble. These also allow conduction of holes and, therefore, act as hole injection layers in OLEDs. However, mechanical rubbing induces scratching and particle contamination, which can be overcome by using non-contact photoalignment techniques. This technique involves the irradiation of the photoalignment film with polarized light, which generates a surface anisotropy in the polymer alignment layer. Alignment materials such as cinnamate and coumarin undergo an anisotropic ac+2 cycloaddition upon irradiation with polarized light, whilst azobenzene side chain polymers undergo *cis-trans* isomerization, which results in anisotropy. The alignment direction can be changed by a corresponding change in the polarization direction of the incident light, allowing photopatterning of the alignment layer. The subsequent deposition of LCPs on top of this alignment layer induces the LCP directors to align along the easy axis of the alignment layer. Potential applications include 3D displays.

Electro-polymerization

Electro-polymerization is another method of tuning the properties of an LCP, such as its electroactivity. For example, a self-assembly of poly(alkylthiophene) structure can be induced through generation of organized multilayers. This can be achieved by organizing a variety of polyions supported by an electrode. Morphological studies of the resulting electro-polymerized thin film indicate large liquid crystalline domains through polarized microscopy and the presence of fibrils using atomic force microscopy. The fibril dimensions indicate that these comprise of bundles of polymer chains which are randomly oriented with respect to the substrate. The films were also doped with ions and the conductivity was observed to increase from 0.53 S cm^{-1} to 2.36 S cm^{-1}. It is believed that the enhanced conductivity is due to the molecular characteristics of the polymer and its consequent self-assembly.

An applicational example of the importance of alignment layer on device performance is the case of polarized electroluminescence. Main chain LCPs (such as arylene vinylene main chain polymers), which are mostly nematic, but some are smectic, can produce polarized electroluminescence. The alignment layer enables a uniform planar alignment of the LCP and strongly influences the quality of polarized light emitted by the LCP. The polarization ratio, R, of the liquid crystalline semiconductor gives the ratio of the measured intensity of polarized light parallel and perpendicular to the alignment direction. The highly anisotropic geometry of the LCP gives rise to correspondingly anisotropic optical response. In particular, extremely high values of *R*, up to 51, have been achieved by LCPs such as β-phase poly(9,9-dioctylfluorene (PFO) [28].

3.2.4 Dielectric Anisotropy of LCPs and Its Applications

Fundamentals of Dielectric Anisotropy in LCs

For non-polar molecules under an applied electric field, the internal charges are polarized. This involves induced polarization comprising of two components- electronic and ionic polarization. Due to the general rod-shaped structure of the LC molecule, the dielectric permittivity of the molecule is resolved into two directions parallel and perpendicular to the director ($\varepsilon_{\|}$ *and* $\varepsilon_{\perp}$ respectively), resulting in an overall dielectric anisotropy.

The dielectric anisotropy, $\Delta\varepsilon$, of the LCs is given by the following equation:

$$\Delta\varepsilon = \varepsilon_{\|} - \varepsilon_{\perp}$$

where $\varepsilon_{\|}$ and $\varepsilon_{\perp}$ are the dielectric constants measured parallel and perpendicular, respectively, to the long molecular axis of the LCs.

This equation is based on the Maier-Meier theory described as follows [7]:

$$\Delta\varepsilon \propto \left[\Delta\alpha + \frac{\mu^2 F}{2k_B T}(3cos^2\beta - 1)\right]$$

where $\Delta\alpha$ is a polarizability anisotropy, μ is a molecular dipole moment, F is parameter depending on the reaction field factor, k_B is a Boltzmann constant, T is temperature and β is the angle between the molecule axis and the direction of the off-axis μ. $\Delta\varepsilon$ depends on molecular properties such as polarizability, dipole angle and macroscopic orientational order [29]. These factors also determine the sign of $\Delta\varepsilon$, which can either be positive or negative. In addition, the magnitude of $\Delta\varepsilon$ is highly dependent on the strength of the dipole, the frequency of the applied electric field and the temperature. As the LC material becomes isotropic, $\Delta\varepsilon$ decreases because the system is becoming less ordered.

In the case of side chain LCPs, it has been concluded that the behavior is similar to that of normal polymers, in which polarizability is affected by the segmental motion of the main chain, while β is greatly affected by the oscillating motion of the side chain mesogenic groups. In the case of main chain LCPs, the dipole is parallel to the main chain of the polymer in which the sum of dipoles on the main chain is proportional to the vector connecting both ends of the polymer chain. Therefore, in the dielectric property measurement, it is possible to observe the movement in which the distance between both ends of the polymer chain varies, that is, how the whole polymer chain moves. In display applications, a fast response time is desirable which is easily obtained in the case of low molecular weight LCs. However, due to the large size of LCPs, it is difficult to switch the molecule where the required time is around 100 ms compared to 30 ns for low molecular weight LCs [30, 31]. However, there are reports on the designing of LCPs having fast response time (10 ms) and low threshold voltage [6, 32].

Since the slower switching times of the LCPs are less favourable for electro-optic applications, it is preferable to utilize their highly insulating nature to applications in electronics packaging, which will be elaborated below.

LCPs as Electronic Packaging

A segment in electronics where LCPs are of value is in electronics packaging, due to their superior mechanical properties. A collaboration between DuPont and Interplex Quantum Leap Packaging (IQLP) is aimed to develop LCP technology for the high frequency electronics segment. Targeted electronics segments, which may utilize the LCP encapsulation technology, include communications of up to 100 GHz, solar photovoltaics and LEDs. Their advantages as encapsulation for high frequency electronics are owing to their toughness, impermeability to air and water vapour and little thermal expansion. Conventionally, LCP casings are produced through injection molding which render them stiff and bulky. As a solution, roll to roll processing is proposed, by depositing a thin layer of LCP over a flexible substrate. Such processes allow the use of LCP as a component in printable electronics circuitry. A point of interest in the role of LCPs as electronics packaging is that whilst most applications rely on the exploitation of the material anisotropy for functionality, the opposite is desired in packaging. Suppression of the anisotropy of the bulk LCP material can be achieved during the fabrication process. For example, during roll-to-roll deposition of the LCP, where extrusion of the molten LCP desires a twist of the LCP director, will reduce its overall anisotropy.

The role of LCPs as a systems level packaging layer for integrated packaging of monolithic microwave integrated circuits (MMICs) and radio frequency microelectromechanical systems (RF MEMS) switches has been demonstrated by Thompson *et al.* [33], as shown in Figure 3.9. A 4 mm thick layer of LCP is deposited on the substrate and subsequent laser micromachining were performed to fabricate the active device regions. The low dielectric constants of the LCPs were useful as the cavity dimensions can be arbitrarily chosen without significantly affecting the RF performance. This work is the first demonstration of the use of an organic (LCP) material as a flexible and low cost organic substrate for system level electronics packaging.

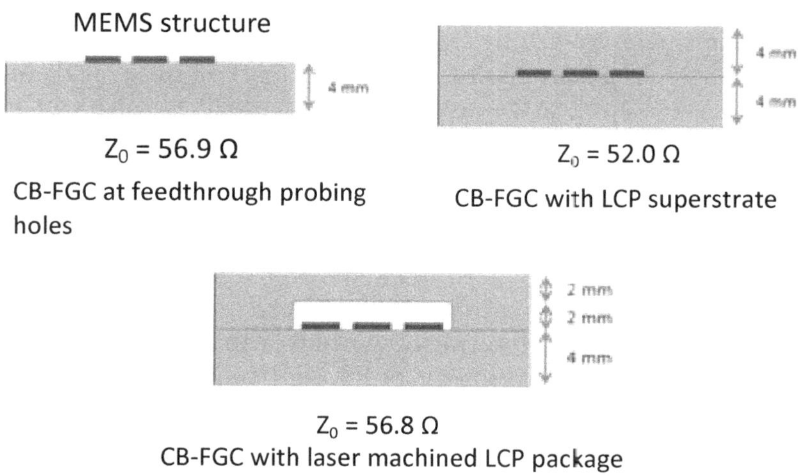

Figure 3.9 Three different conductor-backed finite ground coplanar (CB-FGC) cross-sections present in the measured packaging structures. Impedance values are from HFSS 3D simulator for an CB-FGC with signal width S=200 μm, gap G=120 μm, and r = 3.16. Z_0 only differs by 4 Ω for the three different cross-sections (redrawn from Reference [33]).

Related to the use of LCPs as encapsulation material in biomedical applications, demonstration of LCP as a casing for a radio chip for implantation inside the body has been demonstrated by the Korea Institute of Advanced Sciences, which has been shown to be more durable than other alternatives [34]. A radio-frequency transistor was etched from thin layers of silicon, gold and chromium, and then encased in a biocompatible LCP using thermal pressing. The durability of the chip was demonstrated in terms of its ability to function inside human body for up to two years. In-vitro testing of the chip in a lab rat was carried for up to six weeks, with no detriment to the chip performance or the health of the rat. This finding could be a breakthrough in bionic applications, as it allows control of elements and communication with an external server through its RF communication capabilities.

3.2.5 Optical Effects in LCPs

Birefringence

Birefringence, i.e., the difference in the refractive indices between the long and short axes of the LC, is useful in providing changes in optical polarization of light traversing through the LC device. It can subsequently be exploited as optical waveguides for various applications. The optical response of the liquid crystal director with respect to polarized light is described by its birefringence, as shown in Figure 3.10.

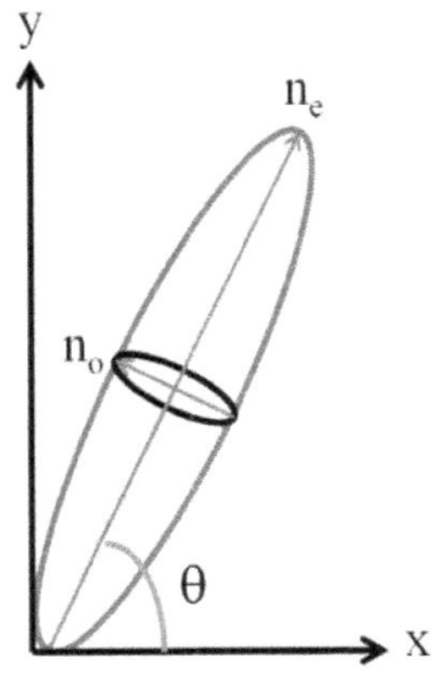

Figure 3.10 Schematic diagram of an LC molecule showing the different refractive index according to the axis.

Given the geometrical anisotropy of the LC molecule, light polarized perpendicular to the director of the molecule experiences a refractive index called the ordinary refractive index, n_o, whereas light polarised parallel to the director is called the extraordinary refractive index, n_e. The birefringence of uniaxial material is given by the following equation [30]:

$$\Delta n = n_e - n_o = n_{\parallel} - n_{\perp}$$

The refractive index of the LC material is given by the following equation:

$$n(\theta) = \frac{n_e n_o}{\sqrt{n_e^2 sin^2\theta + n_o^2 cos^2\theta}}$$

where θ is the angle between the x-axis and the molecule; the light is travelling in the y direction and is polarized in the x direction. The birefringence is also dependent on the wavelength of the incident light and temperature.

Using the refractive index, the optical retardance, Γ, of an LC in a cell with thickness, d and an illuminated light of wavelength, λ, can be calculated as follows:

$$\Gamma = \frac{2\pi d(n(\theta) - n_o)}{\lambda}$$

The phase shift occurs because some of the illuminated light experiences a different refractive index (either ordinary or extraordinary), which results in different velocities through the material. Through manipulation of the birefringence and alignment of the liquid crystal director in a device, the LC device can be designed to act as a linear or circular polarizer or a half wave or circular wave plate, through tuning of its retardance by choice of device thickness and refractive indices of the liquid crystal. For example, a phase shift of $\pi/2$ will result in the liquid crystal device to act as a circular polarizer.

Depending on the alignment of molecules, the birefringence changes significantly. Thus, there are many reports exploring the effect of alignment of LCPs on their birefringence. One of the widely used methods to align the polymer molecule is through photoalignment, where a large difference in birefringence before and after photoalignment has been reported [35]. The change in birefringence was as much as $\Delta n \geq 0.65$ which could be beneficial for optical applications. Another interesting report has proved that it is possible to selectively induce photoalignment under UV irradiation, thus, making it feasible to obtain high degree of molecular orientation at different angles [36]. Photoalignment of LCPs using various techniques has been reported in another study and applied to different device architectures [37]. Another method of alignment is called mechanical alignment in which a uniaxial tension is applied to pre-polymer LCs after which the liquid crystal is polymerized to retain the molecular alignment [10]. According to another study, the molecular weight of LCPs plays an important role in determining the birefringence [11]. The authors found that for a low enough molecular weight, the birefringence exhibited almost isotropic-like behavior. This is in stark contrast with low molecular weight LC which still shows birefringence despite the small molecular size.

LCP Based Polarizers

Liquid crystals can be designed into effective polarizers, given that their strong optical anisotropy can be manipulated to form circular and linear polarizers. Past works have included liquid crystal cells filled with dichroic LC materials, such as Merck ZLI-4714 and polymethacrylate based LC doped with dichroic dye [8]. Myhre *et al.* [38] have demonstrated patterned polarizers for visible wavelengths using dichroic dye in a liquid crystal polymer (LCP) host. A gray dichroic dye mixture with LCP was fabricated to allow absorption in the visible wavelength. A circular polarizer setup consisting of a quarter-wave retarder and a color polarizer were fabricated using bilayer films of this composite as shown in Figure 3.11, allowing applications in interferometry, 3D displays and optical storage. The first layer contains an undoped LCP which acts as a retarder, whilst the second is an LCP layer doped with the dichroic dye to form a circular polarizer. The LCP layer's polarization modulation capabilities can be tuned according to its thicknesses and retardance and concentration of dichroic dyes. The authors were also able to create linear polarizers using bilayer liquid crystal systems comprising of dye embedded in an LCP matrix. For example, a blue polarizer achieved the highest extinction ratio of 40.1 at 633 nm. This polarizer consisted of a dye/LC solution of 30 mg/mL blue dye in $CHCl_3$ and 1 mL of LCP, i.e. near 2% by weight, which is close to its solubility limit. In principle, the extinction ratio can be improved by increasing solubility through the use of other solvents or dichroic materials, thus, indicating a pathway for optimization of the optical properties of LCPs as polarizers by tuning concentration and composition.

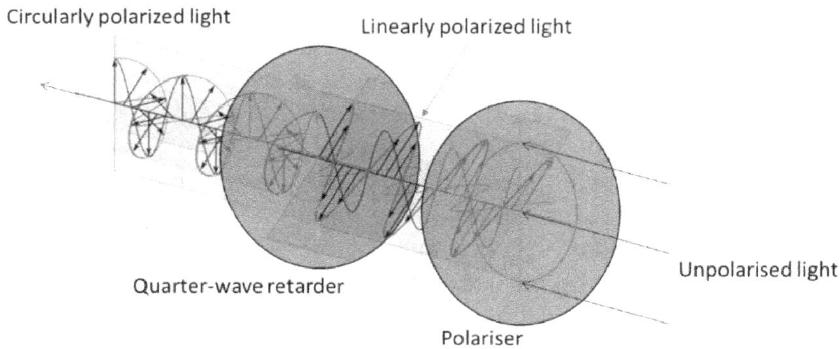

Figure 3.11 A circular polarizer comprising of a combination of a quarter-wave retarder and linear polarizers.

LCP Lasers

Lasing from liquid crystal media is enabled through exploitation of the periodic structure of liquid crystals (such as those in chiral liquid crystal), which gives rise to a photonic bandgap and consequently lasing (Figure 3.12). An advantage of liquid crystals over solid crystals is the spontaneous formation of the period structure without the need for complex fabrication processes. The first liquid crystal laser induced through band edge lasing was observed by Kopp *et al.* [39], Taheri *et al.* [40] and Alvarez *et al.* [41] in dye-doped chiral nematic liquid crystals. Since then, band edge lasing has been exhibited by a range of LCPs [42-46]. Tuning of the band edge lasing through changes in the LCP pitch can be induced by changes in temperature and through influence of ultraviolet illumination, mechanical stress and applied electric fields. Apart from band edge lasing, laser emission can also be induced through the introduction of a defect into the band edge structure and is also known as defect-mode lasing. The location of the mode is in the photonic bandgap and allows the excitation and control of multiple modes [47].

Laser Emission in a Dye Doped Cholesteric Polymer Network

Cholesteric liquid crystals (CLCs), such as those shown in Figure 3.13,

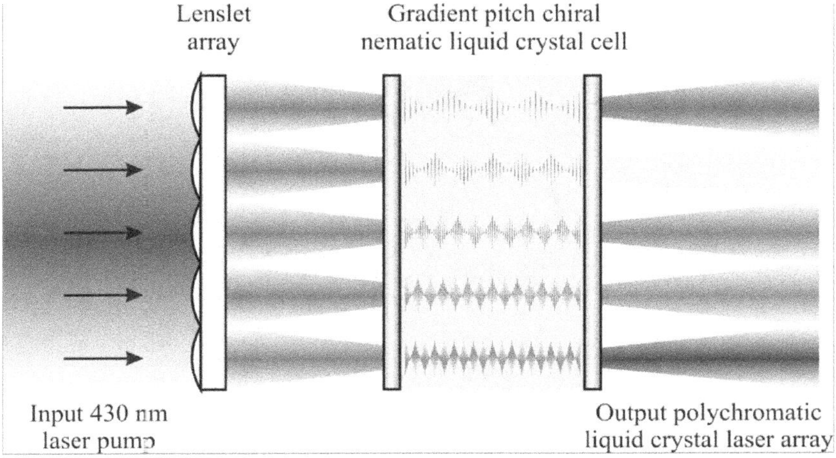

Figure 3.12 Schematic of LCP lasers producing variation of light by tuning arrangement of LCP in liquid crystal cell.

possess an intrinsic helical structure which provides a 1D refractive index modulation formed by rod-shaped molecules that arrange themselves within a helical structure. For a given frequency, CLC allows selective reflection of light for which the polarization has the same handedness as the cholesteric medium. Thus, CLC can be likened as a 1D photonic crystal, which can be used for lasing. An advantage of the CLC is that the lasing frequency is tunable as a function of the cholesteric pitch and its intrinsic self-assembly capability allows spontaneous formation of the structure needed to induce lasing. In order to retain the lasing capabilities, the cholesteric helix is frozen into the required helical configuration, by either cooling it below the glass transition temperature or crosslinking the mesogens. For example, a mixture of the nematic p-pentylphenyl-2-chloro-4-(pentylbenzoyloxy)-benzoate and the cholesteric cholesteryl nonanoate were doped with the laser dye 4-(dicyano-methylene)-2-methyl-6-(4-dimethylamino styryl)-4-H-pyran. The lasing behavior is largely dependent on the liquid crystal mesophase of the resulting composite: in the isotropic phase, the fluorescence spectrum is very broad, but upon entering the cholesteric phase, emission is strongly enhanced at the long wavelength edge or the reflection band.

However, due to the fact that LCs are not as strongly ordered as solid state photonic crystals, liquid-like variation in LCs have some detrimental effect on the overall lasing characteristics. A reduced degree of ordering in some regions and variation of pitch results in the

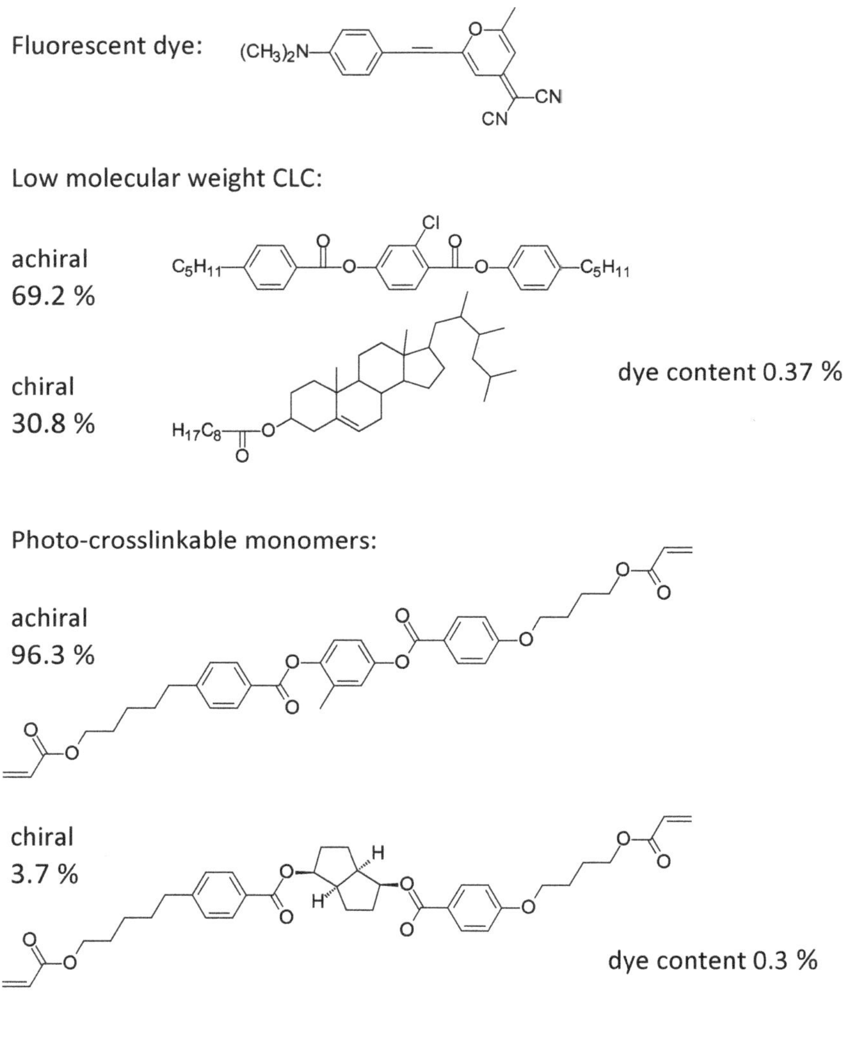

Figure 3.13 Examples of CLC molecular structure. Two types of CLCs are shown: low molecular weight and photo-crosslinkable CLCs with achiral and chiral structures.

absence of clear-cut edges of the reflection band and the presence of several resonance frequencies for laser emission, respectively. LCPs show some advantages over lasers based on lower mass CLCs, where these provide lasing which is one order of magnitude higher compared to a low molar mass CLC with similar optical parameters. LCPs also lend themselves well to further developments, as the resulting crosslinked samples form thick films which can be easily peeled off from substrates and allow further processing through coating or micro-structuring [42].

3.2.6 Ferroelectricity in LCPs

Ferroelectricity is exhibited by materials which exhibit a spontaneous polarization that is reversible through application of an electric field. The term ferroelectricity is analogous to ferromagnetism, where a material possesses a permanent magnetic moment, even though in many cases the material does not contain iron. Ferroelectricity is a subset of pyroelectricity and exhibits spontaneous polarization below a transition temperature known as the Curie temperature. In ferroelectric materials, this spontaneous polarization can be reoriented by application of an electric field, where full reversal of the polarization is called "switching". The direction of the spontaneous polarization is defined by the crystal symmetry of the material and is most significant well below the Curie temperature. A ferroelectric material can be defined as a material that has a nonlinear relation between polarization P and applied electrical field E. This relation is given by a hysteresis loop, which is a function of temperature, defects, mechanical stress and electrical measurement condition imposed on the sample. Since the pyroelectric materials are piezoelectric, ferroelectric materials should inherently possess piezoelectric properties. Thus, the ferroelectric material would produce electrical charge when it is applied with mechanical stress (direct piezoelectric effect) and in contrast, a change in dimension would occur when it is applied with electrical field (converse piezoelectric effect).

A ferroelectric material should have the following features: an electric hysteresis loop, a very high dielectric constant, the presence of spontaneous polarization without an external electric field, temperature dependence of the dielectric temperature and possession of polar and domain structures in their morphology [48]. In the case of LCs, ferroelectricity can be established in a material that satisfies the following three conditions: (i) at least one asymmetric carbon must be included in the compound to induce chirality, (ii) there must exist a component of the dipole moment perpendicular to the long axis of the molecules which induces spontaneous polarization and (iii) the material must show a smectic phase with a nonzero tilt angle. Thus, these materials are also known as chiral smectic liquid crystals (SmC* or SmH*). Examples of types of LCPs able to form the chiral smectic phase are polysiloxanes and polyacrylats, as shown in Figure 3.14:

Polysiloaxanes:

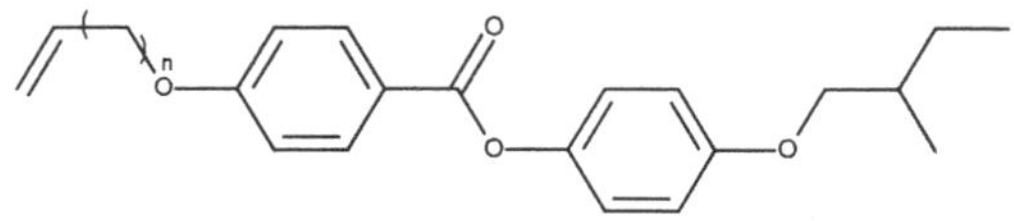

n = 8 (*S*)-4'-(2-methyl-butyloxy)-phenyl-4-(9-decenyloxy)-benzoate
n = 9 (*S*)-4'-(2-methyl-butyloxy)-phenyl-4-(10-undecenyloxy)-benzoate

Polyacrylates:

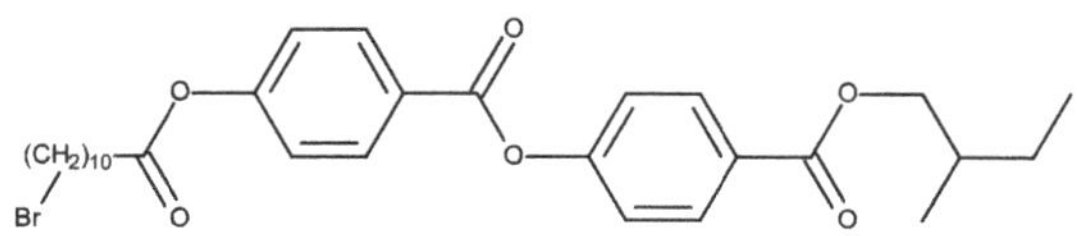

(S)-4'-(2-methyl-butyloxycarbonyl)-phenyl-4-[11-bromo-undecanoyloxy]-benzoate

Figure 3.14 Examples of polysiloaxanes and polyacrylates exhibiting ferroelectric characteristics.

Visualization of the LCP director can be represented as in Figure 3.15 below:

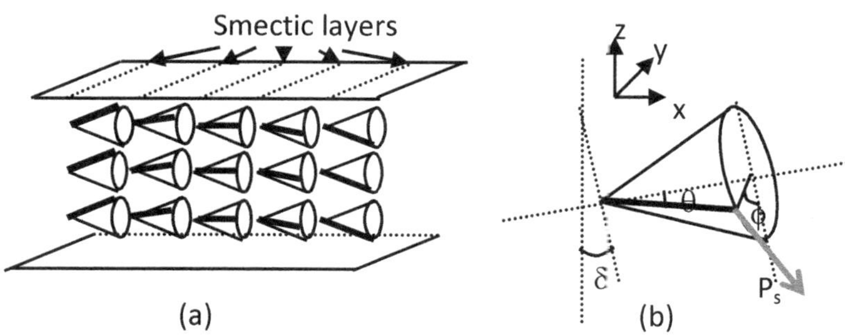

Figure 3.15 a) Illustration of the progression of the chiral smectic C director, with respect to the cell geometry and smectic layers. (b) Close up of one of the chiral smectic C directors, where δ defines the layer tilt angle, θ the smectic cone angle, and ϕ the azimuthal angle of the director.

The spontaneous polarization, P_s is always tangential to the base of the cone and will couple to the electric field applied in the vertical direction (either up or down) respective to the cell. The direction and magnitude of electric field will allow rotation of the director on the cone and hence drive the electro-optic effect of the cell. In the case of ferroelectric LCPs, their response to the applied electric field is significantly slower than FLCs and, thus, unsuitable for electro-optic and displays applications. Instead, these are identified as potential actuators, due to the possible high mechanical strain which accompanies the polarization switching. This can be further explained below, through exploitation of the electroclinic effect at the Sm A* - Sm C* transition, where switching also implies a volumetric change and, hence, a change in mechanical strain.

The electroclinic or soft mode effect is an example of a linear optical response in ferroelectric liquid crystals. In chiral materials, molecular tilt can be induced by applying an electric field perpendicular to the optic axis, as shown in Figure 3.16. This effect was first observed at the SmA*-SmC* transition [49]. Although it is an intrinsic property of all chiral materials, its effect is more marked in the Smectic A* phase, as we can observe a transition from the untilted Sm A* phase to an observed tilt when an electric field is applied. This is due to the manifestation of the local polarization fluctuations along the layers and perpendicular to the optic axis. As the field is applied perpendicular to the optic axis, the permanent dipole moments of the

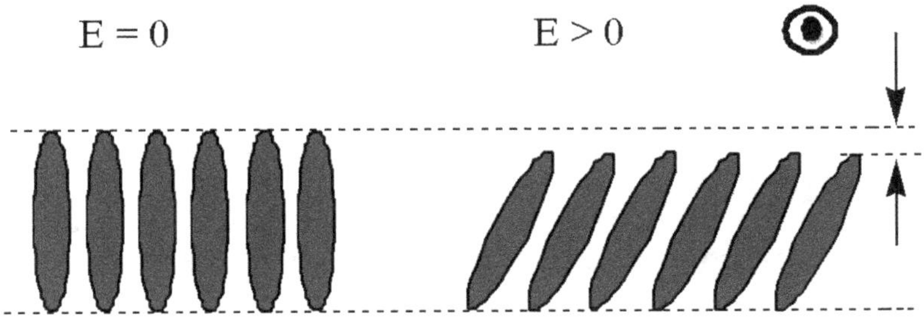

Figure 3.16 An illustration of the elecgtroclinic effect, where an applied electric field causes a tilting of the smectic layers, and consequently, a reduction in the smectic layer spacing.

molecules are forced to align with the electric field and this causes the axis to tilt. Due to the interaction with the polarization components, this effect is also polar, i.e. application of electric field of opposite polarities will result in tilt angles of equal magnitude but in opposite directions. The resultant tilt angle, for a reasonable range of voltages, is proportional to the field. The tilting induced by the electroclinic effect causes a corresponding constriction in the smectic layer spacing and results in a constriction of the smectic layers, which correspondingly results in an electrostrictive effect.

Electrostriction in an electron-irradiated poly(vinylidene fluoride-trifluoroethylene) copolymer exhibited a fast actuation response of strain up to 4%. Further exploitation of the electrostrictive effect has been shown by combining ferroelectric liquid crystalline elastomer thin films of less than 10 nm thickness. The electric field strength needed for a strain of 4% was 100 times lower in FLCE than the next best electrostrictive material. For example, solid state materials such as quartz and lead zirconate titanate demonstrated small strains of less than 0.1%. The electrostrictive response for this material showed a pronounced maximum near the phase transition of the uncrosslinked FLCP from ferroelectric (S_C* phase) to paraelectric (S_A* phase), at the Curie temperature $T_C \approx 61$ °C. This corresponds well to the fact that the electroclinic effect is most pronounced at the boundary between the Smectic A and Smectic C phases. Potentially, the electrostrictive effect in LCPs can be utilized as actuators in applications, such as motors, pumps, artificial muscles and biomedical nano- and micro-robots.

3.2.7 Electrorheological Effect in LCPs and Its Applications

Viscosity in LCs

It is well known that polymers have high viscosity, which correspondingly will have an impact on the physical properties of LCPs. Viscosity is defined as the internal friction of a fluid to deform under a shear stress present due to velocity gradients in the system. For LC material confined in a cell of a few microns thick, measuring the force required to move one substrate while the other is held anchored at a constant speed will give the shear bulk viscosity [50]. The viscosity is defined by the ratio of the shear stress to the shear strain; shear stress is defined as the force F_x per unit across the sectional area and shear strain is defined as the velocity v_x divided by the thickness of the cell. As the thickness becomes small enough, the effect of surface-induced ordering becomes increasingly important. The shear stress is dependent on the direction of the flow field and the configuration of the flow gradient both relative to the director field. This was shown by Miesowicz who found that LCs have three viscosity coefficients, as shown in Figure 3.17 [51].

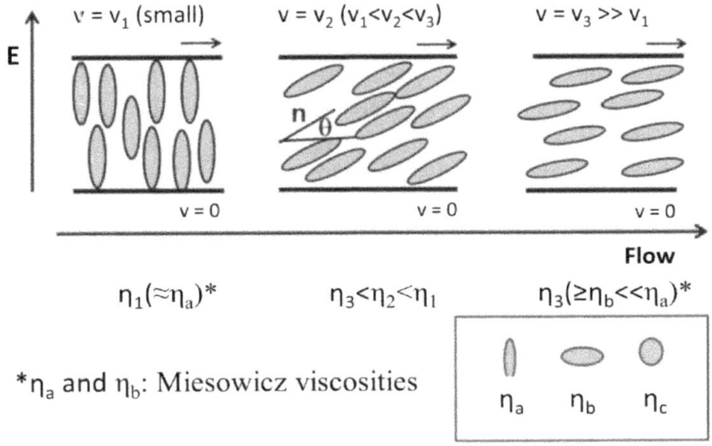

Figure 3.17 Schematic diagram of the measurement to determine the shear viscosity of a nematic LC where a force of F_x with a constant velocity of v_x is exerted on the top plate while the bottom plate is static. The director is a) homeotropic and parallel to the z-axis, b) planar and parallel to the x-axis, and c) planar and parallel to the y-axis (adapted from [52]).

Viscosity opposes the rotational motion of the liquid crystal director when subjected to an electric field required for electrooptical switching. Hence, the relationship between the switching time and viscosity can be described using the following equation:

$$\tau_r = \frac{\eta_i d^2}{\varepsilon_0 |\Delta\varepsilon| V^2 - K_i \pi^2}$$

$$\tau_d = \frac{\eta_i d^2}{K_i \pi^2}$$

where τ_r is the rise time, η_i is the viscosity coefficient, d is the distance between top and bottom electrode, $\varepsilon_0$is the permittivity of free space, $\Delta\varepsilon$ is the dielectric permittivity, V is the applied bias voltage, K_i is the elastic constant and τ_d is the decay time.

The relationship between the viscosity and the electrooptical properties of SLCPs has been investigated by Pranoto *et al.* [53]. It was reported that for homogeneously oriented siloxane, τ_r, τ_d and V_{th} depend greatly on temperature, in which case the values decreased as the temperature is increased. Calculating the splay elastic modulus, K_1 of the siloxane LCP showed similar value to that of low molecular weight LC although η was 100 magnitudes higher than the low molecular weight LC. Another report by Goozner and Finkelmann [54] showed that the bend elastic moduli K_3 and η_3 of homeotropically aligned polymethyl methacrylate were similar to those of the low molecular weight LC. The same group also investigated the effect of terminal polar group and spacer length of various SCLC polysiloxanes [55]. The study concluded that higher polarity and longer spacer length led to the decrease of V_{th}, attributed to the increase in $\Delta\varepsilon$.

The Electrorheological Effect in LCPs

Given the increased switching time in LCPs, these may not be the best candidates for electrooptic switching. An interesting applicational alternative to LCPs is in the field of electrorheology. Electrorheological fluids comprise of fine domains of non-conducting but electrically active particles in an electrically insulating fluid. Upon application of an electric filed, the effective viscosity of these fluids changes reversibly by up to 5 orders of magnitude, i.e. these materials can transform from the consistency of a fluid to gel within the order of miliseconds. LCPs are a particularly interesting manifestation of this effect, given the dielectric anisotropy of the LCPs and their ability to form aggregations which may allow the electrorheological effect to occur, as illustrated in Figure 3.18 [56].

Low molar mass liquid crystal solvents possessing a positive dielectric anisotropy exhibit an electrorheological effect where a strong electric field applied transverse to flow produces a η_{on}, which is larger than η_{off}, where there is no applied field. A liquid crystal polymer (LCP) is able to enhance this effect by an amount $\sigma\eta_{on}$ - $\sigma\eta_{off}$, which is a function of LCP concentration and molecular architecture. Specifically, main chain LCPs are thought to exhibit stronger electrorheological effect compared to side on side chain LCP and end-on side chain LCP. A molecular description is obtained by relating η_{on}, and η_{off}, respectively, to the Miesowicz viscosities η_c and η_b. A hydrodynamical model developed by Brochard deduces that the ratio of the viscosities is proportional to the geometrical anisotropy of the molecule, i.e. $\sigma\eta_c/\sigma\eta_b = R_{\parallel}{}^4/R_{\perp}{}^4$, where $R_{\parallel}$ and $R_{\perp}$ are the end-to-end distances of the chain, parallel and perpendicular to the director. This observation is based on the theory that the electrorheological effect is a balance between hydrodynamic and electric torques experienced by the LCP director and hence the ellipsoid of revolution resulting from the torque balance can be related to the molecular conformation which represents the mass distribution of the LCP director. Thus, the main-chain LCPs are thought to be highly prolate, whilst the side-on side chain LCPs are moderately prolate and the end-on side chain LCPs are quasi-spherical and hence give the least electrorheological response amongst the LCPs. In other words, the highest geometric anisotropy of the LCP molecule as demonstrated by main chain LCP will give rise to the most significant electrorheological effect. Examples of this phenomena are demonstrated by a polysiloxane based LCPs in dimethyl silicone, where LCPs containing mesogenic groups in the main chain produced a large increase in shear stress, up to 1,300 Pa [57]. It is suggested that the strong electrorheological effect between homogenous LCP fluid is related to flexible chain linkage between the crystalline domains.

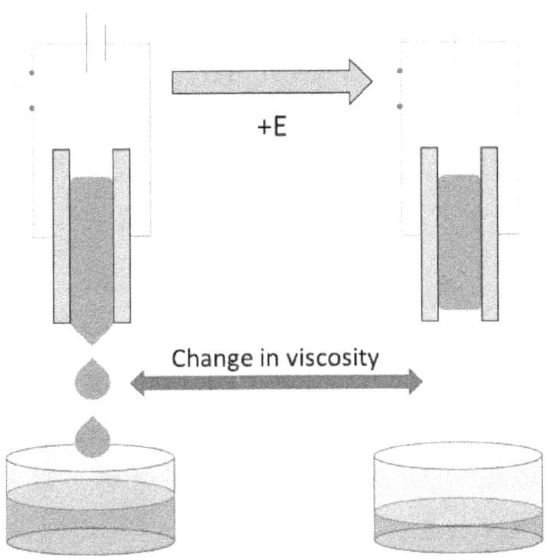

Figure 3.18 Depiction of changes in viscosity of LCP when electric field is applied.

3.2.8 Controlled Drug Delivery

An area where smart, responsive and biodegradable LCPs can find extensive use is in biomedical applications, not just as actuator materials described in the earlier sections, but also as drug delivery vehicles.

Hwang *et al.* [58] first reported the synthesis of cholesterol end functionalized oligo(L-lactic acid). Subsequently, Wan *et al.* [59] and Guo *et al.* [60] demonstrated the synthesis of cholesteryl end-capped aliphatic polycarbonates and dicholesteryl end functionalized block poly(3-caprolactone), as biodegradable polymers. These can be of particular value as encapsulation of drugs for delivery in the human body. Compared with the biodegradable polycarbonates with main chain functionalized cholesteryl mesogenic groups, the polycarbonates based on side functionalized cholesteryls groups have more advantage such as controlled multiple structure and molecular weight. So, it is necessary to synthesize new biodegradable side chain LC polycarbonate derived from cholesterol and study their structure-property relationships and explore the potential applications in drug delivery and tissue engineering templates.

A side chain functionalized cholesterol with hydrophobic moiety was added onto the amphiphilic block copolymers with polycarbonate and mPEG backbone (chiral LC monomers with two, four, six and eight methylene segments) for the introduction of physicochemical functionality. The ability of side cholesterol groups to form mesophase is anticipated to provide a driving force for self-assembly into nanostructures. The following amphiphilic LC polymers on a polycarbonate backbone compounds (as shown in Figure 3.19 and 3.20)

Chiral monomers:

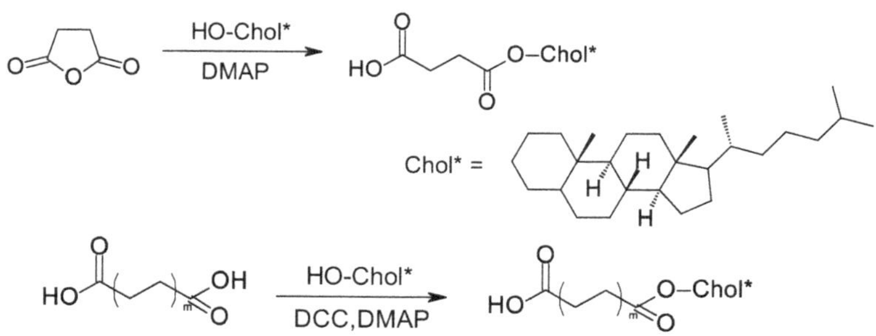

M$_1$:4-cholesteroxy-4-oxobutanoic acid
M$_2$:6-cholesteroxy-6-oxocaproic acid
M$_3$: 8-cholesteroxy-8-oxooctanoic acid
M$_4$: 10-cholestroxyl-10-oxodecanoic acid

m= 2, **M**$_2$; m= 3, **M**$_3$; m= 4, **M**$_4$

Figure 3.19 Schematic of cholesteryl end-capped aliphatic polycarbonates and dicholesteryl end functionalized block poly(3-caprolactone), as biodegradable polymers.

exhibit a mesophase state below body temperature, thus, showing potential for self-assembly into Smectic A (SmA) phase.

From the compounds shown above, **M₂-M₄** exhibited enantiotropic cholesteric phase. Generally, mesophase formation is a function of aspect ratio, the presence of a flexible spacer at the centre, and polarity of the monomers.

Copolymers:

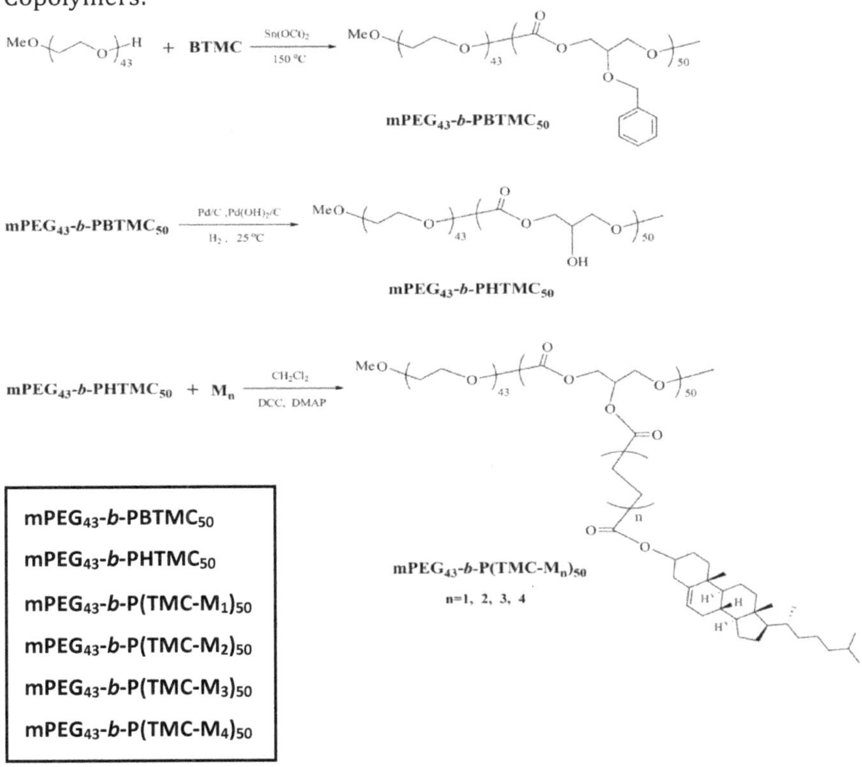

Figure 3.20 Compound **M_2-M_4** in enantiotropic cholestric phase.

Amongst the copolymers listed in Figure 3.20, the copolymer **mPEG$_{43}$-*b*-P(TMC-M$_1$)$_{50}$** demonstrated a high interaction between side chain units which is conducive for generation of cholesteric mesophases, whilst the other copolymers incorporating chiral monomer **M_2-M_4** exhibited SmA phases, suggesting that the polymer chains could hinder the formation of a supramolecular helical structure. Generally, the mesophase temperature ranges of the copolymers were greater than corresponding monomers, indicating the stabilizing effect of the polymers. These compounds are targeted as a self-assembly materials for controlled drug delivery.

3.2.9 Light Activated LCPs

Light activated materials which operate on the principle that an external stimulus, such as light, can change the birefringence of the material. Tuning of molecular alignment through photo-controllability allows control of optical coefficients of electroactive materials, selective light diffraction, production of surface relief structures, photo induced chirality and photomechanical actuation. By extension, these materials are particularly valuable in applications such as self-assembled nanotemplates, photonics, optical data storage, holography and nanomechanics.

Azopolymer and azobenzene systems are model systems which undergo photoisomerization in order to induce a change in birefringence, through either (i) introduction of order, i.e. assembly of the azobenzene moieties or (ii) destruction of order in an aligned mesogenic system through a photochemical phase transition. Azobenezenes are photochromic molecules that are able to change from a rod-like conformation in *trans*- form to a *cis*-form upon irradiation of visible light. In the first case, irradiation of polarized light on azobenzene systems allows self-assembly of its molecular alignment through axis-selective photoreaction. Linear polarized light will cause the alignment of molecules normal to the direction of the polarized light, through photoisomerization of the azobenzene. In the second case, azobenzene LCPs which undergo the nematic-isotropic phase transition demonstrate a significant change in Δn. On homogenous alignment, these materials demonstrate a high Δn; irradiation of light causes a change in birefringence through destruction of the alignment through *cis-trans* isomerization. For example, Okano *et al.* [32] introduced a molecular architecture for LCPs which contain an azotolane moiety in the side chain in order to maximize the responsiveness of the molecule to light by destruction of the alignment by *trans-cis* photoisomerization. An azotolane LCP with two azobenzene units demonstrated the largest change in birefringence.

A particularly interesting case is the formulation of photo-deformable materials which can lead to photoactuation. This phenomenon has been recorded for both photo-isomerizable and photo-crosslinkable molecules. The examples of photo-isomerizable molecules include azobenzene [61-63], diarylethene [64] and spiropyran [65], whilst anthracene [66, 67] is a photo-crosslinkable molecule. The change in molecular conformation which defines the *cis-trans* isomerization can also be accompanied by a bulk volumetric change, as demonstrated by Wermter and Finkelman [68] who showed that the shrinkage ratio can be up to 20% in azobenzene LCPs. The fluid-like properties of the LCP also means that the response time is faster than solid state photoactuators, due to the cooperative molecular reorientation during irradiation. The direction of film shrinkage is defined by the alignment direction of the azobenzene molecules and, hence, can be manipulated during the fabrication process of the material, either through rubbing or stretching processes.

The intrinsic molecular anisotropy of LCPs, which allows for long range self-assembly, can also allow bending of the LCP elastomer, through careful manipulation of its elastic constants. In a thick homogenously aligned azobenzene LCP, irradiated light does not pass to the back of the film, thus, causing photoisomerization to occur only on the side which is irradiated and consequently drives a bending motion towards the light source. Conversely, a homeotropic thick film will initiate a photoinitiated isotropic expansion which will cause the material to bend away from the light source. This bending can be reversed through irradiation with visible light. Permutations of this bending principle, such as a hybrid alignment structure can be employed to tune the response time and degree of bending, as demonstrated by van Oosten *et al.* [69], as shown in Figure 3.21. In this case,

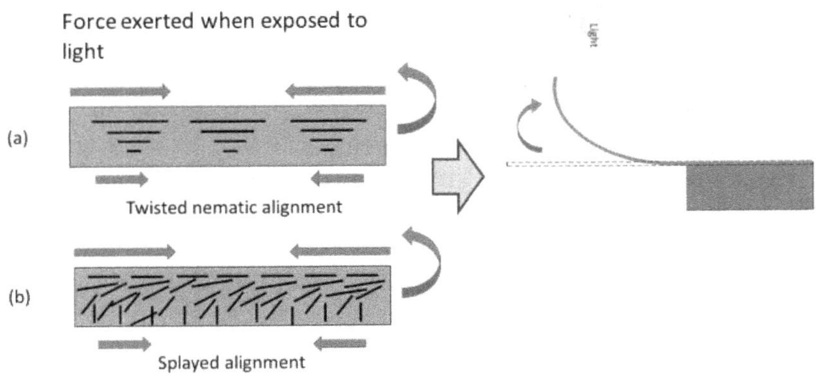

Figure 3.21 (a and b) Schematic illustrations of molecular alignment configurations: (a) twisted nematic (TN) alignment and (b) splayed alignment. Right: model of a photomechanical actuator when exposed to light.

the molecular alignment of the film changes along its thickness, where the molecules near the incident light are chosen to be homogenous, whilst on the opposite side these are chosen to be homeotropic. The concentration of the azobenzene was also tuned in order to render it transparent to light, resulting in an overall bending through shrinking of the incident side and expansion of the opposite side. Such principles have been further extended to devise translational and rotational motions and are particularly useful for actuation, such as biomimicry of muscles [70-72].

Light responsive azobenzene LCPs can be used to form diffractive gratings with anisotropic refractive index modulations through the use of polarization holography [73]. Polarization holography employs two orthogonal polarized light beams to inscribe relief gratings on the azobenzene films. The diffraction characteristics of these gratings are strongly dependent on the polarization direction of the beam, recording energy and resultant relief depth. Demonstration of this concept to produce an in-plane multiplexed high density optical memory showed the ability of the photo-crosslinkable LCP to retain optical information without any crosstalk. [74-76]. The uses of LCPs in anisotropic diffractive gratings have shown potential applications in image holograms and high density memories [77,78]. A high diffraction efficiency was achieved when tolane moieties were copolymerized into azobenzene LCPs, thus, producing large changes in photoinduced refractive index [79]. In particular, employment of photosensitive polymer blends and block copolymers for anisotropic recording media have shown improved diffraction efficiency [80], large photo-induced birefringence and hologram stability [81] and a high degree of repeatability in the read/write process of the optical data [82], thus, impacting advances in information technology through highly functionalized polarization devices.

Extension of the patterning of azobenzene LCP diffraction gratings into the third dimension signal their potential as photonic crystals, as investigated by Yaroshchuk *et al.* [83, 84]. By using two different wavelengths of polarized light, a range of optical anisotropy can be built into the material and may be considered as an alternative to conventional photonic crystals [85].

3.3 Conclusions

In this chapter, we have discussed the unique properties of LCPs which are a result of their polymeric structure and liquid crystal mesophase, which give rise to interesting applications. A strong interdependency of molecular structure and self-assembly of LCPs through a choice of fabrication processes allow the tuning of electro-optic, mechanical and electronic properties. In addition to relatively conventional electro-optic applications such as optical polarizers, lasers and electroluminescence, novel phenomena such as electrorheology, ferroelectrocity and light activation through *cis-trans* isomerization of LCPs give rise to applications such as soft actuators, data storage and drug delivery. Such applications indicate the strong potential of LCPs in a wide range of electronics applications, through exploitation of unique physical characteristics through optimized design of materials and device development.

References

1. Reinitzer, F. (1888) Beiträge zur Kenntniss des Cholesterins. *Monatshefte für Chemie und verwandte Teile anderer Wissenschaften,* **9**, 421-441.
2. Kawamoto, H. (2002) The history of liquid-crystal displays. *Proceedings of the IEEE,* **90**, 460-500.
3. Liu, K., Chen, D., Marcozzi, A., Zheng, L., Su, J., Pesce, D., Zajaczkowski, W., Kolbe, A., Pisula, W., Mullen, K., Clark, N. A., and Herrmann, A. (2014) Thermotropic liquid crystals from biomacromolecules. *Proceedings of the National Academy of Sciences of the United States of America,* **111**, 18596-18600.
4. Louise, K. S. and Winthrop, M. P. (1966) Process for the Production of a Highly Orientable, Crystallizable, Filamentforming Polyamide, US patent, US3287323A.
5. Mitchell, G. R., Davis, F. J., and Ashman, A. (1987) Structural studies of side-chain liquid-crystal polymers and elastomers. *Polymer,* **28**, 639-647.
6. Moore, J. S., and Stupp, S. I. (1987) Orientation dynamics of main-chain liquid-crystal polymers. 2. Structure and kinetics in a magnetic-field. *Macromolecules,* **20**, 282-293.
7. Reck, B., and Ringsdorf, H. (1985) Combined liquid-crystalline polymers - Mesogens in the main chain and as side groups. *Makromolekulare Chemie-Rapid Communications,* **6**, 291-299.
8. Percec, V., and Keller, A. (1990) A thermodynamic interpretation of polymer molecular-weight effect on the phase-transitions of main-chain and side-chain liquid-crystal polymers. *Macromolecules,* **23**, 4347-4350.
9. Hess, L. D. (1984) Book Review of Electro-Optical and Magneto-Op-

tical Properties of Liquid-Crystals by Blinov, L. M. *IEEE Journal of Quantum Electronics,* **20**, 1294-1294.

10. Woon, K. L., Aldred, M. P., Vlachos, P., Mehl, G. H., Stirner, T., Kelly, S. M., and O'Neill, M. (2006) Electronic charge transport in extended nematic liquid crystals. *Chemistry of Materials,* **18**, 2311-2317.
11. O'Neill, M., and Kelly, S. M. (2003) Liquid crystals for charge transport, luminescence, and photonics. *Advanced Materials,* **15**, 1135-1146.
12. Yazaki, S., Funahashi, M., Kagimoto, J., Ohno, H., and Kato, T. (2010) Nanostructured liquid crystals combining ionic and electronic functions. *Journal of the American Chemical Society,* **132**, 7702-7708.
13. McCulloch, I., Heeney, M., Bailey, C., Genevicius, K., MacDonald, I., Shkunov, M., Sparrowe, D., Tierney, S., Wagner, R., Zhang, W., Chabinyc. M. L., Kilne, R. J., McGehee, D., and Toney, M. F. (2006) Liquid-crystalline semiconducting polymers with high charge-carrier mobility. *Nature Materials,* **5**, 328-333.
14. Choi, S. K., Lee, J. H., Kang, S. J., and Jin, S. H. (1997) Side-chain liquid-crystalline poly (1,6-heptadiyne)s and other side-chain liquid-crystalline polyacetylenes. *Progress in Polymer Science,* **22**, 693-734.
15. Manickam, M., Iqbal, P., Belloni, M., Kumar, S., and Preece, J. A. (2012) A brief review of carbazole-based photorefractive liquid crystalline materials. *Israel Journal of Chemistry,* **52**, 917-934.
16. O'Neill, M., and Kelly, S. M. (2011) Ordered materials for organic electronics and photonics. *Advanced Materials,* **23**, 566-584.
17. Bayer, A., Zimmermann, S., and Wendorff, J. (2003) Low molar mass and polymer discotics: Structure, dynamics and opto-electronic properties. *Molecular Crystals and Liquid Crystals,* **396**, 1-22.
18. Marguet, S., Markovitsi, D., Millie, P., Sigal, H., and Kumar, S. (1998) Influence of disorder on electronic excited states: an experimental and numerical study of alkylthiotriphenylene columnar phases," *The Journal of Physical Chemistry B,* **102**, 4697-4710.
19. Seguy, I., Destruel, P., and Bock, H. (2000) An all-columnar bilayer light-emitting diode. *Synthetic Metals,* **111**, 15-18.
20. Heppke, G., Kruerke, D., Lohning, C., Lotzsch, D., Moro, D., Muller, M., and Sawade, H. (2000) New chiral discotic triphenylene derivatives exhibiting a cholesteric blue phase and a ferroelectrically switchable columnar mesophase. *Journal of Materials Chemistry,* **10**, 2657-2661.
21. Kumar, S. (2005) Triphenylene-based discotic liquid crystal dimers, oligomers and polymers. *Liquid Crystals,* **32**, 1089-1113.
22. Boden, N., Bushby, R. J., and Cammidge, A. N. (1995) Triphenylene-based discotic-liquid-crystalline polymers: A universal, rational synthesis. *Journal of the American Chemical Society,* **117**, 924-927.
23. Abeysekera, R., Bushby, R. J., Caillet, C., Hamley, I. W., Lozman, O. R., Lu, Z., and Robards, A. W. (2003) Discotic liquid crystalline triblock

copolymers: Interplay of liquid crystal architecture with microphase separation. *Macromolecules,* **36**, 1526-1533.
24. Kawata, K. (2002) Orientation control and fixation of discotic liquid crystal. *Chemical Record,* **2**, 59-80.
25. Gunduz, B. (2015) Controlling of spectral and optical parameters of the F8T2 liquid-crystalline polymer (LCP) by molarity for optoelectronic devices. *Optik,* **126**, 4566-4573.
26. Le, T. H., Kim, Y., and Yoon, H. (2017) Electrical and electrochemical properties of conducting polymers. *Polymers,* **9**(4), 150.
27. Toyoshima, R., Narita, M., Akagi, K., and Shirakawa, H. (1995) Synthesis and properties of polythiophene derivatives with liquid crystalline substituents. *Synthetic Metals,* **69**, 289-290.
28. Misaki, M., Chikamatsu, M., Yoshida, Y., Azumi, R., Tanigaki, N., Yase, K., Nagamatsu, S., and Ueda, Y. (2008) Highly efficient polarized polymer light-emitting diodes utilizing oriented films of beta-phase poly(9,9-dioctylfluorene). *Applied Physics Letters,* **93**, 023304.
29. Maier, W., and Meier, G. (1961) Eine einfache Theorie der dielektrischen Eigenschaften homogen orientierter kristallinflüssiger Phasen des nematischen Typs. *Zeitschrift für Naturforschung A,* **16**, 262-2671.
30. Vuks, M. (1966) Determination of the optical anisotropy of aromatic molecules from the double refraction of crystals. *Optics and Spectroscopy,* **20**, 361.
31. Fuh, A. Y. G., Tsai, M. S., and Huang, C. Y. (1996) Polymer network formed in liquid crystals: Polymer-network-induced birefringence in liquid crystals. *Japanese Journal of Applied Physics, Part 1-Regular Papers Short Notes & Review Papers,* **35**, 3960-3963.
32. Okano, K., Tsutsumi, O., Shishido, A., and Ikeda, T. (2006) Azotolane liquid-crystalline polymers: huge change in birefringence by photoinduced alignment change. *Journal of the American Chemical Society,* **128**, 15368-15369.
33. Thompson, D., Kingsley, N., Wang, G. A., Papapolymerou, J., and Tentzeris, M. M. (2005) RF Characteristics of Thin Film Liquid Crystal Polymer (LCP) Ppackages for RF MEMS and MMIC Integration. *2005 IEEE MTT-S International Microwave Symposium,* USA, pp. 857-860.
34. Researchers Develop Flexible Nanoelectronics for In-vivo Wireless Biomedical Applications (2013) *Nanowerk.* Online: https://www.nanowerk.com/news2/newsid=30351.php (assessed 19th June 2018).
35. Noel, C., Monnerie, L., Achard, M. F., Hardouin, F., Sigaud, G., and Gasparoux, H. (1981) Orientation of nematic liquid-crystalline polymer in a magnetic-field - Evaluation of magnetic-susceptibility. *Polymer,* **22**, 578-580.

36. Lizuka, E. (2000) Effects of magnetic fields on polymer liquid crystals. *International Journal of Polymeric Materials,* **45**, 191-238.
37. Blinov, L. M. (1983) *Electro-Optical and Magneto-Optical Properties of Liquid Crystals,* John Wiley & Sons, USA.
38. Myhre, G., Sayyad, A., and Pau, S. (2010) Patterned color liquid crystal polymer polarizers. *Optics Express,* **18**, 27777-27786.
39. Kopp, V. I., Fan, B., Vithana, H. K. M., and Genack, A. Z. (1998) Low-threshold lasing at the edge of a photonic stop band in cholesteric liquid crystals. *Optics Letters,* **23**, 1707-1709.
40. Taheri, B., Munoz, A. F., Palffy-Muhoray, P., and Twieg, R. (2001) Low threshold lasing in cholesteric liquid crystals. *Molecular Crystals and Liquid Crystals,* **358**, 73-82.
41. Alvarez, E., He, M., Munoz, A. F., Palffy-Muhoray, P., Serak, S. V., Taheri, B., and Twieg, R. (2001) Mirrorless lasing and energy transfer in cholesteric liquid crystals doped with laser dyes. *Molecular Crystals and Liquid Crystals,* **369**, 75-82.
42. Schmidtke, J., Stille, W., Finkelmann, H., and Kim, S. T. (2002) Laser emission in a dye doped cholesteric polymer network. *Advanced Materials,* **14**, 746-749.
43. Shibaev, P. V., Tang, K., Genack, A. Z., Kopp, V., and Green M. M. (2002) Lasing from a stiff chain polymeric lyotropic cholesteric liquid crystal. *Macromolecules,* **35**, 3022-3025.
44. Matsui, T., Ozaki, R., Funamoto, K., Ozaki, M., and Yoshino, K. (2002) Flexible mirrorless laser based on a free-standing film of photopolymerized cholesteric liquid crystal. *Applied Physics Letters,* **81**, 3741-3743.
45. Shibaev, P. V., Madsen, J., and Genack, A. Z. (2004) Lasing and narrowing of spontaneous emission from responsive cholesteric films. *Chemistry of Materials,* **16**, 1397-1399.
46. Ohta, T., Song, M. H., Tsunoda, Y., Nagata, T., Shin, K.-C., Araoka, F., Takanishi, Y., Ishkawa, K., Watanabe, J., Nishimura, S., Toyooka, T., and Takezoe. H. (2004) Monodomain film formation and lasing in dye-doped polymer cholesteric liquid crystals. *Japanese Journal of Applied Physics, Part 1-Regular Papers Brief Communications & Review Papers,* **43**, 6142-6144.
47. Coles, H., and Morris, S. (2010) Liquid-crystal lasers. *Nature Photonics,* 4, 676-685.
48. Vargha, V., Pozsgay, A. G., and Valentini, P. (1999) Ferroelectric liquid crystal polymers. *Periodica Polytechnica Chemical Engineering,* **43**, 17-33.
49. Dierking, I. (2014) Chiral liquid crystals: Structures, phases, effects. *Symmetry,* **6**, 444-472.
50. Benicewicz, B. C., Johnson, J. F., and Shaw, M. T. (1981) Viscosity behavior of liquid-crystals. *Molecular Crystals and Liquid Crystals,* **65**, 111-131.

51. Miesowicz, M. (1946) The three coefficients of viscosity of anisotropic liquids. *Nature,* **158**, 27.
52. Brehm, T., Pereira, G., Leal, C., Goncalves, C., Borges, J. P., and Cidade, M. T. (2015) Electrorheological characterization of dispersions in silicone oil of encapsulated liquid crystal 4-n-penthyl-4′-cyanobiphenyl in polyvinyl alcohol and in silica. *Physica Scripta,* **90**, 035802.
53. Pranoto, H., Bormuth, F. J., and Haase, W. (1986) Dielectric-properties of a liquid-crystalline siloxane copolymer. *Makromolekulare Chemie-Macromolecular Chemistry and Physics,* **187**, 2453-2460.
54. Goozner, R. E., and Finkelmann, H. (1985) Electric-field effects on a liquid-crystalline side-chain polymer with a negative dielectric anisotropy. *Makromolekulare Chemie-Macromolecular Chemistry and Physics,* **186**, 2407-2414.
55. Ujiie, S., and Kato, T. (2014) Design and synthesis of side-chain liquid crystal polymers. In: *Handbook of Liquid Crystals,* Wiley, USA, doi:10.1002/9783527671403.hlc114.
56. Chiang, Y. C., Jamieson, A. M., Campbell, S., Lin Y., OSidocky, N., Chien, L. C., Kawasumi, M., and Percec, V. (1997) Effect of molecular architecture on the electrorheological behavior of liquid crystal polymers in nematic solvents. *Rheologica Acta,* **36**, 505-512.
57. Inoue, A., and Maniwa, S. (1995) Electrorheological effect of liquid-crystalline polymers. *Journal of Applied Polymer Science,* **55**, 113-118.
58. Hwang, J. J., Iyer, S. N., Li, L. S., Claussen, R., Harrington, D. A., and Stupp, S. I. (2002) Self-assembling biomaterials: Liquid crystal phases of cholesteryl oligo(L-lactic acid) and their interactions with cells. *Proceedings of the National Academy of Sciences of the United States of America,* **99**, 9662-9667.
59. Wan, T., Zou, T., Cheng, S. X., and Zhuo, R. X. (2005) Synthesis and characterization of biodegradable cholesteryl end-capped polycarbonates. *Biomacromolecules,* **6**, 524-529.
60. Guo, J. B., Sun, J., Cao, H., Zhao, D. Y., and Yang, H. (2007) Synthesis and characterization of functionalized triblock polymer: The prepared polymer is cholesteryl terminated and chain-extended PCL. *Journal of Applied Polymer Science,* **105**, 3505-3512.
61. Koshima, H., Ojima, N., and Uchimoto, H. (2009) Mechanical motion of azobenzene crystals upon photoirradiation. *Journal of the American Chemical Society,* **131**, 6890-6891.
62. Nakano, H. (2010) Direction control of photomechanical bending of a photochromic molecular fiber. *Journal of Materials Chemistry,* **20**, 2071-2074.
63. Bian, S. P., Robinson, D., and Kuzyk, M. G. (2006) Optically activated cantilever using photomechanical effects in dye-doped polymer fibers. *Journal of the Optical Society of America B-Optical Physics,* **23**, 697-708.

64. Kobatake, S., Takami, S., Muto, H., Ishikawa, T., and Irie, M. (2007) Rapid and reversible shape changes of molecular crystals on photoirradiation. *Nature,* **446**, 778-781.
65. Athanassiou, A., Kalyva, M., Lakiotaki, K., Georgiou, S., and Fotakis, C. (2005) All-optical reversible actuation of photochromic-polymer microsystems. *Advanced Materials,* **17**, 988-992.
66. Zhu, L. Y., Al-Kaysi, R. O., and Bardeen, C. J. (2011) Reversible photoinduced twisting of molecular crystal microribbons. *Journal of the American Chemical Society,* **133**, 12569-12575.
67. Al-Kaysi, R. O., and Bardeen, C. J. (2007) Reversible photoinduced shape changes of crystalline organic nanorods. *Advanced Materials,* **19**, 1276-1280.
68. Wermter, H., and Finkelmann, H. (2001) Liquid crystalline elastomers as artificial muscles. *e-Polymers,* **1**, doi: 10.1515/epoly.2001.1.1.111.
69. van Oosten, C. L., Harris, K. D., Bastiaansen, C. W. M., and Broer, D. J. (2007) Glassy photomechanical liquid-crystal network actuators for microscale devices. *European Physical Journal E,* **23**, 329-336.
70. Camacho-Lopez, M., Finkelmann, H., Palffy-Muhoray, P., and Shelley, M. (2004) Fast liquid-crystal elastomer swims into the dark. *Nature Materials,* **3**, 307-310.
71. Yamada, M., Kondo, M., Miyasato, R., Naka, Y., Mamiya, J.-i., Kinoshita, M. Shishido, A., Yu, Y., Barett, C. J., and Ikeda, T. (2009) Photomobile polymer materials - various three-dimensional movements. *Journal of Materials Chemistry,* **19**, 60-62.
72. Yamada, M., Kondo, M., Mamiya, J.-i., Yu, Y., Kinoshita, M., Barrett, C. J., and Ikeda, T. (2008) Photomobile polymer materials: Towards light-driven plastic motors. *Angewandte Chemie-International Edition,* **47**, 4986-4988.
73. Nikolova, L., and Ramanujam, P. S. (2009) *Polarization Holography,* Cambridge University Press, UK.
74. Kawatsuki, N., Goto, K., Kawakami, T., and Yamamoto, T. (2002) Reversion of alignment direction in the thermally enhanced photoorientation of photo-cross-linkable polymer liquid crystal films. *Macromolecules,* **35**, 706-713.
75. Ono, H., Emoto, A., Kawatsuki, N., and Hasegawa, T. (2003) Multiplex diffraction from functionalized polymer liquid crystals and polarization conversion. *Optics Express,* **11**, 2379-2384.
76. Ono, H., Hatayama, A., Emoto, A., Kawatsuki, N., and Uchida, E. (2005) Two-dimensional crossed polarization gratings in photocrosslinkable polymer liquid crystals. *Japanese Journal of Applied Physics, Part 2-Letters & Express Letters,* **44**, L306-L309.
77. Todorov, T., Nikolova, L., Stoyanova, K., and Tomova, N. (1985) Polarization holography. 3. Some applications of polarization holographic recording. *Applied Optics,* **24**, 785-788.

78. Ilieva, D., Nedelchev, L., Petrova, T., Tomova, N., Dragostinova, V., and Nikolova, L. (2005) Holographic multiplexing using photoinduced anisotropy and surface relief in azopolymer films. *Journal of Optics A-Pure and Applied Optics,* **7**, 35-39.
79. Ishiguro, M., Sato, D., Shishido, A., and Ikeda, T. (2007) Bragg-type polarization gratings formed in thick polymer films containing azobenzene and tolane moieties. *Langmuir,* **23**, 332-338.
80. Minabe, J., Maruyama, T., Yasuda, S., Kawano, K., Hayashi, K., and Ogasawara, Y. (2004) Design of dye concentrations in azobenzene-containing polymer films for volume holographic storage. *Japanese Journal of Applied Physics, Part 1-Regular Papers Short Notes & Review Papers,* **43**, 4964-4967.
81. Forcen, P., Oriol, L., Sanchez, C., Rodriguez, F. J., Alcala, R., Hvilsted, S., and Jankova, K. (2007) Methacrylic azopolymers for holographic storage: A comparison among different polymer types. *European Polymer Journal,* **43**, 3292-3300.
82. Hackel, M., Kador, L., Kropp, D., and Schmidt, H. W. (2007) Polymer blends with azobenzene-containing block copolymers as stable rewritable volume holographic media. *Advanced Materials,* **19**, 227-231.
83. Yaroshchuk, O. V., Kiselev, A. D., Zakrevskyy, Y., Bidna, T., Kelly, J., Chien, L. C., and Lindau, J. (2003) Photoinduced three-dimensional orientational order in side chain liquid crystalline azopolymers. *Physical Review E,* **68**, 011803.
84. Yaroshchuk, O. V., Dumont, M., Zakrevskyy, Y. A., Bidna, T. V., and Lindau, J. (2004) Molecular structure of azopolymers and photoinduced 3D orientational order. 1. Azobenzene polyesters. *Journal of Physical Chemistry B,* **108**, 4647-4658.
85. Yager, K. G., Tanchak, O. M., Godbout, C., Fritzsche, H., and Barrett, C. J. (2006) Photomechanical effects in azo-polymers studied by neutron reflectometry. *Macromolecules,* **39**, 9311-9319.

Chapter 4

Photoresponsive Main Chain Liquid Crystalline Polymers

Asit Baran Samui[1*] and Kolli Balakrishna[2]

[1]*Institute of Chemical Technology, N. Parekh Marg, Matunga, Mumbai 400019, India*
[2]*School of Technology, Gitam University, Hyderabad, India*
**Corresponding author:* absamui@gmail.com

4.1 Introduction

Organic small molecules, forming liquid crystals, can be incorporated in a polymer structure to make it liquid crystalline polymer (LCP), so that processability is improved along with ample scope of designing the structures with desired properties. Another important advantage in LCP is the formability of thin film, which is useful in numerous devices. Further advancement in the area is the introduction of photoresponsive character, which directly qualifies it to be smart material. Integrating photochromic molecules into LC materials enables one to manipulate unique features such as photoinduced phase transition, photocontrolled alignment and phototriggered molecular cooperative motion leading to innovative applications. As an example, the LCs, in the form of elastomer, show photomechanical and photomobile properties by converting light energy directly into mechanical work. Being responsive to light, these smart materials have advantage over other forms of energy, as light is used to harness the smart characteristics as a clean energy. Due to anisotropic nature and self-assembling capabilities, LCPs can be applied in many areas such as engineering plastics and optical and electrooptical devices, etc. The performance of LCPs as functional material depends on LC properties and supramolecular structure of LC polymers. Therefore, phase structures and transitions of LC polymers have become an important area of research. Various designs using main chain, side chain, their combinations and hyperbranched polymers have been developed and the inventory of photoactive smart materials has been enriched with variety of properties.

Liquid Crystalline Polymers, edited by Vikas Mittal

4.1.1 Commonly Used Chromophores in Photoactive Liquid Crystalline Polymers

A photochromic compound has the ability to alternate between two different chemical forms having different absorption spectra during irradiation by light of appropriate wavelengths. During the photoisomerization, various physicochemical properties, such as refractive index, dielectric constant, oxidation/reduction potential and geometrical structure change together with the absorption spectra. These molecular property changes can be gainfully utilized in various photonic devices, such as erasable optical memory media and photooptical switch components, 2-D and 3-D actuators. Few chromophores are given in Figure 4.1.

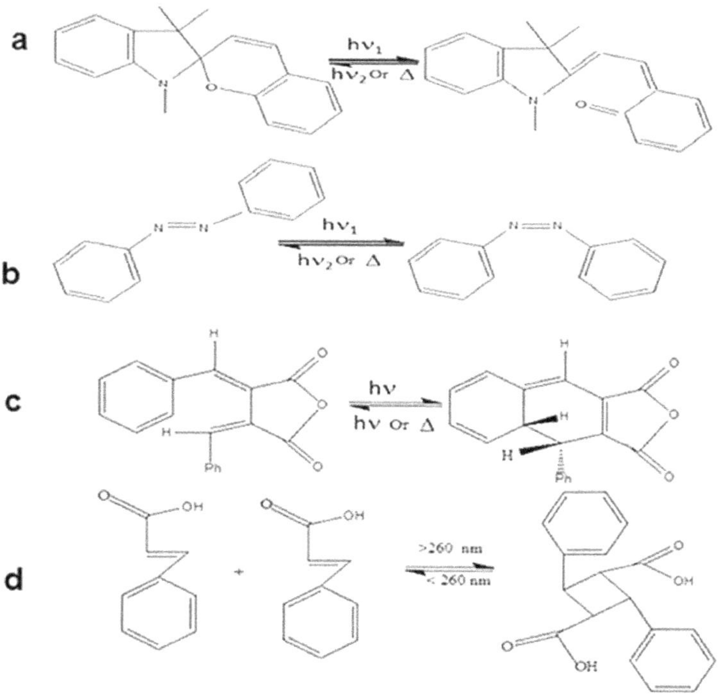

Figure 4.1 Common chromophores and their photoactivity: (a) photoisomerization of fulgides via ring closure and back; (b) photoisomerization of azobenzene from *trans* to *cis* and back; (c) photoactivity of spiropyrans via ring opening and back; (d) photodimerization of cinnamic acid derivative via cyclobutane ring formation.

4.1.2 Liquid Crystalline Polymers (LCPs)

Applications of LCs boosted the research to reach a level that the low molecular mass LCs become a part of high molecular mass LCPs in early 1970s. The design of various architectures of LCPs can be seen in Figure 4.2.

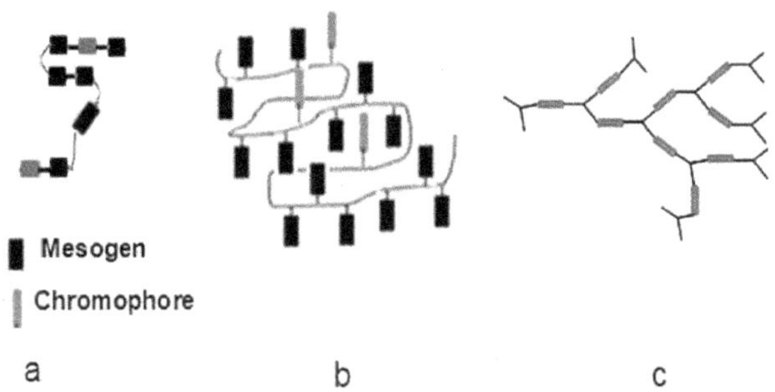

Figure 4.2 Photoactive liquid crystalline polymers: (a) main chain, (b) side chain and (c) hyperbranched.

The initial development was not so beneficial, as rigid-chain polymers are characterized by exceptionally high melting temperatures, and the hypothetical temperatures corresponding to the transition from crystalline state to the LC phase lie within the temperature range of their thermal degradation. Based on the mesogenic moieties attachment in the polymer, LCPs can be categorized into main chain, side chain and hyperbranched LCPs. Small flexible aliphatic chain (spacer) introduction along with mesogen in the polymer, in both main and side chain of the polymers, decreases the rigidity of the polymer which subsequently decreases various thermal transition temperatures of the polymer. Thus, the thermal stability as well as processability is improved by lowering the transition temperature. Further, due to spacer induced flexibility, the orientation of mesophases (LC phase) is also facilitated, which is expressed in higher LC order.

During 1890-1902, the observations of nematic as well as smectic LC phases for the first time motivated the researchers towards synthesizing variety of LCs. By integrating photochromic molecules into LC materials, the resulting materials are made photo-addressable, which have common application in LC display that is based on interactions of light and LCs. With further progress, the features such as photoinduced phase transition, photocontrolled alignment and phototriggered molecular cooperative motion reached a stage at which these can be photomanipulated. The materials are finding important applications in displays [1] along with application in smart devices such as artificial muscles which can respond to electric fields [2], heat [3] and light [4].

Photosensitive LC polymers are generally designed by three principal methods. The first approach involves simple physical blending of photochromic dopant with liquid crystal polymer. In this case, the polymer functions as matrix, or 'host' and the additive (photochromic dopant) as 'guest'. Such mixtures have the advantage of independent design of LCPs and control on the properties of the guest component by light irradiation without any changes in the matrix itself. The second method is the incorporation of both mesogenic and photoactive moieties via covalent bonding to the polymer. In the third method, photochromic dopants are loosely bound to the polymer matrix via non-covalent interactions, such as hydrogen bonding or ionic bonding.

Generally, the optical and electrical properties are dependent mostly on orientation and morphology of the polymers [5]. Thus, the device performance is optimized by the nature and intensity of supramolecular organization of the material [6]. Further, the spacer length variation causes periodic changes in the transition properties based on whether there are an even or odd number of units in the LC polymer [7].

Liquid crystalline elastomer (LCE) can change dimensions under low stresses through phase transitions while remembering the original network [8]. The memory is very much helpful for various applications such as actuators and shape memory devices [9]. The photochromic moiety undergoes structural change under photoirradiation with change in molecular ordering of mesogens, which can be amplified to large mechanical deformation. Under normal condition (before irradiation), the polymer chains in LCE have an elongated chain conformation together with orientational ordering of mesogens. By irradiating with UV radiation, the structure of photochromic moiety changes and as a result the orientational ordering of mesogen is destroyed to attain isotropic phase.

Bisbenzylidene group comprise another very important photoactive series of materials having liquid crystalline nature [10]. A large number of bisbenzylidene based architectures have been developed over the years due to its dual nature and reversible photocrosslinking ability via $2\pi + 2\pi$ cycloaddition reaction. Isomeric structures of bisbenzylidene are presented in Figure 4.3.

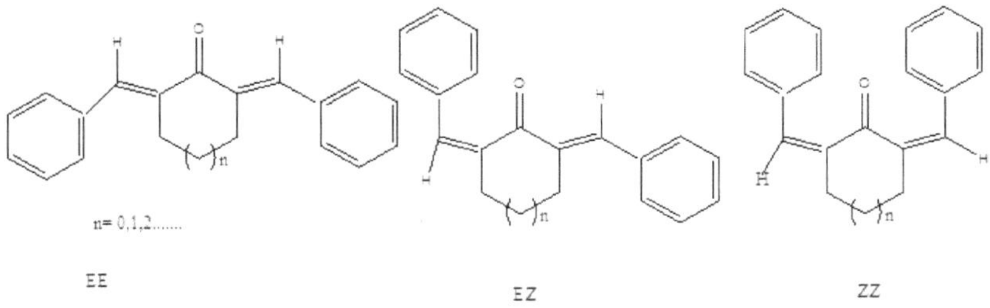

Figure 4.3 Isomeric structures of bisbenzylidene cycloalkanone.

Photoactive LCP (PLCP) blends such as polymer dispersed LC (PDLC) and composites have also been developed for integrating with advanced devices. The blends and composites provide an excellent prospect to combine the properties to suit the application and processing. PDLC are used for a variety of electro-optic applications ranging from switchable windows to projection displays. Further, dye/PDLC combination enhances the transmission (~50%) and makes the switching responses faster [11].

For main-chain and side-chain LC polymers, various strategies to harness the LC properties have been adopted by proper choice of mesogenic groups for incorporation in the polymer along with their location and the placement of flexible spacers with variable lengths [12,13]. The role of spacer length is important as can be seen with main chain LCP; for example, the smectic layer thickness increases with the spacer length.

The extension of main chain LCP can be a combination of the chemical features of main chain LCP and side chain LCP that result in a hybrid structure called as main-chain/side-chain LCP [14]. The structural specialty leads to their broad LC-phases and polymorphism making them interesting for various applications. Interestingly, the main chain and side chain mesogens orient parallel to each other and interact cooperatively to form the LC-phase. This synergism broadens the LC phase temperature range as compared to that of the corresponding LC main chain and LC side chain polymers. Photoresponsive natures are introduced by incorporating photoresponsive moieties such as azobenzene.

The interest in liquid crystalline elastomers (LCE) arises from the competition of the liquid-crystalline order and mechanical properties. The most fascinating property of LCE is to couple the liquid crystalline order of the mesogenic groups by macroscopically orienting the whole sample by the application of a mechanical stress [15]. This results in a polydomain-to-monodomain transition, in which the director of the mesophase becomes uniformly aligned throughout the whole sample with respect to the direction of the external stress [16]. This order can be locked in by using cross-linking strategy, which results in stable monodomains with almost perfectly oriented mesogens. These species called as liquid single-crystal elastomers (LSCEs) are stable materials with interesting properties suitable for various potential applications such as artificial lenses and muscles [17].

Block copolymers with incompatible polymer blocks have the ability to self- assemble into a variety of ordered morphologies. The blocks being linked together via chemical bonds, phase separation at a macroscopic level is not possible; rather, it occurs at nano-scale. The properties of two completely different blocks can be combined in this way with development of nanostructures. This type of architectures can be designed with azobenzene moiety as mesogen, which will exhibit photochromism [18].

Shape memory polymers (SMP) have the unique ability to come back to the original or permanent shape after being changed to different shape with external stimulus. Unlike amorphous counterpart, photoresponsive LC networks exhibit fast and large deformation in the alignment direction of mesogens. LC networks can be functionalized with photoresponsive molecules, which have properties of LCs, elastomers as well as photoisomerization ability [19].

The functional capabilities of LCs can be enhanced by designing self-assembled structures from the nano- to the macro-scale [20]. The concept of supramolecular chemistry has enlarged the scope of newer designs of LCs. In fact, the supramolecular LC phases with well-defined structures have been formed by exploiting intermolecular hydrogen bonding [21]. These dynamic LC assemblies can also be built by using other molecular interactions, such as ionic and charge-transfer interactions.

Physical gelation of liquid crystals with low molecular weight gelators results in a new class of anisotropic gels, which are found to be excellent materials for applications as optical, electrical and photofunctional materials. LC physical gels are considered composites, which comprise liquid crystals and a small amount of self-assembled fibers [22]. The incorporation of azobenzene moieties into the gelator leads to the preparation of the photoresponsive liquid crystalline gels, which are used for rewritable information recording media [23].

Non-linear optical (NLO) polymers, having application in integrated optics and telecommunication, are the materials of choice to replace the inorganic crystals. However, the stability of poled dipoles for application towards second harmonic generation is not good. Introduction of the LC phase enhances the NLO property by increasing the extent of alignment of dipoles during poling and extending the poling derived orientation stability of the poled polymer [24].

Microcapsules are microcontainers, which can encapsulate gases, liquids, and solids and release the contents in controlled manner, while the coating material can be designed to suit particular application and respond to certain stimulus. The release of core (microencapsulated material) by light irradiation has become popular for application in agricultural and cosmetic field. UV and visible light-sensitive microcapsules are best suited for applications, while near-IR absorbing capsules are required for applications related to biological systems. Generally, the light-sensitive polymers, functional dyes, and metal nanoparticles are incorporated to make the microcapsules sensitive to light [25].

4.2 Photochromic LC Polymers

4.2.1 Main Chain LC Polymers

The liquid crystalline polymers have one common feature: these comprise mesogen units connected via spacers (Figure 4.4). The spacers allow the freedom of rearrangement of mesogens during phase formation, phase change, stretching, etc. In an alternative type of LC polymer, the mesogenic groups may be attached to the polymer chain via either ionic interaction among the components or hydrogen bonding among main chain (or side chain) and the mesogenic fragments. Hyperbranched architectures also result in LC polymers. The photoresponsive nature is derived from introduction of photoactive moiety in the LC structure or as a guest. Photochromic LCs exhibit unique properties in response to light as their deformation can be easily controlled by irradiating at specific wave length of light. Photo-induced structural change of photochromic moiety has the ability to change the molecular ordering of mesogens. The ordered mesogens and photochromic moiety exist together in ordered molecular arrangement. During irradiation, the molecular ordering is destroyed, which is then amplified to large deformation.

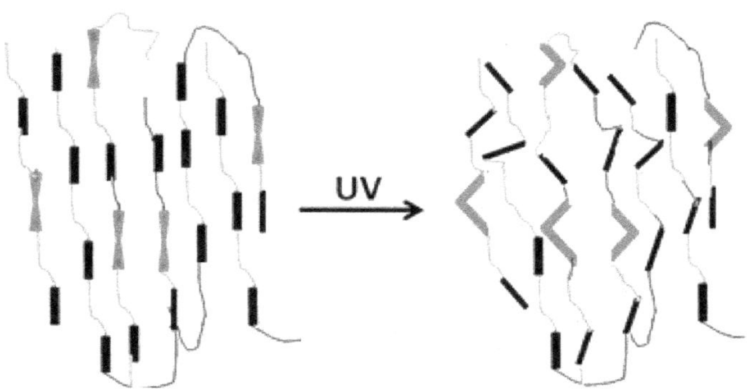

Figure 4.4 Schematic representation of main chain photoactive liquid crystalline polymer.

As an example, the azo molecules undergo isomerization from *trans* to *cis* state during irradiation of light. The *trans* state, having rod like structure fits in the nematic phase of LC molecules without disturbing the arrangement. During irradiation, the *trans* form of azo transforms to bend *cis*-form. During this process, the whole system gets disorganized together with the reduction of the size of azo molecule. This induces contraction on the side of the film exposed to light. The light cannot penetrate to other side of the film. Thus, only one side of the film experiences contraction, while the other side remains unchanged. This causes bending of the film, which is due to photomechanical effect.

Photoinduced bending and unbending of LC film are possible by alternative irradiations of polarized UV and unpolarized visible light [26]. Although side-chain LC polymers containing azobenzene groups are commonly used as photoresponsive materials, the main-chain LC polymers show higher actuator performance due to improved coupling of ordered structure of the main chain and its network strain as compared to its side-chain analogues. Due to the disadvantages, such as high crystal to nematic transition temperature, a solvent swollen state or high temperature is required to exhibit photo-induced deformation. Also, the designed LCEs are generally crosslinked elastomers, which is not desirable in respect of processability. More efficient materials, such as main-chain LC polymer with photoisomerizable azobenzene side chains, can be designed from photochromic LC monomer having a photochromic diazobenzene moiety, a phenol benzoate nematic core and two polymerizable olefin groups (Figure 4.5) [27]. The polymer film can undergo up to 22% photo-induced contraction at room temperature.

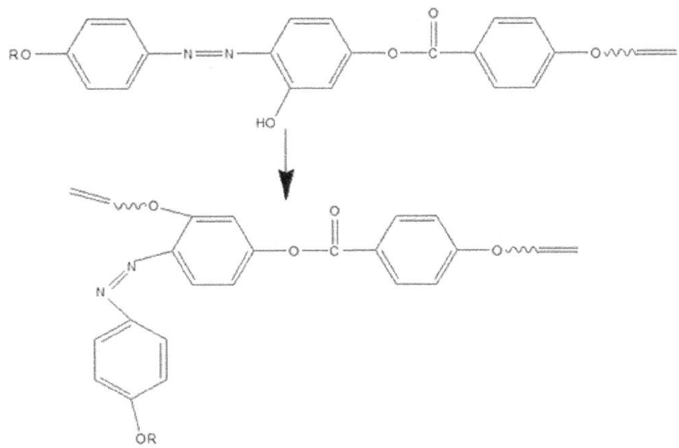

Figure 4.5 Schematic representation of synthesis and polymerization of photochromic LC monomer.

The presence of the azobenzene units in the polymer structures imparts important properties with practical applications in optics and optoelectronics. In command layer of LC cell, azobenzene dissolved in a polymer layer or attached as side chain of a polymer are commonly used, so that it can change the orientation of LC component. Although, the azo group is placed in a main polyether chain, photoisomerization takes place during irradiation with UV light [28]. The chemical structure of the polymer influences the isomerization rate; more rigid the chain, less is the isomerization rate.

A polymer network carrying azobenzene groups can stabilize liquid crystal together with altering its orientation through irradiation. In contrast with surface alignment, a polymer network offers "bulk" effects, which has the advantage of aligning liquid crystals in relatively thick films. Also, the alignment of liquid crystals via irradiation of the polymer network provides selectivity for the orientation that surface treatment cannot. As the alignment of azobenzene is confined only in areas exposed to irradiation, the orientation of liquid crystal molecules occurs only in these areas. Thus, combined with surface alignment, an irradiation-regulated orientation provides polymer stabilized LC (PSLC) materials more possibilities for design of electrooptical properties. In addition to PSLC, the approach can be utilized for designing electrooptical materials, for which the basis of applications is often the induction and control of birefringence. Diacrylate monomer having an azobenzene moiety as the central core is used for thermal polymerizations in the presence of liquid crystal, which results in the formation of a homogeneous polymer network stabilizing the liquid crystal [29]. The structure of diacrylate monomer is shown in Figure 4.6.

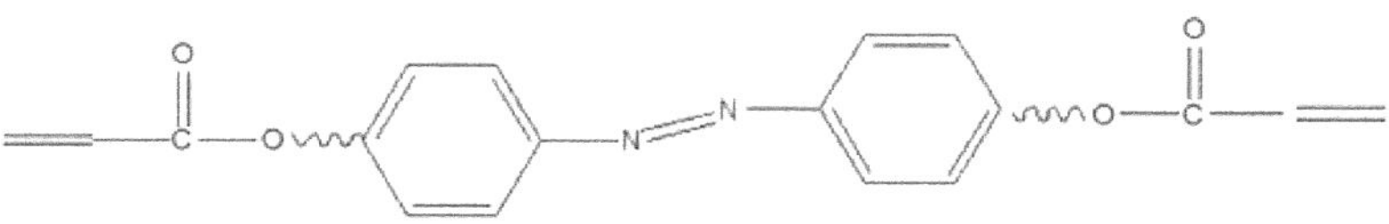

Figure 4.6 Schematic representation of diacrylate monomer with azobenzene moiety.

The effect of alignment of azo moiety on LC molecules can be evaluated by dichroism study. The composite film exhibits strong perpendicular dichroism for the bands at 2227 (CN), 1606 and 1494 (phenyl) cm^{-1} after irradiation with UV radiation. It indicates that the orientation of liquid crystal (E7) molecules is perpendicular to the UV polarization. Thus, the alignment of azobenzene groups in the network dictates the alignment of director fields of liquid crystal molecules and orients them along the same direction. The change of orientation of azo moiety in response to UV light is fast. However, the changes in the director fields are rather slow. In any case, the bulk orientation is faster than the surface alignment. In fact, the orientation of LC molecules surrounding the azo moiety as immediate neighbor changes first, which activates the next layer and so on. Therefore, at high network density, all the LC molecules cannot be oriented; the farthest LC molecules remain unaffected. The induced orientation decays very fast during first few days after irradiation. Later, the decreased orientation is maintained over long time. This is possible due to network formation, which is able to hold the diminished orientation for long time.

Nematic liquid crystalline polymers can be synthesized from 4′-hydroxyphenyl(4-hydroxy-3-allyl)azobenzene as photoresponsive mesogenic unit [30]. The allyl-substituent disturbs the packing of the polymer chains, which results in lowering crystal melting temperature to great extent. However, the nematic phase remains stable to be detected. The allyl unit is quite stable as identical calorimetric curves are exhibited during repeated heating and cooling cycles. Thermal crosslinking does not occur under mild conditions. It is possible only by using an activator agent like tert-butylperoxibenzoate. The rate of isomerization of azobenzene chromophores in the amorphous solid state is slower than that in solutions and does not occur in the nematic liquid crystalline phase. This occurs due to the presence of strong J-aggregation (antiparallel arrangement) and molecular constraints that minimizes the free volume in the polymer matrix.

Efficient synthesis of a series of azo main-chain LCPs, with variable spacer length and having reactive secondary amino groups in the polymer backbones, can be performed under mild reaction conditions using Michael addition polymerization [31]. These main-chain azo polymers are thermally stable, have low glass transition temperatures, a broad temperature range of smectic C LC phase and reversible photochemical behaviors. In addition, the presence of secondary amino group in the backbone makes it highly reactive precursors for various modifications and crosslinking. Due to the unique characteristics of these main chain azo based LCPs, supramolecular hydrogen-bonded LCP fibers can be fabricated by the simple melt spinning method, which have high mesogen order along the fiber axis. Further, fast and reversible photoinduced bending and unbending behaviors with controllable three-dimensional photomobile directions are common during irradiation of UV and visible light even near ambient temperature. The unbending of bent fiber under application of thermal energy is somewhat slower as compared to that by visible light irradiation. The fatigue resistance of such fibers is also very high as confirmed by repetitive bending and unbending. The stress generated by the fiber during UV irradiation can reach very near to that of human muscle (300 kPa).

4.2.2 Combined Main-Chain/Side-Chain Liquid Crystalline Polymers

Main chain/side chain LCPs (MCSCLCP) exhibit more complex behaviors as compared to conventional main chain or side chain LCPs. The interplay between main-chain and side-chain of MCSCLCP leads to versatile LC properties and supramolecular structures. The common synthetic method for MCSCLCP is the condensation polymerization and the general architecture is the main-chain having either rigid rod or alternative structure of rod-like mesogenic groups and flexible spacers, and the side-chain comprising mesogenic group linked to the main-chain repeating unit through a flexible spacer. In another strategy, mesogenic side groups are attached as lateral substituents onto the mesogenic moieties of a semi-flexible main-chain polymer. Hydrogen bonding can also be used to generate MCSCLCP. Synergistic stabilization of the LC phase between the main-chain and side-chain mesogenic units in MCSCLCP acts in widening the mesophase temperature range as compared to the corresponding main or side chain LCP [32]. The orientational order parameter may approach values of 0.9. In fact, orientation of some of the polymers is possible in weak external fields such as 1 T magnetic fields. The re-orientational motion of the mesogens about their short axis, having coupled to dielectric and electro-optical properties, has very low activation energy and a strong anisotropy of the dielectric constant despite a relatively small concentration of dipoles.

Mesogen-Jacketed Liquid-Crystalline Polymers

Although the mesogen-jacketed liquid-crystalline polymer (MJLCP), having mesogenic units laterally attached to the polymer backbone via without or only short spacers, is considered a type of SCLCPs, it exhibits special liquid crystalline property of main chain LCPs. The spatial requirement of the bulky and rigid mesogenic units allows the main chain of MJLCP to adopt an extended conformation and the "jacket" formation takes place by the arrangement of mesogenic units around the backbone. The specific liquid crystal characteristic of MJLCPs is that the main chain decides the director of the LC phase. The mesogen-jacketed model describes the close wrapping of laterally attached mesogenic side groups around the flexible polymer backbone, which forces the polymer chain to stretch along the main-chain direction [33]. The whole polymer chain acts similar to a supramolecular rod and its molecular characteristics can be tuned easily as the rod length can be controlled by varying the molecular weight of the polymer. The rod diameter can also be controlled by varying the lengths of both the rigid core and the flexible tails in the mesogenic side chains. Among XRD, FTIR, POM, DSC, WAXD techniques for analyzing phase behavior of MJLCPs, 2D WAXD is mostly used for analyzing the phase structure. Mostly, the MJLCPs form columnar phases together with smectic phases in few cases. Rod-like MJLCP chains self-organize into columnar LC phases, whose type is dependent on the molecular properties including the shape, size and flexibility of side chains. The columnar nematic phase is commonly found in many systems. More ordered columnar phases, such as 2D hexagonal columnar and rectangular columnar phases, are also observed [34]. Due to flexible nature of the soft tails at the ends of the mesogenic side groups and rigid nature of mesogens, a core–shell cylinder can be realized by using relatively long tails. This amounts to the possibility of the surface chemistry adjustment with a variation in the tail structure. Thus, an MJLCP can be considered as a semi-rigid polymer chain, which is used to constitute rod–coil block copolymers. Together with shape tunability, the MJLCPs are easily designed by controlled/living free radical polymerization with molecular weight up to 400,000 g mol^{-1} and good solubility in most solvents.

In MJLCP blocks containing block copolymers, the most common morphology is the lamellar morphology, and the liquid crystallinity in the lamellae is maintained by MJLCPs, when the molecular weight of the MJLCP block reaches the critical value. Solution behavior of a typical rod-coil block copolymer, e.g., polystyrene-2,5-bis[(4-methoxyphenyl) oxycarbonyl]styrene copolymer (PS-b-PMPCS) presents interesting behavior. On cooling the dilute solution, the copolymer chains self-assemble into a core–shell nanostructure having PMPCS and PS segment as the core and shell respectively. With increasing solution concentration, the calculated core radius remains constant while shell thickness rises. Core radius is mostly dependent on the length of the PMPCS block. In case of more chains inserted into the core–shell structure, the coil PS blocks in the shell are stretched to accommodate more chains.

Photoinduced anisotropy and photochemical phase transition can be observed in MJLCP containing azobenzene moiety. During exposure to UV-radiation (λ= 365 nm) at low temperatures, the thin films of polymer having single azobenzene per repeating unit lose birefringence under POM due to complete conversion to isotropic state [35]. The behavior is different with attachment of two azobenzene groups per repeating unit. Even the POM images of thin film under UV irradiation at 80 °C cannot convert it to isotropic state as the original grainy texture changes only to schlieren texture. Thus, the remaining LC texture does not belong to azobenzene as it is converted to bend *cis* state. As the main chain in MJLCP acts as mesogen, the LC texture originates from it.

4.2.3 Photocrosslinkable Liquid Crystal Main-Chain Polymers

Main-chain liquid crystal elastomers respond to changes in temperature especially in the vicinity of a phase transition. Mostly, these crosslinked materials are synthesized in single step reaction leading to macroscopically aligned elastomer. For making thin films, the polymer solution is kept in-between two glass slides and selectively irradiated to crosslink. The copolycondensation reaction of a mesogenic monomer and a functionalized benzophenone photocrosslinker can yield main chain LCP. The polymer solution is also electrospinned together with photocrosslinking to obtain LC nanofiber. The LC elastomeric film can be obtained by spin coating of polymer solution, which is followed by drying and UV irradiation. Also, the lightly photo-crosslinked film is stretched till it becomes transparent and subsequently exposed to UV light to complete the photocrosslinking. These elastomers possess exceptionally high mechanical properties such as a strong load-dependent thermoelastic effect and a nonlinear stress–strain relation.

4.2.4 Photoresponsive Block LC Polymers

Block copolymers with LC segments are ideal models for semicrystalline block copolymers that can be employed as highly versatile interfacially active additives and viscosity modifiers, which are capable of providing enhanced optimization of material performance. By using azo macroinitiators, block copolymers comprising amorphous/main-chain LC blocks [36], and both main-chain and side-chain LC blocks can be prepared [37]. For the first copolymer system, the chemically different blocks being mutually immiscible, produce multidomain structures in which the two blocks are strongly segregated. The latter class of block copolymer, the two chemically different blocks are partially phase-separated in the glassy and LC phases as can be confirmed by thermal and dynamic-mechanical data. However, the glass transition and the smectic-nematic and nematic-isotropic phase transition parameters of the main-chain block are different from those common to structurally analogous homopolymer, which is due to the existence of specific interactions between the main-chain and side-chain blocks. However, the main chain/side chain block copolymers, having side-chain LC polymethacrylate block with side-chain mesogenic unit and a main-chain LC polyester block made up by two mesogenic p-oxybenzoyl diads alternatively interspaced by aliphatic chains of five and ten methylene groups and connected to the mesogenic cores by two ester and ether linking groups respectively, exhibit lower stability than the corresponding homopolymers [38]. The phase separation in block copolymers is being gainfully utilized for broad applications in nanotechnology because of the possibility of controlling their size, shape and periodicity by changing the structural parameters such as molecular weight, composition and chain connectivity [39]. Furthermore, by using external stimuli such as light, the morphology of phase separation can be manipulated by incorporating photoresponsive component into a polymer segment. By incorporating photoresponsive moiety such as azobenzene, the phase-separated structures in LC block copolymers can be controlled. The photoisomerization induced change in the molecular alignment of an azobenzene containing polymer can be fixed below glass transition temperature. Thus, the use of polarized light can reversibly control the alignment of a nanostructure. Despite containing a similar block mesogen, an amorphous polymer is not able to generate phase-separated structures. This is due to the fact that the cooperative effect of the amphiphilic mesogen in the LC phase is necessary for nanophase separation [40].

4.2.5 Bisbenzylidene Based Main Chain LCPs

The bisbenzylidene cycloalkanone (BBCA) molecule, a stilbenoid class of organic molecules, is mesogenic in nature and has distinct photo-active properties. Linear LCPs containing photocrosslinkable groups, besides their common use in making anisotropic networks, have attracted considerable attention in areas like information storage, non-linear optics, permeation of gases and drugs through the aligned membrane, etc. Bisbenzylidene can be incorporated into different kinds of architecture to generate main chain LCPs [10]. Bisbenzylidene containing main chain LC polycarbonates can be synthesized by interfacial polycondensation of benzylidene derivatives of cycloalkanones or ketones with phosgene or bischloroformates in the presence of sodium hydroxide. Bis(4-hydroxybenzylidene) cycloalkanone and various aliphatic dicarboxylic acid chlorides react to form bisbenzylidene containing polyesters [41]. Two kinds of photoreactions, i.e. photoisomerization and photo-crosslinking occur during photolysis of the polyesters. The former reaction disrupts the parallel stacking of the chromophores. The extent of each reaction depends on the mobility of the polymer chains. If the photolysis is performed below T_g, the photo-crosslinking dominates over photoisomerization. Above T_g, the photoisomerization occurs and is followed by photo-crosslinking. With increasing the size of the cycloalkanone ring, the photo-sensitivity of the polymers decreases. Linear benzylidene polyesters containing aromatic diacid components are synthesized by solution polycondensation of bis(4-hydroxybenzylidene)-4-phenyl cyclohexanone and terephthaloyl chloride in the presence of pyridine (Figure 4.7) [42].

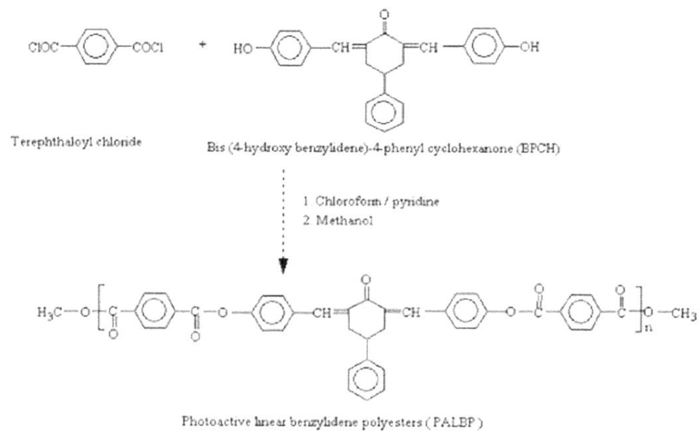

Figure 4.7 Preparation of linear poly(BBCA) ester by solution polycondensation of bis(4-hydroxybenzylidene)-4-phenyl cyclohexanone and terephthaloyl chloride in the presence of pyridine. Reproduced from Reference 42 with permission from John Wiley and Sons.

The presence of conjugated rings makes the polymers fluorescent in nature and the fluorescence intensity increases with time of UV irradiation. The UV irradiation causes the photocrosslinking via formation of cyclobutane ring which, in turn, increases the rigidity of structure. The increase in rigidity is conducive for enhancing the fluorescence of the benzylidene main chain LCP. The viscosity increases with UV irradiation time because of intermolecular photocycloaddition of double bonds in benzylidene moieties. By polycondensation reactions of α,ω-bis(chloromethyl) polydimethylsiloxanes with varying molecular weights and 2,6-bis(4-hydroxybenzylidene) cyclohexanone, BBCA polyethers are obtained [43]. The presence of highly flexible siloxane moieties induces a thermotropic smectic mesophase, unlike other types having only a nematic phase.

Series of main chain photoactive liquid crystalline polyethers containing bisbenzylidene, variable substitution type and flexible methylene spacers of variable length are synthesized by polycondensation of bisbenzylidene diols and dibromoalkanes [44]. Variation of spacer length affects the LC phases, which are either smectic A (fine texture) or a combination of smectic A and nematic droplets. There is a reduction in width of LC phase with increasing the number of methoxy substituents on mesogenic unit. Polymers with two substituents on benzylidene diol unit and without a substituent are stable up to 400 °C, whereas polymers with four substituents are stable up to only 260 °C. The steric hindrance of substituents prevents the chromophores to move close to each other leading to a decrease in the rate of photocycloaddition. In fact, the rate of photocycloaddition decreases with increasing number of the methoxy substituents on the rigid units (Figure 4.8).

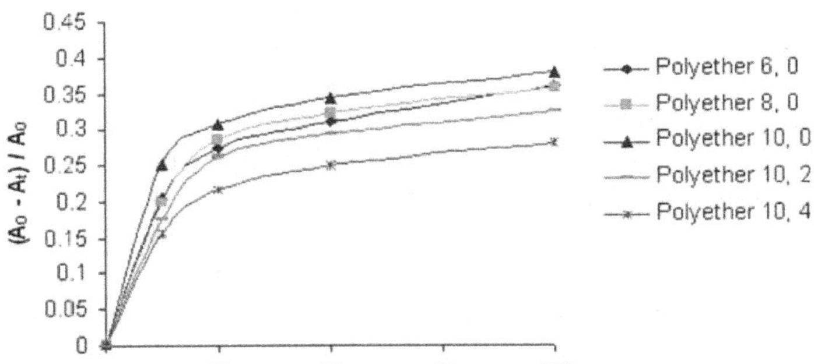

Figure 4.8 Effect of substitution and spacer length on photocrosslinking of substituted benzylidene polyether solutions. Reproduced from Reference 44 with permission from John Wiley and Sons.

There is an increase in photoviscosity of all the polymer solutions with increasing irradiation time as there is progressive increase of crosslinking (Figure 4.9). The least substituted polymers exhibit faster increase of photoviscosity due to faster photocrosslinking.

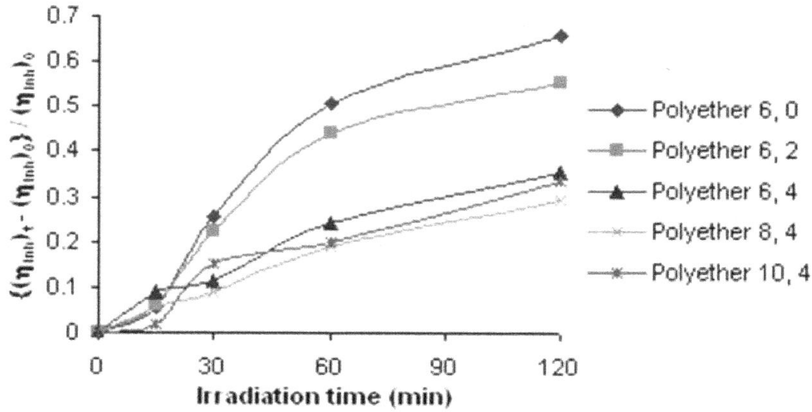

Figure 4.9 Effect of substitution and spacer length on photoviscosity of substituted benzylidene polyether solutions. Reproduced from Reference 44 with permission from John Wiley and Sons.

With increasing spacer length, the potential energy of the cross-linked dimer does not change much, while it increases with increasing the number of substituents in the polymer chain, which has a direct correlation with steric hinderance arising from the presence of substituents.

Other main chain LCP can be designed with specific polymer classes such as poly(BBCA) phosphate esters, poly(BBCA) azomethine ethers, poly(BBCA) sulfonates and poly(BBCA) phosphoramides. The photoresponsive nature, thermal and chemical stability make them useful for a variety of applications. Bisbenzylidene derivatives exhibit distinct features in second harmonic generation (SHG) as a non-linear optical material. These types of materials are more efficient than conventional urea in non-linear application and also satisfy the criteria required for practical applications such as, phase matching, cut-off wavelength, stability and laser-power tolerance. Also, the damage threshold of these materials is much higher than urea.

Poly(BBCA) sulphides, derived from 2,5-bis-(p-chloro-m-nitrobenzylidene) cyclopentanone and sodium sulfides in N-methlpyrrolidone at higher temperatures, exhibit an electrical conductivity of about 2.3x 10^{-12} S cm^{-1} [45]. Intramolecular charge transfer exists in the molecule in which the nitro and the 2-oxo-1,3-cyclopentanediylidene moieties act as acceptor and donor respectively. Further, the presence of conjugation is responsible for the conduction mechanism. Similarly, a series of BBCA polymers, having photoresponsive nature, exhibit electrical conductivity due to presence of conjugation present in the polymer backbone.

One of the important series of main chain LCP is the elastomers, possessing photoisomerization and photocrosslinking ability along with the high flexibility. In this respect, poly(benzylidene) ether elastomers can be synthesized from cyclopentanone, 4,4'-diformyl-α,ω-diphenoxy alkene and 4,4'-diformyl-2,2'-dimethoxy-α,ω-diphenoxy alkane in 20% aqueous alkali at 80 °C [46]. Wide LC phase region, presence of crystalline phase together with photoresponsive nature make these materials ideal for application as smart elastomers. The degradation temperatures of such type of elastomers are quite high at around 300 °C. Poly(BBCA) urethanes are prepared via step growth solution polymerization, by reacting arylidene monomers, like 2,5-bis(4-hydroxybenzylidene) cyclopentanone, 2,5-divanillylidene cyclopentanone, 2,6-bis(4-hydroxybenzylidene) cyclohexanone, 2,6-divanillylidene cyclohexanone or 2,7-bis(4-hydroxybenzylidene) cycloheptanone with diisocyanates such as toluene diisocyanate and methylene bis(4-phenylisocyanate), etc. [47]. The presence of –O–CO–NH– (urethane group) along with –C=C– generates crystalline domains. Siloxane-benzylidene sulfone segmented copolymers containing unsaturated units and ester groups are synthesized by the polycondensation of chlorine-terminated benzylidene-sulfone macro-monomers and sodium salt of 1,3-bis(sebaco methyl) tetramethyl disiloxane at 140 °C [48]. Incompatibility between the hard units (arylidene sulfone) and the soft segments (aliphatic siloxane) results in phase separated morphology. High thermal stability, photoresponsive nature and two-phase morphology promotes application as impact modifier and the presence of the internal unsaturation qualifies these type of elastomers for multipurpose application. Heat curable BBCA polymers belong to amides, imides, macrocylic polymer, etc. Unsaturated heat curable polyamides and polyimides are prepared from 2,6-bis(3-aminobenzylidene) cycloheaxnone by reacting with terephthaloyl chloride and pyromellitic dianhydride or benzophenone tetracarboxylic anhydride respectively [49]. The unsaturated polymers are cross-linked through thermal curing at 265 °C for 27 h. Macrocyclic poly-(2-vinylnaphthalene), containing a single 1,4-benzylidene unit, exhibits negligible absorption and emission of 1,4-benzylidene [50].

4.3 Oligomers

4.3.1 Photoresponsive Main-Chain Oligomers

The properties and functions of polymers/oligomers depend on the chain length and chemical structure. However, the polymers/oligomers are polydisperse because of common synthesis routes. Thus, the dependence of physical properties on the chain length and chemical structure cannot be measured correctly. Precise synthesis of polymers with monodisperse molecular weight, chemical structure and topologies is required to be achieved to eliminate the uncertainty. Preparation of precise oligomers/polymers containing azo-functional groups is performed by synthetic approach using acyclic diene metathesis (ADMET) technique, which leads to the introduction of azobenzenes into the olefin structures repetitively for the generation of saturated and unsaturated azo oligomers [51]. The azo (-N=N-) bond survives both the metathesis and the subsequent hydrogenation chemistry. The azo oligomers, forming well-defined aggregates with an average hydrodynamic radius in the range of 40-60 nm in THF, exhibit molecular weight dependent growth with time; the larger molecular weight oligomers form higher aggregation sized micelles. The aggregate sizes of saturated azo-oligomers are bigger than their unsaturated counterparts and the growth of the former is more rapid than the latter. The hydrodynamic radius of the micelles decreases by approximately 14% during UV irradiation. This is due to increased polarity induced by isomerization of the azobenzene from its planar rod-shaped form to the non-planar bent-shaped *cis* form, which enhances the solubility in polar solvents.

4.3.2 Macrocycles Containing Azobenzene Main Chain Oligomers

Together with chain length, the polymeric topologies usually influence the properties and functions of polymers to considerable extent. Due to their chain-end free topology, the cyclic polymers often display exceptional performance as compared to their linear counterparts. Considering the influence of chain length and cyclic topology, the molecularly-defined cyclic polymers provide a good model for precisely exploring the structure–property relationship [52]. By incorporating one or more azobenzene moieties in a cyclic structure, the topological constraint imposed by the ring strain can trigger a much larger impact on the photoswitching properties of azobenzene, which emphasizes the superior performance of the cyclic compounds as compared to non-cyclic structures [53]. Molecularly-defined linear and cyclic oligomers, containing alternate azobenzene and tetraglycol moieties in the main chain, can be prepared by CuAAC "click" chemistry and stepwise chain growth approach [54]. As compared to linear analogues, the cyclic topologies have faster photo-isomerization from the *trans*-form to *cis*-form and a slightly faster reversible relaxation from the *cis*-isomer to *trans*-isomer. This behavior originates from more stable conformation for cyclic *cis*-azobenzene relative to the linear analogues, particularly for the small ring. Also, the linear polymers being more flexible than the cyclic one, due to the absence of strain and the presence of swinging chain ends, experience higher transition resistance of isomerization than cyclic ones. The first factor plays a dominating role in the faster *trans*-to-*cis* photo-isomerization for the cyclic products. With increased chain length, progressive change of the *cis* isomer percentage and different self-quenching effect of fluorescence emission are common for both linear and cyclic oligomers.

4.4 Liquid Crystalline Elastomers

Liquid crystalline elastomers (LCEs) are weakly cross-linked elastic polymer networks containing rigid mesogens bonded to the polymer chains. The mesogens can exhibit spontaneous orientational ordering similar to low molar mass liquid crystals. However, the mesogens, being chemically bound to the cross-linked polymer network, any change in their orientation will be coupled to changes in the mechanical properties of the bulk material. If the overall preferred direction for the mesogens is same throughout the material, then this network is termed aa a mono-domain LCE. During phase transition of the mesogens from the liquid crystalline state to the isotropic state, mono-domain LCEs exhibit a change of their macroscopic shape. The transition between the ordered and disordered states is controlled by the thermodynamic equilibrium, which indicates that the shape change is fully reversible. The transition energies are available from either of electrical, magnetic and photochemical sources. The synthesis of main chain LCE can be achieved by using rod like mesogen and flexible spacer in the main chain, along with a flexible crosslinker up to 10 mol% concentration respectively

[55]. This amounts to the number of mesogens on average in the polymer chains (degree of polymerization) as 4, 8 and 16 respectively with 10, 5 and 2.5 mol% crosslinker.

4.4.1 Photocrosslinking of Main Chain LCE

Main-chain LCEs, showing high response in the vicinity of a phase transition, are crosslinked in a single step reaction, thus, resulting in a macroscopically aligned elastomer in only macroscopic scale. Thin films can be prepared by synthesizing photocrosslinkable nematic main-chain polymers containing ethylene oxide units, LC units and benzophenone as UV crosslinker [56]. The thin films are first produced and oriented, which is followed by crosslinking by UV light to generate liquid single crystal main chain elastomers. These elastomers possess exceptional mechanical properties. Electrospinning of the polymer can be performed from solution under UV irradiation leading to highly oriented fibers, which are insoluble and exhibit a stable monodomain structure.

4.4.2 Nanoparticle-Liquid Crystalline Elastomer Composites

In most LCE systems, the large deformations can be realized by heating the material so that the LC state changes into the isotropic phase. However, the polymeric LCEs, having low thermal conductivity, do not allow quick delivery of heat to all mesogens in the material. This delays the LC-isotropic phase transition, so that the response time of thick and large LCE films or other shapes indicates a drawback for possible applications. The mechanism of switching between the LC phase and the isotropic phase being the underlying principle, any external source acting towards such phase transition can be used to trigger the actuation. Photosensitive materials have limitation of the penetration depth, which allows only thin films for quick actuation. Various nanoparticles can absorb irradiation and convert it into heat for triggering phase transition. Polymer-functionalized single-walled carbon-nanotubes (SWNTs) and liquid crystal elastomers (LCEs) nanocomposites can exhibit reversible IR-induced strain (~−30%) at very low SWNT loading levels (≤0.2 wt%) [57]. SWNTs can efficiently absorb IR light and convert it into thermal energy, thereby serving as a nanoscale heat-source and making a thermal-conduction pathway. The absorbed thermal-energy induces a nematic–isotropic phase-transition, which results in generating a shape change in the nanocomposite film.

Besides absorbing visible and near-IR light, CNTs also show absorption within the microwave, terahertz and megahertz frequency ranges [58]. High loading and alignment of CNTs also makes LCE composites anisotropic at these frequencies.

4.5 Supramolecular Assembly

Supramolecular polymerization changes the phase organization, viscosity, optical features, etc., such as the interaction of non-mesogenic molecules may form supramolecular species having LC behavior [59]. It is possible that selective interaction occurs only from complementary species, which may lead to recognition at molecular level. The most common molecules forming the thermotropic LC are designed by using a rigid core with flexible unit at two ends. As the complementary monomers are mixed, initiation and chain growth occurs, while chains of different compositions are formed side by side without any cross-coupling. Chain growth is initiated after the recognition groups are set free by using external stimulus like light. Similar to normal polymerization, chain length can be altered or reduced by using single recognition groups as chain terminator.

Specific intermolecular interactions, such as hydrogen bonding, play a major role in the supramolecular assembly of liquid crystals and induce dynamic functions to the assemblies [60]. Although, hydrogen-bonded LC materials have been known earlier, they were not considered useful for advanced technologies until supramolecular concept involving hydrogen bonding was introduced into the design of liquid crystals in 1989 [61].

The supramolecular liquid crystals are designed for the first time by formation of a complex via one hydrogen bond, which exhibits smectic and nematic phases [62] and the complex with three hydrogen bonds shows a columnar phase (Figure 4.10) [61].

Nanosegregation of the incompatible section of amphiphilic or block copolymers is one of the key driving forces for LC formation. The morphology of segregation at nanometer scale has direct relation with the micro-segregation behavior of block copolymers [63]. The rigid rod-like structure of azobenzene molecules makes them suitable candidates for exhibiting liquid crystallinity and their photoisomerization characteristics provide the possibility of molecular motion in response to light. Stoichiometric mixture of a proton acceptor and a proton donor is heated in an oil bath at temperatures above their melting points with vigorous magnetic stirring and is allowed to remain in the melt state for about 1 min before it is air-cooled to room temperature. In the polymeric complex, the proton acceptor is locked by two dibenzoic acids via hydrogen bonds. The hydrogen bonding results in the formation of less flexible pseudo-polymer chain due to conformational constraints as well as due to being in the polymer main chain. The *trans–cis* photoisomerization is, therefore, reduced substantially [64].

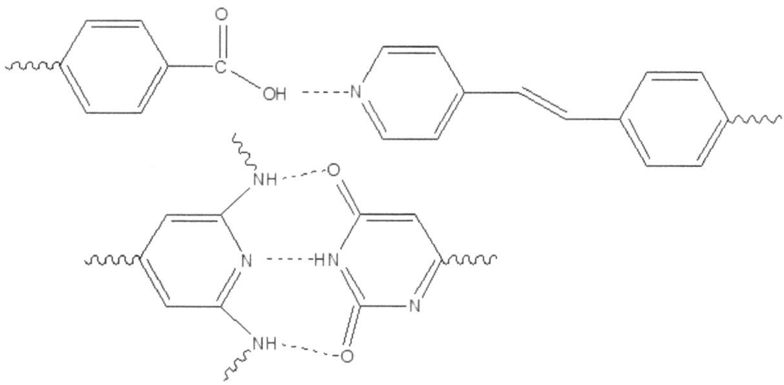

Figure 4.10 Supramolecular liquid crystals complex with hydrogen bonds.

4.6 Typical Synthesis Methods for Main Chain LC Polymers

In a typical method, LC diol is reacted with vinyl side group containing diester via melt polycondensation by using tetraisopropylorthotitanate in diglyme as the catalyst in the temperature range of 140-150 °C [65]. α,ω-dihydro-oligo(dimethylsiloxane) is employed to crosslink the main chain LCP in presence of platinum catalyst overnight at 90-100 °C, which is followed by evaporation of solvent over two days.

The synthesis of the elastomer network film can be achieved by spin-casting of a mixture of LC polyether end-capped with vinyl-functionalized monomer and cyclosiloxane in absolute toluene and platinum catalyst during 24 h at 6000 rpm [66]. The Si-H moieties of cyclosiloxane react with vinyl functionalities to produce crosslinked main chain elastomers.

Photoactive liquid crystalline main chain linear benzylidene polyester epoxy with pendant hydroxyl groups is synthesized by polyaddition of photoactive bis benzylidene cyclopentanone biepoxy monomer with terephthalic acid in the presence of tetra-n-butyl ammonium in dimethyl formamide at 90 °C, under nitrogen over a period of 24 h (Figure 4.11) [67].

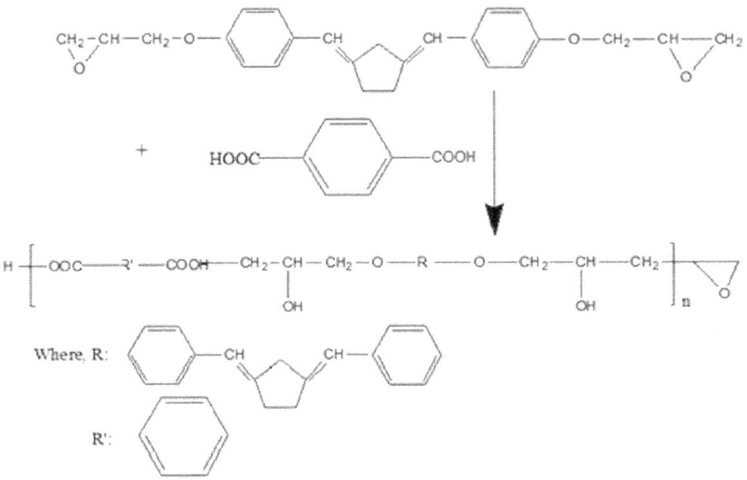

Figure 4.11 Schematic representation of synthesis of photoactive linear benzylidene polyester epoxy.

The benzylidene based main chain polyester epoxy with pendant hydroxyl group provides additional reactivity to the polymer, which can be utilized for desired crosslinking or further modification.

Allyl 4-((4-(allyloxy) phenyl)diazenyl)benzoate, dry toluene and Karstedt's catalyst mixture is heated to 50 °C and 1,1,3,3-tetramethyl disiloxane (0.16, 1.82 mmol) is added drop wise to it (Figure 4.12) [68]. The reaction temperature is raised to 90 °C and reaction continued till Si-H bond disappears. The molecular weight remains low at around 3000-4000 g/mol. Low molecular weights of the polymers are considered for fast *trans-cis* isomerization, which can be utilized in photoactation.

Synthesis of condensation type LC polymers from isolate diene monomers, with built-in aromatic type mesogenic unit bearing the cholesterol pendant through oxydodecyloxy spacer or monomesogenic terephthalate ester unit bearing the cholesterol pendant through a spacer, can be performed in both solution and bulk [69].

By this method, reasonably high molecular weight polymers can be prepared at a low temperature under very mild conditions. Further, the reaction condition being mild, there is no interference with catalyst activity.

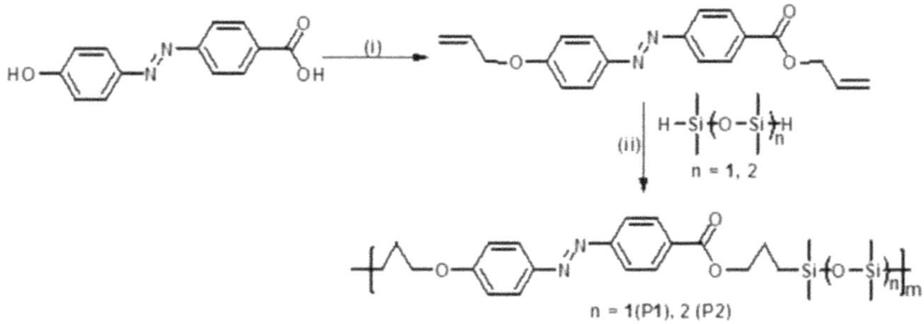

i) Allyl bromide, K_2CO_3, DMF, 80 °C; ii) karstedt's catalyst, toluene, 90 °C, 48h

Figure 4.12 Schematic representation of synthesis of azo-dialkene and its hydrosilylation polymerization. Reproduced from Reference 68 with permission from Royal Society of Chemistry.

The SmA LC main-chain copolymers are based on soluble polymalonates and a laterally brominated biphenyl [70]. The incorporation of crosslinkable groups into the polyester backbone can be achieved via two routes: the first one comprises the incorporation of phenols into the polyester. The less reactive phenol groups do not participate in the transesterification reaction during condensation polymerization. These can be esterified after the polymer is generated by using acrylates. Thermally or photochemically created radicals perform the crosslinking. In the second route, benzophenone is incorporated as a side group, which carries out photochemical crosslinking. Crosslinked fibers (monodomains) have the potential to be used as smectic LC main-chain elastomers based actuators.

4.7 Opto-mechanical Effects

4.7.1 Shape Memory Polymers

Shape-memory polymers are dual-shape materials, having the ability to change from one shape to other and the change can be triggered by heat, light, electric field, alternating magnetic field, immersion in water, etc. The shape-memory effect is not an intrinsic property, i.e. the polymers do not display this effect spontaneously. In fact, this type of behavior originates from a combination of polymer morphology and specific processing procedures, which is akin to polymer functionalization. By conventional processing techniques such as, extrusion or injection molding, the polymer is formed into the permanent shape. The polymer is then deformed under stress to fix into the temporary shape. Upon application of an external stimulus, the polymer recovers its initial permanent shape and the process of deformation and recovery can be repeated several times. Light-induced stimulation of shape-memory polymers can be designed by incorporating reversible photoreactive molecular switches or fillers [71]. The strategy for incorporating photoactive groups into polymer environment has transferring effects from the molecular level into effects that are macroscopically visible. Also, the temperature increase required for inducing shape memory in polymers can be provided by IR light illumination [72]. Molecular switches and permanent "netpoints" constitute the photoresponsive shape-memory polymers. Together with tailored transition temperatures, such as isotropization temperature for liquid crystalline elastomers (LCE), the shape memory polymers require specific mechanical properties. Liquid crystalline elastomers with varying compositions can be used for providing both properties. Main-chain smectic-C LCEs have the exceptional capabilities to be used as shape memory elastomers having low transition temperatures. As an example, the benzoate-based mesogenic groups, 1,4-bis[4-(4-pentenyloxy) benzoyl] hydroquinone and 2-tert-butyl-1,4-bis[4-(4-pentenyloxy) benzoyl] hydroquinone, can be coupled with hydride-terminated poly(dimethylsiloxane) spacers to produce main chain LCE [26]. After the polymer is heated above the clearing transition of the initial polydomain smectic-C primary shape, the network loses its smectic ordering and becomes highly deformable. A secondary shape can then be given to the network by applying stress, such as stretching or twisting, which is fixed by cooling below the clearing transition (I-SmC). The secondary state possesses a highly ordered smectic-C structure, which is energetically stable and the shape is retained over long period of time. If and when required, the secondary shape can return to primary one via second heating above the clearing transition, which results in "melting" of the smectic-C phase and release of the deformed network toward its equilibrium configuration. As the clearing temperature determines the shape fixing and recovery, the polymer can be designed with mixed-mesogen according to applications. Although, the modulus is not high, it has immense applications in biomedical field. The siloxane chemistry is the preferred choice for synthesizing main chain LCE. The linear and non-functional/functional polyhydrosiloxane chain is reacted with the mesogenic groups and a crosslinking agent in one step. This reaction is platinum catalyzed, which results in the attachment of mesogens and crosslinking moieties to polyhydrosiloxane backbone [30].

4.7.2 Liquid Crystalline Elastomers as Actuators and Sensors

Liquid crystalline elastomers (LCEs) are materials that combine the elastic properties of polymeric elastomers with the self-organization behavior of liquid crystals. The most interesting property of LCEs is the ability to change their shape reversibly after the application of a certain external stimulus. By irradiating a photoresponsive LCE film with unpolarized UV light (366 nm for azobenzene), the bending of the film occurs towards the irradiation direction [73]. This is followed by unbending to initial stage by exposing the bent film to unpolarized visible light. The bending and unbending behaviors are repeatable by alternate irradiation of UV and visible light. Further, the bending of the LCE films occurs anisotropically, only in the direction parallel to the rubbing direction of the alignment layers. During irradiation of photoresponsive LCE films, the *trans–cis* photoisomerization occurs near the surface, which triggers the nematic-isotropic phase transition only in the surface region, leaving the bulk undisturbed. The isomerization derived *cis* state has smaller dimension than *trans* state, which results in contraction of azobenzene. This is exhibited by bending motion of the film.

Crosslinked Systems and Their Properties

Generally, the synthesis route of LCEs comprises the build-up of the LC polymer having mesogenic groups and polymer chains along with the crosslinking of the polymer chains to form an infinitely large network. Silicone based LCP can be crosslinked by using hydrosiloxane and tetraethoxy/tetramethoxy silanes, respectively. Liquid crystalline polymer having free functional groups can be mixed with a bi- or multifunctional crosslinking agents to react selectively with the functional groups under controlled reaction conditions, leading to network formation, as also shown in the reaction scheme in Figure 4.13.

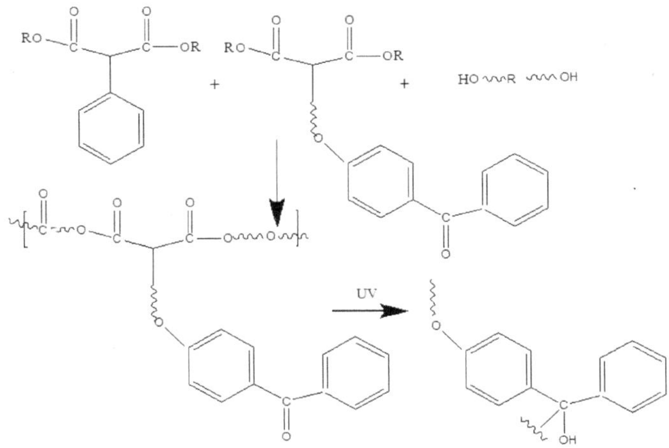

Figure 4.13 Schematics representation of synthesis of polyester and subsequent UV mediated crosslinking via benzophenone moiety attached as side group.

The crosslinking is mostly based on isocyanate reaction, click chemistry, hydrosilylation, etc. The main chain LCP is also synthesized by a melt polycondensation of a mesogenic diol and a malonic acid ester, which is partially functionalized with a benzophenone derivative for subsequent crosslinking by UV irradiation. In another strategy, LC monomer with polymerizable moieties is mixed with multifunctional crosslinker and initiator and polymerized by either thermal or UV-irradiation conditions. LC monomer with a vinyl and a mercapto group ends is mixed with a multifunctional crosslinker (two vinyl and two mercapto groups) and a radical photoinitiator and subjected to UV irradiation to yield main-chain elastomer [74].

In a monodomain nematic single crystal elastomer in which all the polymer chains are, on average, parallel to each other and crosslinked, the molecular level shape change translates to the macroscopic scale [74]. In fact, the strongest shape changing effects have been observed mostly in main chain systems. Also, by using a main-chain polymer as a crosslinker for a side-chain system yielding a combined system, the shape changing effect can be increased to large extent as compared to the pure side-chain system [75].

Between smectic and nematic main-chain polymers having almost similar structure, the former shows stronger shape variation (210%) compared to the nematic (80%) because of higher order parameter of the smectic phase [76].

Photopolymerization of a nematic monomer (thiol-ene) can be carried out along with a crosslinker with the help of a photoinitiator and finally the molten product is molded in pillar shape in a PDMS soft mold, prepared by replica molding [74]. After initiating the reaction in the temperature range corresponding to iso phase, it is cooled down to temperature corresponding to LC phase, while using a permanent magnet to align the nematic director. As the main chain nematic LC polymer is heated above the nematic to isotropic transition temperature (around 170 °C), the pillars experience an ultra-large contraction up to 400%.

The use of heat as stimulant is impractical. Electric field can be used to trigger the actuation. However, mostly the energy gained by the interaction of a nematic phase with an electric field is quite small as compared to the energy needed for deformation [77].

The response of well oriented photoactive main chain LCEs, with azo derivatives as co-monomers attached to the load cell, can be measured during irradiation and during relaxation process when the irradiation is terminated. The nematic photoactive main chain LCE containing symmetric azo derivative show exponential growth and exponential decay of stress upon turning the UV-light on and off respectively. Further, the clearing temperature for higher crosslink density sample is lower than the isotropization temperature of the lower crosslinked sample, which is due to the isotropic crosslinker acting as an impurity. This results in some disorder that lowers the transition temperature [78].

Sample with higher mechanical response is expected for the higher crosslinked sample and, in fact, the sample with 10 mol% of crosslinking density is about 6 times more mechanically-efficient than the corresponding sample with 2.5 mol% crosslinking density (Figure 4.14) [55]. At lower temperature, the mechanical response increases with crosslink density, while at higher temperature, the increase is not significant. It is also true that it is much more efficient than the side chain LCE.

On superimposing the normalized curves of the samples, it is observed that slightly faster photo- and thermal-isomerization processes occur in the sample with higher crosslinking density. This is possible as any change in the polymer conformation is transmitted more efficiently when the polymer chains are shorter or more nodes are present in the polymer network. The presence of smectic layer in the main chain LCE is responsible for higher Young's modulus as compared to nematic LCEs. However, the smectic photoactive main chain LCE can recover only 55% of the maximum stress during the first thermal-isomerization process. This is possible as after first photo-isomerization process, the smectic layers are partially broken and the correlation is lost. Thus, the smectic LCE shows an irreversible contraction, if free, which is due to the shape memory effect. Then, the starting point of next cycles is the increased initial value of the stress (higher than the applied pre-load) corresponding to the final value reached at the end of the first cycle. The maximum stress value during the irradiation processes is more than six times than that for the nematic photoactive main chain liquid crystalline elastomers.

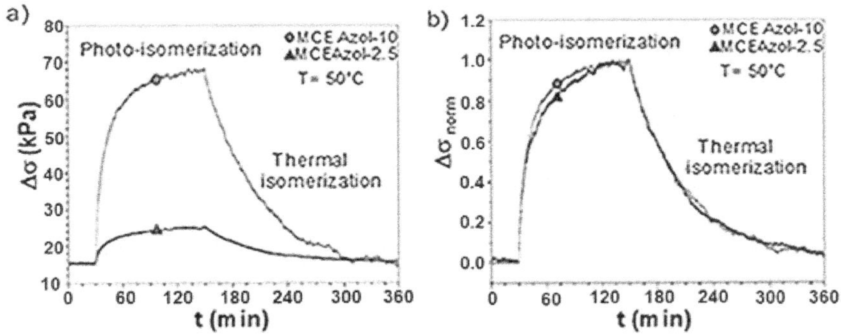

Figure 4.14 a) Opto-mechanical experiment on the nematic main-chain liquid-crystalline elastomers MCEAzo1-10 and MCEAzo1-2.5 at 50 °C and λ = 380 nm. b) Normalized curves during the photo- and thermal-isomerization processes for the two nematic photoactive MCLCEs. Reproduced from Reference 55 with permission from SPIE.

The activation energy in the photo-isomerization process depends on the nature of the azobenzene compound together with the nature of the macromolecule. The value of activation energy for highly crosslinked elastomer is 63 kJ mol $^{-1}$, while the corresponding value for the low crosslinked system is only 36 kJ mol $^{-1}$ which matches with that of side-chain elastomers [79]. The stress-avoiding isomerization is due to the short length and rigidity of the polymer backbone, acting as a spring between crosslinking points, which delay the photo-isomerization process for the highly crosslinked main chain elastomer.

By incorporating superparamagnetic nanoparticles into an LCE film and applying a rapidly changing electromagnetic field, vibration derived heat can be used to raise the temperature to clearing temperature. Photoisomerizable groups (mainly azo) can change the state of LC phase during *trans-cis* isomerization. On irradiation with UV light, the isomerization is mostly concentrated on the surface. Therefore, a bilayer can be designed to form a workable system. Motor based on an LCE material transforms light energy into motion [80]. An LCE-covered ribbon can be placed around two pulleys and irradiated on the right side of the belt with UV light for *trans–cis* isomerization and on the left side with visible light for *cis–trans* isomerization. Contraction occurs on one side while expansion on the other, which leads to a rotation of the pulleys. The alternating bending and stretching upon UV and visible light irradiation can induce a lateral motion on a solid substrate. In another arrangement, one edge of an LCE covered film is made sharp and the other end round, which on irradiation with changing wavelength displays an inchworm like motion of the film.

Opto-mechanical Effect in Liquid Single Crystal Elastomers

Liquid single crystal elastomers (LSCE) have sufficient network anisotropy consistent with a macroscopically uniform director alignment, which is realized by synthetic design. The LSCE will be stable under this condition without any external fields in the LC state. Due to uniform director orientation without defects, the LSCE has excellent transparency suitable for application for optical elements. The LC moieties can lead to optimum non-linear optical property and the rubber elasticity is responsible for electromechanical effects such as piezoelectricity of chiral smectic C-LSCE [81]. As UV light falls on LSCEs, besides maximum stress that the system is able to generate, the time required by the network to produce its maximum mechanical response and that necessary to recover the initial state thermally are vital for complete performance evaluation of LSCEs for artificial muscle-like photo-actuation. The UV-irradiation process of an LSCE is a bidirectional step, since the *trans*-to-*cis* photoisomerization reaction competes with the thermal *cis*-to-*trans* isomerization simultaneously.

$$trans \underset{k^{th}}{\overset{k^{ph}}{\rightleftarrows}} cis$$

Both individual isomerization processes are observed to follow a first order profile for low-molecular weight azo derivatives in isotropic and nematic solution as well as in dense polymer matrices [82,83].

Stimulus-induced Deformations in Various Geometries

Liquid-crystal networks can be designed to undergo deformations in various geometries in response to light. Photoresponsive moieties such as azobenzene incorporated LC polymer films exhibit only bending motions, while photo-isomerization remains slow. The core can be modified or substituents can be added to the azobenzene moiety so that substantial changes in photophysical and photochemical properties can be realized [84]. This helps in avoiding the use of a complex set-up that involves multiple light sources, lenses or mirrors. By incorporating azobenzene derivatives with fast *cis*-to-*trans* thermal relaxation into liquid-crystal networks, photoactive polymer films can be generated, which exhibit continuous, directional and macroscopic mechanical waves under constant light illumination [85]. Azo derivatives are copolymerized with a mixture of LC methacrylate and acrylate. The monomer mixture is first aligned in a splayed configuration over the cross-section of a thin film, with homeotropic alignment at one surface and planar alignment at the other, and then polymerized using photoinitiator to LC network. The presence of hydrogen bond enhances the push–pull effect, which results in a decrease in the half-life as observed with thermal *cis-trans* relaxation. In these samples, the bending and relaxation are instantaneous due to increase of temperature of the film beyond glass transition temperature during irradiation and the decrease of half-life to very low value.

As a strip of the thin LC network film is placed on a glass surface, enabling free movement without any constraints, the polymeric film exhibits a caterpillar-like motion. On clamping the film at one end, oscillations are observed. By gluing together two edges of the film, a tube formation occurs, which buckles and continuously deforms during irradiation. If the two edges of the film are glued separately, continuous travelling wave is generated during irradiation. By using continuous light source, the initiated wave continuously regenerates and propagates in a repeating and snake-like motion. By this motion, small object can be successfully transported over a certain distance.

4.7.3 LC Anisotropic Gels

LC anisotropic gels are oriented LCs having long-range molecular orientation permanently retained by the network. The interesting feature is that the LC orientation disappears in the isotropic phase and reappears when the gel is cooled to the LC phase temperature range. This memory effect present in the network suits well for the design of attractive applications such as electrically switchable mirrors made from cholesteric anisotropic gels [86]. To prepare anisotropic gels, the conventional method comprises dissolution of a reactive monomer in a nonreactive LC, together with a photoinitiator. The homogeneous mixture is aligned using surface orientation layers in the LC phases and photopolymerization is carried out to form the network. Optical alignment can also be performed to prepare anisotropic gels without the use of surface orientation layers. LC anisotropic gels can be obtained by polymerization of optically aligned azobenzene monomer in the isotropic phase of the LC, which, in turn, induces and stabilizes the long-range LC orientation [87]. Formation of anisotropic network in the isotropic phase of the LC compound has the advantage of using high monomer concentrations without any concern about the temperature corresponding to LC phases. However, at higher monomer concentrations, the miscibility between the LC molecules and monomer may decrease. With monomer concentration up to 10%, better orientation is achieved. By irradiation with linearly polarized light during cooling and the liquid to nematic transition, the anisotropic network formed in the isotropic phase is able to set long range LC orientation. On heating the network again to isotropic phase and cooling back in the absence of irradiation, the LC orientation is completely or partially recovered.

4.7.4 Ferroelectric Liquid Crystals

Ferroelectric liquid crystals (FLCs) have a chiral smectic-C phase, S^*_C, having each layer with permanent electric polarization confined in the plane of the layer and perpendicular to the director tilted by an angle from the layer normal. The chirality in a bulk FLC prompts the rotation of the directors of the successive layers and the polarization vectors are cancelled by induced helical structure over a pitch length. Therefore, the spontaneous polarization of FLCs can be made practically useful by suppressing the helical structure in the S^*_C

phase by aligning molecules in one direction. The incentives for developing FLC technologies are that FLCs have important advantages over other liquid crystals (LCs) for use in electronic and photonic devices with respect to fast switching speed of FLCs in an electric field. The surface-stabilized FLCs can be generated by using rubbed surface [88]. However, the fabrication process and difficulty of inducing and sustaining a uniform alignment are the critical components in the development, which need betterment. An azobenzene containing monomer is polymerized in FLC host and irradiated with linearly polarized light, which results in photoalignment in an anisotropic polymer network that achieves both the induction and stabilization of the FLC orientation in the S^*_C phase [89]. The advantages in this strategy can be listed as:

- The photoalignment of the azobenzene monomer or network is a bulk process and can be performed uniformly along the thickness direction, which indicates the possibility of alignment and stabilization for samples thicker than the helical pitch.
- As no rubbing is required, the difficulties associated with the rubbing process are eliminated.
- The polymer network makes it favourable for improving the shock resistance of FLC cells, which is of great interest for studies on polymer-stabilized FLCs.

Few disadvantages are also there:

- The polymer network reduces the spontaneous polarization of the FLC that slows down the switching speed to some extent.
- The network may also decrease the homogeneity of the cell and reduce the contrast ratio.

On dossolving chiral dimethacrylate and diacrylate monomers in a FLC host and polymerizing under linearly polarized irradiation, the monomers are able to induce a bulk alignment of FLC in the SC* phase, with a minimum concentration of 1% [90]. With higher percentage (~10%), no bulk alignment is possible as the polymer network is too dense to allow alignment. If azobenzene moieties in main chain are connected via shorter spacers, their mobility is reduced, which exhibits less efficient photoinduced orientation. Generally, the chirality on monomers affords stronger interaction between the azobenzene groups and FLC in the chiral S^*_C phase, which facilitates bulk alignment of the latter via the action of the former.

4.7.5 Light-Sensitive Microcapsules Based on Liquid Crystalline Polyester

Microcapsules are known to encapsulate various kinds of substances, such as gases, liquids and solids, which are called as core materials. The coating substances, known as shell, can also be selected from a wide variety of natural or synthetic polymers, depending on the requirement of particular applications. By proper selection of core and shell material, it is possible to prepare microcapsules with a variety of functions, such as adhesives, coatings, food additives, catalysts, dyes, drugs, perfumes, etc. Microcapsules are designed to release their contents in response to a stimulus such as mechanical damage, temperature, pH, ionic strength, electric field, magnetic field, irradiation with light, etc. The light as stimulant finds applications in the fields of intelligent microstructures and microsystems, phase-change materials, self-healing composites, etc. The possibility of triggering release by light irradiation has specific demand for application in agricultural and cosmetic field and in this respect, UV and visible light-sensitive microcapsules are preferred. For applications in biological systems, near-IR absorbing capsules are more suitable.

It is obvious that microcapsules can be made sensitive to light by incorporation of light-sensitive polymers, functional dyes and metal nanoparticles [91]. LCPs are known to form efficient chain packing, which imparts excellent barrier properties to the LCP film. This state can be changed reversibly by disturbing the orientation to isotropic phase, thus, resulting in the change of permeability of the film. The strategy for making the film smart is to incorporate photoresponsive groups as guest or as a part of the polymer chain to use clean energy. Of course, the heat can also be used as a trigger for LC phase transformation. Thus, azo containing monomer is used, along with trifunctional crosslinker, to react with diamine in interfacial polycondensation of an oil-in-water dispersion to generate lightly crosslinked main chain LC polyamide microcapsules [92]. In this approach, toluene or concentrated solution of naphthalene or β-carotene are incorporated as core material of the microcapsules that have diameter lying in the range 30-120 μm. By beaming UV light continuously on the microcapsules, β -carotene is released quickly to reach the maximum within 5 min.

Poly(α-methylstilbenesebacoate-co-α-methylstilbeneisophthalate), containing chloroform or chloroform/vanillin as a filler, is sprayed on methanol so that precipitation of polymer occurs immediately to form the microcapsule [93]. The condition of microcapsule formation is maintained such that the polymer remains in *trans*-state. During UV irradiation, the roughness of the microcapsule decreases by about 24%. The surface skewness, which describes the asymmetry of the height distribution, changes from a negative value to a positive one during UV irradiation. The negative value indicates a surface-porous sample in which the valleys dominate over the peak regimes, while the positive value indicates a lack of surface porosity and smoothening of the microcapsule surface [94].

The smoothening of the surface is due to photoisomerization of α-methylstilbene moieties. The moieties are in the E form with a molecular length of 9.1 Å and after UV irradiation, these convert into Z isomers having a length of 6.8 Å. This leads to the substantial modification and smoothening of the film morphology, in response to the contraction of the polymer main chain. The release of the core liquid is accelerated to a large extent during the irradiation of ultraviolet light.

After microencapsulation, nanoencapsulation approach has been developed, which enables producing particles with size in the nano region. Nanosizing of the capsules results in a dramatically increased interfacial area with the surrounding environment, which offers significant advantages in pharmaceutics and material science by assisting in controlling the release kinetics. Along with more efficient drug detoxification and intracellular uptake, the nanoparticles can be made biocompatible with the tissue and cells, when synthesized from biocompatible or biodegradable materials. The UV-responsive nano-sized capsules based on a lightly cross-linked polyamide containing azobenzene moieties in the main chain are prepared by oil-in-water miniemulsion interfacial polycondensation [95]. By applying intense shear forces, miniemulsions can be obtained that leads to fairly stable suspensions of highly distributed nanosized droplets. The polymer nanocapsules act as phototriggered reservoir for the controlled release of active agents. The UV responsive polyamide nanocapsules find important application in the encapsulation and release of photo-protective agents that can protect cells from damage caused by UV-induced reactive intermediates. Also, the nanocapsule-based film/coatings can be used in packaging to increase the shelf life of perishable substances. On UV light exposure, there is substantial increase in average diameter of the particles. The smaller capsules exhibit faster release rates and the photoresponsiveness of the capsule systems can be effectively designed by proper selection of emulsion and processing parameters.

4.7.6 Photoresponsive Membranes

Smart membranes are being used in many advanced fields of biotechnology, including drug delivery, biosensors, microfluidics, light-powered molecular machines, molecular shuttles, data storage, etc. In responsive polymer membranes, which sense the changes in their environment acting as a stimulus and make desired response, either a controlled porosity or a texture and chemical composition are coupled with properties such as pH-, ionic strength-, thermo-, light-, electric-, and magnetic-response and molecular-recognition.

LCP based nanoporous materials are interesting because of their uniform nanometer pore size, high surface area and permeability, which enhances the adsorption capacity and kinetics. Nanoporous LC networks can be used as nanoporous membranes, which can be controlled by using stimuli so that the nanopore dimensions are varied. Azobenzene changes the dimension during isomerization from *trans* to *cis* isomer and vice versa. Thus, its incorporation in a host molecule can vary the pore dimension that will control the selectivity toward molecules of different sizes. Gas permeation LC membranes can switch between an impermeable ordered state and a permeable disordered state [96], while azobenzene containing smectic LCs exhibit anisotropic ion conduction and photoswitchable ion conduction [97].

Photoresponsive nanoporous LC materials can be generated by using photoresponsive azobenzene cross-linker for the hydrogen-bonded smectic liquid crystalline polymer network [98]. The nanopores are created by treating the polymer network with a base leading to the breakup of the hydrogen bonds, thus, resulting in a photoresponsive smectic nanoporous material. During exposure to light, the lamellar spacing decreases due to the *trans-cis* isomerization that alters the pore size. In fact, after UV irradiation the *d*-spacing decreases to 3.3 nm, which is 0.2 nm smaller than the original value. The change in molecular length of the azobenzene crosslinker under UV exposure causes the decrease. The number of binding sites in this material can also be controlled, which leads to light controlled adsorption of cations and cationic dyes. Polymer films are treated with an alkali and acid before use to erase the initial molecular ordering. The films are then immersed in buffer solution and equilibrated for a day, which is followed by irradiation with UV light at room temperature. The hydrogen bonds are converted to the polymer salt during UV exposure, which is caused by the change in dipole moment of the isomerized azobenzene crosslinker. As a result, a change in the pKa of the acid moieties promotes formation of a nanoporous polymer salt film.

4.8 Nonlinear Optical (NLO) Polymers

Electric field (E_a), such as an applied dc field, always induces displacements of the electronic cloud. If the applied field strength is low, the magnitude of induced polarization will be proportional to the applied field. However, as the field strength is increased to high value, the polarization exhibits non-linearity. Only non-centrosymmetric photochromic molecules have second-order NLO properties. Molecules possessing the largest second order susceptibility ($\chi^{(2)}$) contain donor and acceptor connected in push-pull fashion via π-backbone linkage. A common method used to induce non-centrosymmetry is to "pole" the polymer film by heating it above glass transition temperature (T_g) and applying a strong electric field. The poled film is cooled to room temperature while the voltage is on, so that the orientation of the NLO chromophores is frozen in the direction of the poling field. The most common applications include communication and optical signal processing. The LC polymers can be easily poled at room temperature to exhibit significant $\chi^{(2)}$ in the mesophase. During the application of field, the $\chi^{(2)}$ shows decreasing trend as the temperature increases, which reaches zero at the LC-isotropic transition. As the sample is cooled back to temperature at which the mesophase appears, the $\chi^{(2)}$ reappears and increases with decreasing temperature [99]. Thus, the orientational order parameter of the LC phase and $\chi^{(2)}$ vary in the same direction. In normal NLO molecules, the poled dipoles can relax back to decrease $\chi^{(2)}$, whereas these cannot relax back in the mesophase with "closely packed" parallel orientation of neighboring rods. This amounts to retardation in the motion of NLO dipoles and maintenance of $\chi^{(2)}$.

The molecular structure of the liquid crystal polymers is non-centrosymmetric by nature. The molecules can be oriented to obtain solid structure with broken symmetry, when these are processed in the liquid crystalline temperature range. The advantages of main chain liquid crystalline polymer (MCLCP) in respect to NLO application are: (1) accumulation effect of individual units due to head-to-tail bonding, (2) spontaneous orientation of the chain molecules, (3) high thermal and mechanical stability and (4) non-durability [100].

Main-chain NLO polymers have good thermal stability of dipole alignments. However, solubility issue hinders widespread use of MCLCP for making films for NLO applications. Main chain polymers having second order NLO properties can be synthesized by self-condensing monomer having –OH at one end and –COOH at the other. Electron pulling –CN is attached as side group, which is connected to the conjugated LC system in push-pull arrangement [101]. The nematic to isotropic transition occurs at reasonably high temperature (141°C), which indicates very good stability of poled dipoles. The second order NLO response is quite effective with very low scattering characteristics.

Copolymers can be synthesized by melt condensation of 2-hydroxy-6-naphthoic acid (HNA) with 4-hydroxy-benzoic acid (PHB) in varying proportions (Figure 4.15) [102].

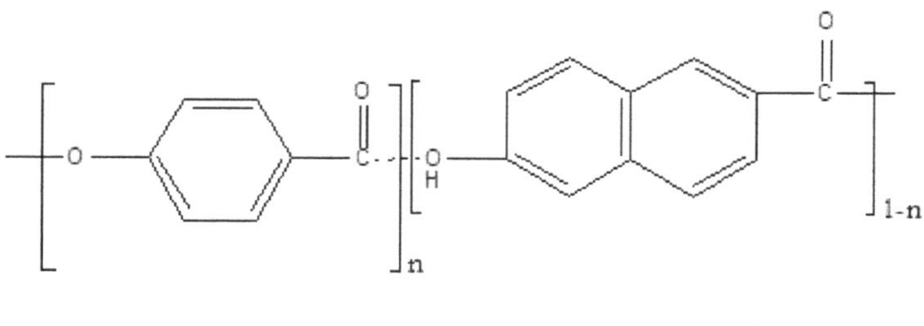

Figure 4.15 Schematic representation of molecular structure of PHB-HNA main chain NLO copolymer.

The film can be processed by heating roller extension (HR-3) and instant two-opposite way extension method, generally called mechanical processing. The molecules assume orientation parallel to the machine direction. Large second harmonic generation (SHG) value of the polymers originates from the additive nature of individual units arranged in head-to-tail fashion. The processing conditions affect the SHG of the polymer film. The LC Polymer, heated into liquid crystalline state and quickly stretched by two sets of rollers along two-opposite-way, exhibits much higher SHG intensity. During measurement, the direction of electric vector of incident light needs to coincide with the direction of molecular long axis to exhibit SH. The mechanically oriented samples do not have uniform polarization direction. After application of high voltage and temperature (12kV/cm; 290 °C), the sample attains an electrically oriented monodomain-like state within 2-5 min. The SHG for the sample, using horizontally (transversely) applied electric field, is about 2-3 times that of mechanically aligned samples. Corona poled samples exhibit lower SHG values due to insufficient alignment.

The alignment of periodically inverted-polarized domains can fashion quasi-phase matching and often shows high SH efficiency. Periodically inverted-polarized domain can be easily attained by stacking the films so that the machine directions of the films change alternatively, maintaining directions with each other. The system requires 4-5 sheets to improve the SH efficiency. By inserting refractive index matched thin liquid crystal films between the oriented polymer films and on the surfaces, the SH efficiency increases by more than 4-5 times. After withdrawal of the electric field, generating highest SH efficiency, the high SH efficiency is held to large extent. In fact, about 60% of the SH efficiency is retained after the termination of electric field [100]. It can be mentioned that 100% efficiency indicates 12-15 times SH efficiency as compared to single sheet of film. By varying the direction of electric field via rotation of the pair of the electrodes, the SH efficiency can be varied from 0 to 100%.

Along with various strategies to restrict the relaxation of oriented dipoles in NLO polymers, attaching NLO chromophores in transverse alignment to the main chain having part of the NLO moiety embedded in the polymer main chain and its dipole moment extending transverse to the backbone makes an intelligent system. Main chain LC polyester with azo side group exhibits LC behavior due to the stabilization of azobenzene mesogenic side groups by the terminal nitro group, whereas, the nitro group replacement by tricyanovinyl group cannot stabilize the LC phase [103]. The replacement of nitro by tricyanovinyl group disturbs the anisotropic ordering due to its nonlinear structure [104]. Nitro group containing polymer has higher SHG coefficient (d_{33}) as compared to the polymer having tricyanovinyl group due to the LC nature of the former. Also, during poling at LC phase transition temperature region, larger NLO activity is possible. The SH activity of the polymers possesses better stability for a longer storage time duration.

In MCLCP, azo groups are incorporated in the main chain of LC polyester [105]. Two types of transitions can be visualized in such types of polymers: the *trans-cis* isomerization and LC-isotropic phase transition. The d_{33} value increases when the poling temperature is around 50 °C and falls in the LC phase temperature range of 200–250 °C. The SHG activity of the polymer remains stable without any loss after storage at 100 and 200 °C for 4-5 days.

Main chain polyesters with one fixed mesogen and other variable mesogen unit can be generated in polyester chains [106]. *Trans-cis* and *cis-trans* photoisomerization in MCLCP occurs at 275 s and 360 s respectively. Fast optical switching processes of the polymers in short time scales provide an opportunity for non-linear applications. In fact, the Z-scan technique with a 532 nm Nd-YAG laser confirms that azo polymer possesses the negative non-linear refractive index together with fast photo-responses and change in optical properties.

4.9 Conclusions

Photoresponsive main chain liquid crystalline polymers have been discussed in the chapter by including various designs, synthesis procedures, characteristics and applications. With a combination of the properties of both crystalline and pure liquid (fluidity) states, the behavior of liquid crystalline polymer (LCP) can be easily modulated by applying external stimulus such as temperature, electromagnetic fields, solvent and pH. The perturbations, generating changes in the organizations in molecular and supramolecular level, can be translated into a range of smart applications. By designing LCPs with chemically linked photoresponsive moieties or physically blended photoresponsive molecules, the light can act as the stimulus, which is a clean energy and can be used in non-contact mode. This imparts new dimension to the concept of smart polymers. Main chain LCPs are also known to possess higher mechanical properties as compared to side chain liquid crystalline polymers.

In the chapter, the general information regarding photoresponsive main chain LCPs has been presented. Although main chain LCPs are very few as compared to side chain LCP, the varieties of their structures and associated properties have created a special place in research and industrial developments. Various photoresponsive molecules, together with their photoinduced changes, etc., have been discussed in brief. Photochromic LCPs have been discussed, particularly the introduction of azo molecules and their characteristics when incorporated in the main chain. Azobenzene molecules have very prominent photoisomerization ability and can make the matrix photoresponsive when combined with LCPs. This gives ample possibilities for design and harnessing the resultant properties. Combined main chain and side chain photoresponsive LCPs, photocrosslinkable LCPs and block copolymers have been discussed with special attention on various structure specific properties. This type of architecture has distinct effect on the properties, particularly the LC phase temperature range. Jacketed mesogens can also be incorporated as a unique possibility to design. Main chain oligomers and macrocycle based oligomers have been discussed as a special type. These types present scopes for studying the basic characteristics of the structures, which can be correlated with small molecules as well as polymers. These materials offer variety and design flexibility for material development. The LC elastomers (LCEs) are discussed along with their photocrosslinking behavior and nanocomposites for photothermal effects. The elastomers are considered ideal for exhibiting photomechanical effects. Few typical synthetic methods of photoresponsive main chain LCPs have also been discussed in brief. It gives an idea about specificity of the methodology to arrive at particular structures and provides inputs for further development of strategies. The LCEs have been discussed for their use as actuators and sensors. Opto-mechanical effect in liquid single crystal elastomers, being an important development in LCE, has been incorporated as a specific type. Stimulus-induced deformations in various geometries, like caterpillar motion, rotation of wheels, etc., are also incorporated and discussed. The strategies for the LCE developments are rapidly coming to a stage for taking off as material for industrial applications. Shape memory polymers are considered an important class, which has been incorporated considering their immense potential for application in biomedical field. The application of non-linear optical polymers having various architectures is being studied. The main goal being the stability of poled dipoles in second order non-linear application, thus, the rigid LC molecules have been incorporated to impact the NLO behavior. Bisbenzylidene, a polymer building block, is also used for generating main chain photoresponsive LC polymers. These have been incorporated as polymers with special architecture, which can provide enhanced thermal and orientational stabilities. The LC anisotropic gels and ferroelectric liquid crystals based on main chain polymers have been discussed to understand their specific behavior and properties. Light-sensitive microcapsules based on liquid crystalline polyesters and photoresponsive membranes have also been discussed to highlight smart behaviors for practical applications.

References

1. Hikmet, R. A. M. (1990) Electrically induced light-scattering from anisotropic gels. *Journal of Applied Physics*, **68**(9), 4406-4412.
2. Urayama, K., Honda, S., and Takigawa, T. (2006) Deformation coupled to director rotation in swollen nematic elastomers under electric fields. *Macromolecules*, **39**(5), 1943-1949.
3. Yusuf, Y., Ono, Y., Sumisaki, Y., Cladis, P. E., Brand, H. R., Finkelmann, H., and Kai, S. (2004) Swelling dynamics of liquid crystal elastomers swollen with low molecular liquid crystals. *Physical Review E*, **69**(2), 021710.
4. Yu, Y., Nakano, M., and Ikeda, T. (2003) Photomechanics: Directed bending of a polymer film by light. *Nature*, **425**, 145-146.
5. Bobrovsky, A., Shibaev, V., Elyashevich, G., Rosova, E., Shimkin, A., Shirinyan, V., and Cheng, K.-L. (2010), Photochromic composites based on porous stretched polyethylene filled by nematic liquid crystal mixtures. *Polymers for Advanced Technologies*, **21**, 100-112.
6. Cipparrone, G., Pagliusi, P., Provenzano, C., and Shibaev, V. P. (2010) Polarization holographic recording in amorphous polymer with photoinduced linear and circular birefringence. *Journal of Physical Chemistry B*, **114**(27), 8900-8904.
7. Ruan, J., Ge, J. J., Zhang, A., Jin, S., Wang, S.-Y., Harris, F. W., and Cheng, S. Z. D. (2002) Polymorphous structures and their phase relationships in a main-chain/side-chain liquid crystalline polyester. *Macromolecules*, **35**(3), 736-745.
8. Ikeda, T., Mamiya, J., and Yu, Y. (2007) Photomechanics of liquid-crystalline elastomers and other polymers. *Angewandte Chemie-International Edition*, **46**(4), 506-528.
9. de Gennes, P.-G. (1997) A semi-fast artificial muscle. *Comptes Rendus de l'Academie des Sciences Series IIB Mechanics Physics Chemistry Astronomy*, **324**(5), 343-348.
10. Murali, M., and Samui, A. B. (2010) Bisbenzylidene cycloalkanone: A versatile molecule as a polymer building block. *Journal of Materials Chemistry*, **20**(14), 2714-2737.
11. Malik, P., Kumar, P., and Raina, K.K. (2006) Guest-Host Polymer Dispersed Liquid Crystal Display Device: Role of Dichroic Dye. *9th Asian Symposium on Information Display*, India, pp. 172-175.

12. Maughon, B. R., Weck, M., Mohr, B., and Grubbs, R. H. (1997) Influence of backbone rigidity on the thermotropic behavior of side-chain liquid crystalline polymers synthesized by ring-opening metathesis polymerization. *Macromolecules*, **30**(2), 257-265.
13. Ranganathan, T., Bhoje Gowd, E., Ramesh, C., and Kumar, A. (2005) Main chain thermotropic liquid crystalline polyurethanes containing biphenyl mesogens based on novel AB-type self-polycondensation route: FT-IR and XRD studies. *Journal of Polymer Science, Part A: Polymer Chemistry*, **43**(9), 1903-1912.
14. Zentel, R., and Brehmer, M. (1996) Combined LC main chain/side chain polymers. *Acta Polymerica*, **47**(4), 141-149.
15. Donnio, B., Wermter, H., and Finkelmann, H. (2000) A simple and versatile synthetic route for the preparation of main-chain, liquid-crystalline elastomers. *Macromolecules*, **33**(21), 7724-7729.
16. Clarke, S. M., Terentjev, E. M., Kundler, I., and Finkelmann, H. (1998) Texture evolution during the polydomain-monodomain transition in nematic elastomers. *Macromolecules*, **31**(15), 4862-4872.
17. Finkelmann, H., Disch, S., and Schmidt, C. (1996) In: *Polymeric Materials Encyclopedia*, Salamone, J. C. (ed.), 5th volume, CRC Press, USA, pp. 3794-3801.
18. Okano, K., Mikami, Y., and Yamashita, T. (2009) Liquid-crystalline polymer with a block mesogenic side group: Photoinduced manipulation of nanophase-separated structures. *Advanced Functional Materials*, **19**(23), 3804-3808.
19. Finkelmann, H., Nishikawa, E., Pereira, G. G., and Warner, M. (2001) A new opto-mechanical effect in solids. *Physical Review Letters*, **87**(1), 015501.
20. Kato, T. (2002) Self-assembly of phase-segregated liquid crystal structures. *Science*, **295**(5564), 2414-2418.
21. Beginn, U. (2003) Thermotropic columnar mesophases from N-H...O, and N...H-O hydrogen bond supramolecular mesogens. *Progress in Polymer Science*, **28**(7), 1049-1105.
22. Kato, T., Mizoshita, N., Moriyama, M., and Kitamura, T. (2005) Gelation of liquid crystals with self-assembled fibers. *Topics in Current Chemistry*, **256**, 219-236.
23. Moriyama, M., Mizoshita, N., Yokota, T., Kishimoto, K., and Kato, T. (2003) Photoresponsive anisotropic soft Solids: Liquid-crystalline physical gels based on a chiral photochromic gelator. *Advanced Materials*, **15**(16), 1335-1338.
24. Sandhya, K. Y., Pillai, C. K. S., Sato, M., and Tsutsumi, N. (2002) Nonlinear optical properties and liquid crystalline behaviour of new polyesters with dipole moments aligned transverse to the backbone. *Macromolecular Chemistry and Physics*, **203**(8), 1126-1134.
25. Esser-Kahn, A. P., Odom, S. A., Sottos, N. R., White, S. R., and Moore,

J. S. (2011) Triggered release from polymer capsules. *Macromolecules*, **44**(14), 5539-5553.

26. Rousseau, I. A., and Mather, P. T. (2003) Shape memory effect exhibited by smectic-C liquid crystalline elastomers. *Journal of American Chemical Society*, **125**(50), 15300-15301.
27. Lee, M.-H., Choi, H. J., Jeong, K.-U., and Chien, L.-C. (2008). A photochromic main-chain liquid crystalline polymer and its photostimulated actuating properties. In: *Emerging Liquid Crystal Technologies III*, Chien, L.-C. (ed.) Editor, SPIE, USA, doi: 10.1117/12.769009.
28. Strat, M., Delibas, M., Strat, G., Hurduc, N., and Gurlui, S. (1998) Optic and spectral characteristics of some aromatic copolyethers embedded in polymer matrices. *Journal of Macromolecular Science, Part B: Physics*, **37**(3), 387-400.
29. Corvazier, L., and Zhao, Y. (1999) Induction of liquid crystal orientation through azobenzene-containing polymer networks *Macromolecules*, **32**(10), 3195-3200.
30. Acierno, D., Amendola, E., Bugatti, V., Concilio, S., Giorgini, L., Iannelli, P., and Piotto, S. P. (2004) Synthesis and characterization of segmented liquid crystalline polymers with the azo group in the main chain. *Macromolecules*, **37**(17), 6418-6423.
31. Fang, L., Zhang, H., Li, Z., Zhang, Y., Zhang, Y., and Zhang, H. (2013) Synthesis of reactive azobenzene main-chain liquid crystalline polymers via Michael addition polymerization and photomechanical effects of their supramolecular hydrogen-bonded fibers. *Macromolecules*, **46**(19), 7650-7660.
32. Endres, B. W., Ebert, M., Wendorff, J. H., Reck, B., and Ringsdorf, H. (1990) Combined main chain/side chain polymers. A new class of liquid crystalline polymers with unusual structural, thermodynamic and dynamic properties. *Liquid Crystals*, **7**(2), 217-239.
33. Zhou, Q. F., Li, H. M., and Feng, X. D. (1987) Synthesis of liquid-crystalline polyacrylates with laterally substituted mesogens. *Macromolecules*, **20**(1), 233-234.
34. Chen, X.-F., Shen, Z., Wan, X.-H., Fan, X.-H., Chen, E.-Q., Ma, Y., and Zhou, Q.-F. (2010) Mesogen-jacketed liquid crystalline polymers. *Chemical Society Reviews*, **39**, 3072-3101.
35. Xie, H.-L., Wang, S.-J., Zhong, G.-Q., Liu, Y.-X., Zhang, H.-L., and Chen, E.-Q. (2011) Combined main-cChain/side-chain liquid crystalline polymer with main-chain on the basis of "jacketing" effect and side-chain containing azobenzene groups. *Macromolecules*, **44**(19), 7600-7609.
36. Angeloni, A. S., Bignozzi, M. C., Laus, M., Chiellini, E., and Galli, G. (1993) Hybrid liquid-crystalline block copolymers based on polystyrene and polyester blocks. *Polymer Bulletin*, **31**(4), 387-392.
37. Galli, G., Chiellini, E., Laus, M., Bignozzi, M. C., Angeloni, A. S., and

Francescangeli, O. (1994) Synthesis and thermal behavior of liquid-crystalline block copolymers containing both main-chain and side-chain mesomorphic blocks. *Macromolecular Chemistry and Physics*, **195**(6), 2247-2260.

38. Laus, M., Bignozzi, M. C., Angeloni, A. S., Francescangeli, O., Galli, G., and Chiellini, E. (1995) LC block copolymers containing side-chain and main-chain blocks. *Molecular Crystals and Liquid Crystals*, **261**(1), 393-404.
39. Park, C., Yoon, J., and Thomas, E. L. (2003) Enabling nanotechnology with self assembled block copolymer patterns. *Polymer*, **44**(22), 6725-6760.
40. Okano, K., Mikami, Y., Hidaka, M., and Yamashita, T. (2011) Photoresponsive liquid-crystalline polymers containing a block mesogenic side chain: Systematic studies on structural parameters for nanophase-separated structures. *Macromolecules*, **44**(14), 5605-5611.
41. Gangadhara, K., and Kishore, K. (1995) Synthesis and characterization of photo-crosslinkable main-chain liquid-crystalline polymers containing bis(benzylidene)cycloalkanone units. *Polymer*, **36**(9), 1903-1910.
42. Murali, M., and Samui, A. B. (2006) Synthesis, photochemical and phase behavior of linear and hyperbranched photoactive benzylidene liquid-crystalline polyesters. *Journal of Polymer Science, Part A: Polymer Chemistry*, **44**(13), 3986-3994.
43. Racles, C., Airinei, A., Cozan, V., Cazacu, M., and Sajo, I. E. (2003) New arylidene-siloxane polyethers: Liquid-crystalline and photosensitive properties. *Journal of Applied Polymer Science*, **90**(11), 3093-3099.
44. Srinivasa Rao, V., and Samui, A. B. (2009) Structure–property relationship of photoactive liquid crystalline polyethers containing benzylidene moiety. *Journal of Polymer Science, Part A: Polymer Chemistry*, **47**(8), 2143-2155.
45. Abd-Alla, M. A., and El-Shahawy, A. S. (1992) Arylidene polymers: 11. Synthesis of poly[2,5-bis(m-nitrobenzylidene)-cyclopentanone sulphide] and some theoretical studies on its monomer unit. *Polymer*, **33**(7), 1525-1529.
46. Aly, K. I., and Hammam, A. S. (2000) Liquid crystalline polymers: I. Main chain thermotropic poly (arylidene-ether) s containing cyclopentanone moiety linked with polymethylene spacers. *European Polymer Journal*, **36**(9), 1933-1942.
47. Aly, K. I. (1999) New polymer syntheses. Part 7. New unsaturated polyurethanes: synthesis and characterization. *Polymer International*, **48**(9), 773-780.
48. Racles, C., Cozan, V., Marcu, M., Cazacu, M., and Vlad, A. (2000) New siloxane–arylidene sulfone segmented copolymers. *European Poly-*

mer Journal, **36**(9), 1951-1957.

49. Mikroyannidis, J. A. (1992) Unsaturated heat-curable polyamides and polyimides derived from 2,6-bis(3-aminobenzylidene)cyclohexanone. *European Polymer Journal*, **28**(4), 439-448.
50. Nossarev, G. G., and Hogen-Esch, T. E. (2002) Synthesis and thermal properties of macrocyclic poly(2-vinylnaphthalene) containing single 1,4-benzylidene or 9,10-anthracenylidene groups. *Macromolecules*, **35**(5), 1604-1610.
51. Appiah, C., Woltersdorf, G., and Binder, W. H. (2017) Synthesis of photoresponsive main-chain oligomers with azobenzene moieties via ADMET oligomerization and their micellization properties. *Polymer Chemistry*, **8**(18), 2752-2763.
52. Binauld, S., Hawker, J. C., Fleury, E., and Drockenmuller, E. (2009) A modular approach to functionalized and expanded crown ether based macrocycles using click chemistry. *Angewandte Chemistry-International Edition*, **48**, 6654-6658.
53. Reuter, R., and Wegner, H. A. (2013) Switchable 3D networks by light controlled π-stacking of azobenzene macrocycles. *Chemical Communications*, **49**(2), 146-148.
54. Jiang, X., Lu, J., Zhou, F., Zhang, Z., Pan, X., Zhang, W., Wang, Y., Zhou, N., and Zhu, X. (2016) Macrocycles containing azobenzene main-chain oligomers Molecularly-defined macrocycles containing azobenzene main-chain oligomers: modular stepwise synthesis, chain-length and topology-dependent properties. *Polymer Chemistry*, **7**(15), 2645-2651.
55. Sanchez-Ferrer, A. (2011) Light-induced disorder in liquid-crystalline elastomers for actuation. In: *Nano-Opto-Mechanical Systems (NOMS)*, Esteve, J., Terentjev, E. M., and Campo, E. M. (eds.), SPIE, USA, pp. 1-8.
56. Krause, S., Dersch, R., Wendorff, J. H., and Finkelmann, H. (2007) Photocrosslinkable liquid crystal main-chain polymers: Thin films and electrospinning. *Macromolecular Rapid Communications*, **28**(21), 2062-2068.
57. Yang, L., Setyowati, K., Li, A., Gong, S., and Chen, J. (2008) Reversible infrared actuation of carbon nanotube-liquid crystalline elastomer nanocomposites. *Advanced Materials*, **20**(12), 2271-2275.
58. Mizuno, K., Ishii, J., Kishida, H., Hayamizu, Y., Yasuda, S., Futaba, D. N., Yunura, M., and Hata, K.(2009) A black body absorber from vertically aligned single-walled carbon nanotubes. *Proceedings of the National Academy of Sciences of USA*, **106**, 6044-6047.
59. Lehn, J.-M. (2002) Supramolecular polymer chemistry-scope and perspectives. *Polymer International*, **51**, 825-839.
60. Kato, T. (2005) In: *Supramolecular Polymers*, Ciferri, A. (ed.), 2nd edition, Taylor & Francis, UK, pp. 131-152.

61. Brienne, M. J., Gabard, J., Lehn, J.-M., and Stibor, I. (1989) Macroscopic expression of molecular recognition. Supramolecular liquid crystalline phases induced by association of complementary heterocyclic components. *Journal of the Chemical Society, Chemical Communications*, 1868-1870.
62. Kato, T., and Frechet, J. M. J. (1989) A new approach to mesophase stabilization through hydrogen bonding molecular interactions in binary mixtures. *Journal of American Chemical Society*, **111**(22), 8533-8534.
63. Tschierske, C. (2001) Micro-segregation, molecular shape and molecular topology – partners for the design of liquid crystalline materials with complex mesophase morphologies. *Journal of Materials Chemistry*, **11**(11), 2647-2671.
64. Toh, C. L., Xu, J., Lu, X., and He, C. (2008) Synthesis and characterisation of main-chain hydrogen-bonded supramolecular liquid crystalline complexes formed by azo-containing compounds. *Liquid Crystals*, **35**(3), 241-251.
65. Hanus, K.-H., Pechhold, W., Soergel, F., Stoll, B., and Zentel, R. (1990) Phase behavior and elastic properties of a slightly crosslinked liquid crystalline main-chain polymer. *Colloid and Polymer Science*, **268**(3), 222-229.
66. Bergmann, G. H. F., Finkelmann, H., Percec, V., and Zhao, M. (1997) Liquid-crystalline main-chain elastomers. *Macromolecular Rapid Communications*, **18**(5), 353-360.
67. Murali, M., Srinivasa Rao, V., and Samui, A. B. (2007) Liquid crystalline photoactive hyperbranched and linear benzylidene polyester with terminal epoxy and pendant hydroxyl groups. *Journal of Polymer Science, Part A: Polymer Chemistry*, **45**(14), 3116-3123.
68. Samui, A. B., Pandey, S., and Mishra, S. P. (2015) Main chain photoresponsive liquid crystalline polymer synthesized through hydrosilylation. *RSC Advances*, **5**(84), 68351-68355.
69. Joo, S.-H., Yun, Y.-K., Jin, J.-I., Kim, D.-C., and Zin, W.-C. (2000) Synthesis of liquid crystalline polyesters of various types by acyclic diene metathesis polymerization. *Macromolecules*, **33**(18), 6704-6712.
70. Beyer, P., Braun, L., and Zentel, R. (2007) (Photo)crosslinkable smectic LC main-chain polymers. *Macromolecular Chemistry and Physics*, **208**(22), 2439-2448.
71. Jiang, H., Kelch, S., and Lendlein, A. (2006) Polymers move in response to light. *Advanced Materials*, **18**(11), 1471-1475.
72. Lendlein, A., and Langer, R. (2002) Biodegradable, elastic shape-memory polymers for potential biomedical applications. *Science*, **296**(5573), 1673-1676.
73. Yu, Y., and Ikeda, T. (2004) Alignment modulation of azobenzene-containing liquid crystal systems by photochemical reactions.

Journal of Photochemistry and Photobiology C: Photochemistry Reviews, **5**(3), 247-265.

74. Yang, H., Buguin, A., Taulemesse, J.-M., Kaneko, K., Bergeret, A., and Keller, P. (2009) Micron-sized main-chain liquid crystalline elastomer actuators with ultralarge amplitude contractions. *Journal of American Chemical Society*, **131**(41), 15000-15004.
75. Tajbakhsh, A. R., and Terentjev, E. M. (2001) Spontaneous thermal expansion of nematic elastomers. *European Physical Journal E*, **6**, 181-188.
76. Sanchez-Ferrer, A., and Finkelmann, H. (2009) Thermal and mechanical properties of new main-chain liquid-crystalline elastomers. *Molecular Crystals and Liquid Crystals*, **508**, 348-356.
77. Ohm, C., Brehmer, M., and Zentel, R. (2010) Liquid crystalline elastomers as actuators and sensors. *Advanced Materials*, **22**(31), 3366-3387.
78. Lebar, A., Kutnjak, Z., Zumer, S., Finkelmann, H., Sanchez-Ferrer, A., and Zalar, B. (2005) Evidence of supercritical behavior in liquid single crystal elastomers. *Physical Review Letters*, **94**, 197801.
79. Sanchez-Ferrer, A., Merekalov, A., and Finkelmann, H. (2011) Opto-mechanical effect in photoactive nematic side-chain liquid-crystalline elastomers. *Macromolecular Rapid Communications*, **32**, 671-678.
80. Yamada, M., Kondo, M., Mamiya, J. I., Yu, Y. L., Kinoshita, M., Barrett, C. J., and Ikeda, T. (2008) Photomobile polymer materials: towards light-driven plastic motors. *Angewandte Chemie-International Edition*, **47**(27), 4986-4988.
81. Finkelmann, H. (1995) Liquid single crystal elastomers (lsce)-mechanical optical and electric properties. *Macromolecular Symposia*, **98**(1), 327-327.
82. Eisenbach, C. D. (1978) Effect of polymer matrix on the cis-trans isomerization of azobenzene residues in bulk polymers. *Macromolecular Chemistry and Physics*, **179**(10), 2489-2506.
83. Mahimwalla, Z., Yager, K. G., Mamiya, J.-I., Shishido, A., Priimagi, A., and Barrett, C. J. (2012) Azobenzene photomechanics: prospects and potential applications. *Polymer Bulletin*, **69**, 967-1006.
84. Bandara, H. M. D., and Burdette, S. C. (2012) Photoisomerization in different classes of azobenzene. *Chemical Society Reviews*, **41**(5), 1809-1825.
85. Gelebart, A. H., Mulder, D. J., Varga, M., Konya, A., Vantomme, G., Meijer, E. W., Selinger, R. L. B., and Broer, D. J. (2017) Making waves in a photoactive polymer film. *Nature*, **546**, 632-636.
86. Hikmet, R. A. M. and Kemperman, H. (1998) Electrically switchable mirrors and optical components made from liquid-crystal gels. *Nature*, **392**, 476-479.
87. Zhao, Y., and Chenard, Y. (2000) Liquid crystalline anisotropic gels

prepared from optical alignment: Polymer network formed in the isotropic phase. *Macromolecules*, **33**(16), 5891-5896.

88. Clark, N. A., and Lagerwall, S. T. (1980) Submicrosecond bistable electro-optic switching in liquid crystals. *Applied Physics Letters*, **36**(11), 899-901.
89. Zhao, Y., and Paiement, N. (2001) Optically aligned ferroelectric liquid crystals. *Advanced Materials*, **13**(24), 1891-1894.
90. Zhao, Y., Paiement, N., Sevigny, S., Leclair, S., Motallebi, S., Giguere, M., and Bouchard, L. (2003) Azobenzene-polymer-stabilized ferroelectric liquid crystals: photoalignment and morphology. In: *Liquid Crystal Materials, Devices, and Applications*, SPIE, USA, doi: 10.1117/12.473841.
91. Esser-Kahn, A. P., Odom, S. A., Sottos, N. R., White, S. R., and Moore, J. S. (2011) Triggered release from polymer capsules. *Macromolecules*, **44**(14), 5539-5553.
92. Tylkowski, B., Pregowska, M., Jamowska, E., Garcia-Valls, R., and Giamberini, M. (2009) Preparation of a new lightly cross-linked liquid crystalline polyamide by interfacial polymerization. Application to the obtainment of microcapsules with photo-triggered release. *European Polymer Journal*, **45**(5), 1420-1432.
93. Bogdanowicz, K. A., Tylkowski, B., and Giamberini, M. (2013) Preparation and characterization of light-sensitive microcapsules based on a liquid crystalline polyester. *Langmuir*, **29**(5), 1601-1608.
94. Raulio, M., Jarn, M., Ahola, J., Peltonen, J., Rosenholm, J., Tervakangas, S., Kolehmainen, J., Ruokolainen, T., Narko, P., and Salkinoja-Salonen, M. (2008) Microbe repelling coated stainless steel analysed by field emission scanning electron microscopy and physicochemical methods. *Journal of Industrial Microbiology & Biotechnology*, **35**(7), 751-760.
95. Marturano, V., Cerruti, P., Carfagna, C., Giamberini, M., Tylkowski, B., and Ambrogi, V. (2015) Photo-responsive polymer nanocapsules. *Polymer*, **70**, 222-230.
96. Glowacki, E., Horovitz, K., Tang, C. W., and Marshall, K. L. (2010) Photoswitchable gas permeation membranes based on liquid crystals. *Advanced Functional Materials*, **20**(17), 2778-2785.
97. Nardele, C. G., Dhavale, V. M., Sreekumar, K., and Asha, S. K. (2015) Ionic conductivity probed in main chain liquid crystalline azobenzene polyesters. *Journal of Polymer Science, Part A: Polymer Chemistry*, **53**(5), 629-641.
98. van Kuringen, H. P. C., Leijten, Z. J. W. A., Gelebart, A. H., Mulder, D. J., Portale, G., Broer,D. J., and Schenning, A. P. H. J. (2015) Photoresponsive nanoporous smectic liquid crystalline polymer networks: Changing the number of binding sites and pore dimensions in polymer adsorbents by light. *Macromolecules*, **48**(12), 4073-4080.

99. Ulrich, D. R. (1990) Polymers for nonlinear optical applications. *Molecular Crystals and Liquid Crystals Incorporating Nonlinear Optics*, **189**(1), 3-38.
100. Asada, T. (1999) Optical second-harmonic generations of liquid crystalline main-chain polymers and their application for EO devices. *Molecular Crystals and Liquid Crystals Science and Technology. Section A. Molecular Crystals and Liquid Crystals*, **331**(1), 391-398.
101. Tsibouklis, J., Richardson, P. H., Ahmed, A. M., Richards, R. W., Feast, W. J., Martin, S. J., Bradley, D. D. C., Warner, M. (1993) A new main-chain thermotropic liquid-crystalline polymer based on a substituted cyanostilbene: synthesis, thermo-optic observations and linear electro-optic effect measurements. *Synthetic Metals*, **61**(1-2), 159-162.
102. Asada, T. (1998) Nonlinear optical properties of the main-chain type liquid crystalline polymers and electric fields. *Molecular Crystals and Liquid Crystals Science and Technology. Section A. Molecular Crystals and Liquid Crystals*, **318**(1), 39-58.
103. Sakai, W., Chiga, K., and Tsutsumi, N. (2001) Nonlinear optical (NLO) polymers. IV. Second-order optical nonlinearity of NLO polyurea and copolyurea with NLO dipole moments aligned transverse to the main backbone. *Journal of Polymer Science, Part B: Polymer Physics*, **39**(2), 247-255.
104. Sandhya, K. Y., Pillai, C. K. S., Sato, M., and Tsutsumi, N. (2002) Nonlinear optical properties and liquid-crystalline behavior of new polyesters with dipole moments aligned transverse to the backbone. *Macromolecular Chemistry and Physics*, **203**(8), 1126-1134.
105. Sandhya, K. Y., Pillai, C. K. S., Sato, M., and Tsutsumi, N. (2003) Highly stable rigid main-chain nonlinear optical polymers with nematic phase: Effect of liquid-crystalline phase on nonlinear optical response. *Journal of Polymer Science, Part A: Polymer Chemistry*, **41**(10), 1527-1535.
106. Subramanian, B., Said, S. M., Rahman, A., Abdullah, N., Murugesan, V. S., and Dhandapani, S. (2016) Main-chain liquid crystalline copolymers: synthesis and comparative study of the mesomorphic properties and non-linear optical properties. *Liquid Crystals*, **43**(12), 1791-1803.

Chapter 5

Synthesis and Liquid Crystalline Properties of 1,3,4-Oxadiazole Derivatives with a Terminal (*R*)-1-methylheptyloxy Group

Jie Han,* Fu-Li Wang, Juan-Juan Sun, Yan-Mei Wang

College of Chemistry and National Demonstration Center for Experimental Chemistry Education, Nankai University, 94 Weijin Road, Tianjin 300071, China
**Corresponding author*: hanjie@nankai.edu.cn

5.1 Introduction

In the past decade, the 1,3,4-oxadiazole moiety has been widely used as an important synthon to prepare emissive liquid crystals (LCs), which exhibit many outstanding advantages such as high photoluminescence quantum yield, good thermal and chemical stabilities, and very rich mesophases [1]. To date, quite a lot of mesogenic 1,3,4-oxadiazoles including rod-like [2] and discotic shapes [3] have been reported for fundamental theory research and potential applications in various fields. Among these materials, the chiral LCs have drawn special interest, as they exhibit various chiral mesophases such as chiral nematic (N*), chiral smectic C (SmC*), twist grain boundary phase with semectic C* blocks (TGBC*) and blue phases (BPs), and might have practical applications in new technologies.

Parra *et al.* [4] first reported a series of chiral liquid crystalline 1,3,4-oxadiazoles **OXD1-OXD4** (Figure 5.1), most of these compounds exhibited enantiotropic N* and SmC* mesophases. It was noted that **OXD4** displayed an additional monotropic blue phase with a narrow temperature range about 2 K. In order to investigate the effect of linking groups, Parra *et al.* [4c] also synthesized chiral 1,3,4-oxadiazole Schiff bases and amides, but most of them are not mesogens. Recently, Chiang *et al.* [5] synthesized two chiral 1,3,4-oxadiazole-based LCs **OXD5** and **OXD6** (Figure 5.1) that display broad ranges of blue phase III (34 and 7 K, respectively). Notably, the electro-optical response of both of these LCs was

Liquid Crystalline Polymers, edited by Vikas Mittal

much faster than those of single-component blue-phase LCs reported previously.

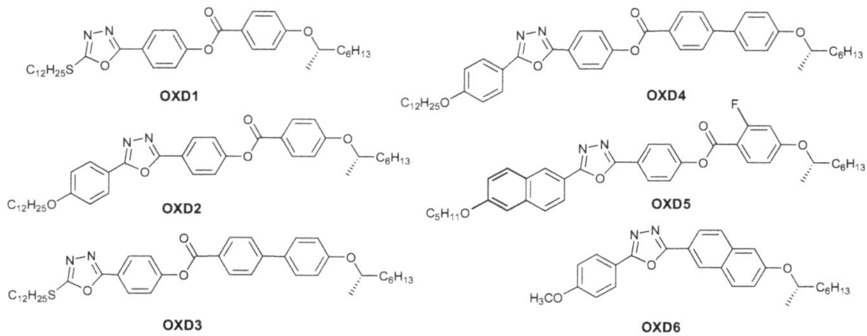

Figure 5.1 Molecular structures of **OXD1-OXD6** [4,5].

Cholesterol has been widely used as an important building block to prepare LCs with various chiral mesophases such as BPs, N*, TGB and SmC*. Taking advantage of the unique properties of cholesteric LCs and the high photoluminescent yields of 1,3,4-oxadiazole derivatives, we recently synthesized a series of liquid crystalline compounds **OXD7** (Figure 5.2) [6]. These compounds exhibited a cholesteric mesophase with very wide mesomorphic temperature ranges. The compounds also displayed an intense absorption band (λ_{max} = 330 nm) and a strong blue fluorescence emission (λ_{max} = 404 nm) with good photoluminescence quantum yields. Majumadar *et al.* [7] also reported the cholesterol based liquid crystalline dimers **OXD8** (Figure 5.2) comprising a mesogenic 1,3,4-oxadiazole unit. These dimers exhibited enantiotropically TGBC* and N* phases and one of them displayed additional blue phases. It is worthy to mention that **OXD8** (R = $C_{18}H_{37}$) exhibited TGBC* over a very wide temperature range (~100 K).

In order to investigate the relationship between the molecular structures and properties, a new series of chiral 1,3,4-oxadiazole based LCs **Ia-Ie** (Scheme 5.1) have been reported in this work. These compounds have two terminal groups: one is a chiral (*R*)-1-methylheptyloxy group and the other is an alkoxy chain with different length. The biphenyl moiety is chosen to construct the conjugated mesogenic core as such kind of liquid crystal materials have many advantages such as high dielectric anisotropy, low melting points and low viscosity coefficient [8]. It is quite interesting to find

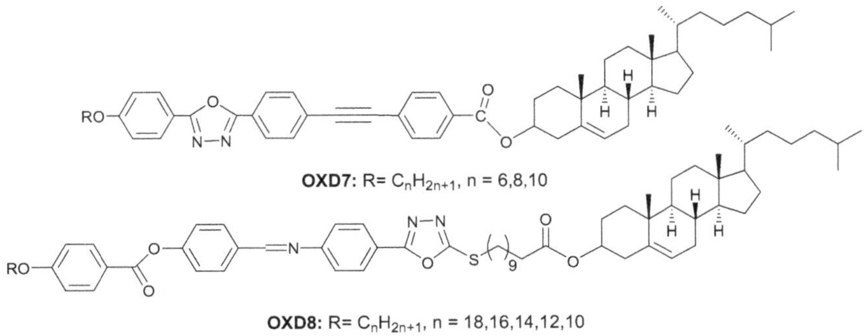

Figure 5.2 Molecular structures of **OXD7** and **OXD8**.

that only **Ia** and **Ib** with a short alkoxy chain exhibited monotropic smectic A phase, while the other homologous compounds **Ic–Ie** with a longer alkoxy chain didn't show any liquid crystalline properties. The structure-property relation has been discussed briefly.

5.2 Results and Discussion

5.2.1 Synthesis

Scheme 5.1 shows the synthetic route for **Ia-Ie**. The intermediate 1,3,4-oxadiazoles **3a-3e** were prepared in good yields according to the standard procedures, which were carried out by refluxing the hydrazines **2a-2e** in anhydrous thionyl chloride. Methyl 4-(4-(5-(4-(alkoxy)phenyl)-1,3,4-oxadiazol-2-yl)phenyl)benzoates **4a-4e** were synthesized in high yields by reactions of **3a-3e** and *p*-(methoxycarbonyl)phenylbornic acid according to the typical Suzuki cross-coupling reaction. Hydrolysis of **4a-4e** and subsequent neutralization of the carboxylate afforded the carboxylic acids **5a-5e**, which were used directly in the next step without purification. Reflux of **5a-5e** in thioyl chloride provided the corresponding acid chlorides, which were further reacted with (*R*)-2-octanol to generate the final products **Ia-Ie** in good yields. The new compounds were characterized by means of proton nuclear magnetic resonance (^{1}H NMR), ^{13}C NMR and high-resolution mass spectrometry.

5.2.2 Liquid Crystalline Properties

The thermotropic liquid crystalline behaviors of **4a-4e** and **Ia-Ie** are

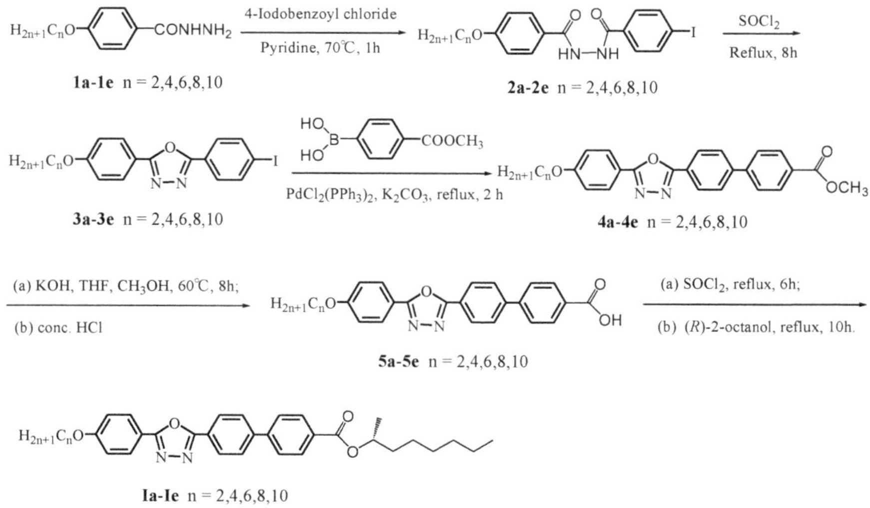

Scheme 5.1 Synthetic route for **Ia-Ie.**

studied by polarized-light optical microscope (POM) and differential scanning calorimeter (DSC). Phase transition temperatures observed through thermal microscopy were in reasonable agreement with the corresponding DSC thermograms. The mesophases were identified according to the texture observed under POM and the classification system reported by Dierking [9] and Demus and Richter [10]. The phase transition temperatures and associated enthalpy changes derived from DSC measurements are presented in Table 5.1. Selected texture images of mesophases are shown in Figure 5.3.

As seen in Table 5.1, all compounds **4a-4e** exhibited enantiotropically liquid crystalline properties. Both **4a** and **4b** exhibited a smectic A and an enantiotropic nematic phase. As the terminal alkoxy chain length increased, the nematic phase disappeared and only smectic A phase was formed in the case of **4c**. In the case of **4d** and **4e**, both semectic A and Smectic C phases were displayed owing to the longer terminal alkoxy chain. This result is consistent with the common rule that the longer terminal chain the compounds have, the higher order mesophases tend to form [11].

As for the final products, both **Ia** and **Ib** exhibited a monotropic smectic A phase in the cooling process of the isotropic liquids and the mesomorphic temperature ranges were much narrower than those of the corresponding compounds **4a** and **4b**. Other compounds **3c, 3d**, and **3e** did not display any liquid crystalline properties both in heating and in cooling cycles. The only difference in molecular structures of **Ia-Ie** is that the terminal alkoxy chain length is different from each other. The experimental phenomena that the compound with longer alkoxy chain does not facilitate the formation of mesophase are contrary to common rule mentioned above. Presumably, the methyl branch present in the chiral chain results in a deviation from linearity for the molecules and disturbs the formation of stable mesophases. However, the definite reason is not clear at present.

Table 5.1 Thermal properties of **4a-4e** and **Ia-Ie** compounds derived from DSC data

Compound	Phase transition[a] T[°C][b](ΔH.kJmol^{-1}])
4a (n=2)	$Cr_1$203.7 (36.7) SmA 227.9 (2.0) N253.9 (0.7) Iso
	Iso253.0 (-0.6) N 226.6 (-1.8) SmA 188.6 (-19.5) Cr_1
4b (n=4)	Cr_1 175.2 (34.6) SmA 235.0 (2.7) N 238.9 (0.3) Iso
	Iso 237.6 (-0.6) N 233.1 (-2.8) SmA170.7 (-15.9) Cr_2
4c (n=6)	Cr_1 161.1 (22.3) Cr_2 168.9 (13.2) SmA 233.2 (7.4) Iso
	Iso 224.4 (-3.5) SmA 161.6 (-9.9) Cr_3
4d (n=8)	Cr_1 87.2 (7.0) Cr_2 159.6 SmC 167.0 (37.7[c]) SmA 228.5 (4.9) Iso
	Iso222.4 (-5.6) SmA 161.4 (-12.5) SmC 65.7(-12.8) Cr_3
4e (n=10)	Cr_1 114.3 (9.2) $Cr_2$145.9 (28.7) SmC 163.9 (12.9) SmA 220.4 (6.04)Iso
	Iso 217.7 (-5.2) SmA 161.2 (-12.9) SmC 72.1 (-27.1) Cr_3
Ia (n=2)	Cr_1 149.5 (28.3) Iso
	Iso 137.1 (-1.2) SmA 133.0 (-7.4) $Cr_2$114.1 (-12.6) Cr_3
Ib (n=4)	$Cr_1$133.9 (10.2) Iso
	Iso 131.7 (-2.0) SmA126.7 (-6.7) Cr_2
Ic (n=6)	Cr_1 135.8 (15.0) Iso
	Iso 126.5 (-15.2) Cr_2
Id (n=8)	Cr_1 125.3 Cr_2 133.0 (28.2 [c]) Iso
	Iso 125.5 (-15.6) Cr_3
Ie (n=10)	Cr_1 128.2 (47.4) Iso
	Iso 121.5 (-13.8) Cr_2 88.7 (-30.8) Cr_3

[a] Cr_n = crystal phase(*n*th); SmA = smectic A phase; SmC = smectic C phase; N = nematic phase; Iso = isotropic liquid

[b] Determined by DSC at a scan rate of 5 °C

[c] Total enthalpy changes of two phases which cannot be integrated separately.

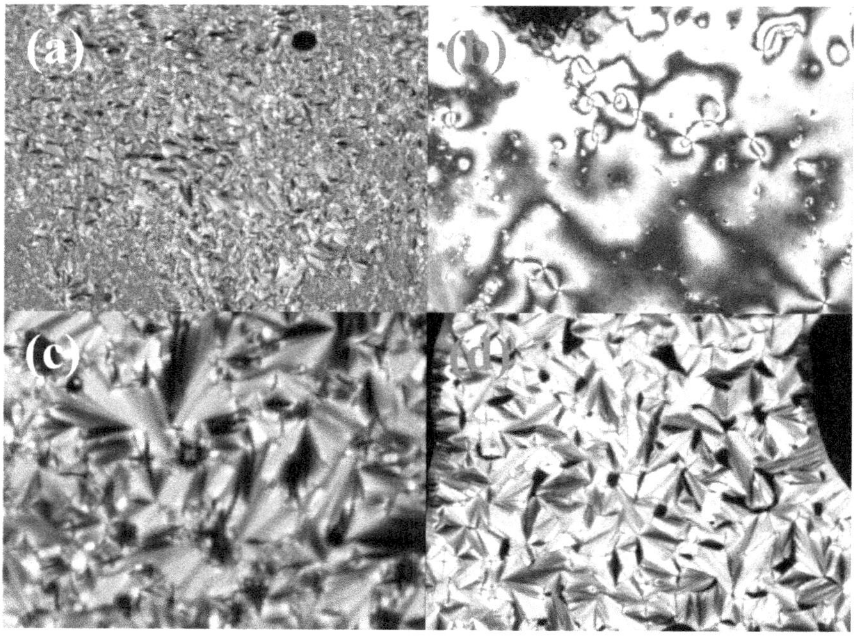

Figure 5.3 POM images (magnification ×200) showing (a) smectic A mesophase with polygonal texture for **4a** at 227.8°C in the cooling cycle; (b) nematic mesophase with thread-like texture for **4a** at 260.1°C in the heating cycle; (c) smectic A mesophase with fan-shaped texture for **4b** at 198°C in the cooling cycle; (d) smectic A mesophase with fan-shaped and focal conic textures for **Ib** at 136.1 °C in the cooling cycle.

5.3 Conclusions

A new class of 1,3,4-oxadiazole-based liquid crystalline compounds **Ia-Ie** with a terminal (R)-1-methylheptyloxy group have been synthesized and characterized by ^{1}H and ^{13}C NMR, MS and high-resolution mass spectra (HRMS) techniques. The phase behaviors were investigated by means of polarized optical microscopy (POM) and differential scanning calorimetry (DSC). The results showed that the liquid crystalline behaviors were influenced greatly by the length of the alkoxy chain. Both **Ia** and **Ib** with a shorter alkoxy chain exhibited a monotropic smectic A phase, while **Ic-Ie** did not display any liquid crystalline properties perhaps due to the longer terminal alkoxy chain. Though there is a chiral terminal chain, none of these chiral compounds exhibit any chiral mesophases.

Experimental

General

Solution ^{1}H NMR and ^{13}C NMR spectra were recorded on Bruker AV400 spectrometer with tetramethylsilane as the internal standard. Electrospray-ionization (ESI) mass spectra were recorded on a Finnigan LCQ Advantage spectrometer and high-resolution mass spectra (HRMS) were obtained with a Finnigan MAT 95 mass spectrometer. The phase transition temperatures and enthalpy changes were measured on NETZSCH DSC 204 differential scanning calorimeter at a heating rate of 5 °C/min and calibrated with a pure indium sample. The texture image of mesophase was recorded using a polarized-light optical microscope (OLYMPUS BX51) equipped with a temperature-controlled hot stage. The starting materials **1a**-**1e** were prepared according to the literature methods [12]. All other chemicals were purchased commercially and used as received.

General Procedure for Synthesis for 2a-2e

4-Iodobenzoic acid (2.5g, 10 mmol) was added to thionyl chloride $SOCl_2$ (10 mL) and refluxed for 8 h to generate 4-iodobenzoyl chloride. Excessive $SOCl_2$ was removed by vacuum distillation and the reaction mixture was cooled to room temperature. *p*-Alkoxybenzoic hydrazide (10 mmol) dissolved in pyridine (15 mL) was added dropwise to the as-formed 4-iodobenzoyl chloride. The reaction mixture was stirred overnight at room temperature and further heated at 70 °C for 1 h. Crude solid was precipitated by pouring the reaction mixture into iced water (50 mL). The crude solid was washed with distilled water and recrystallized from ethanol to generate the intermediate hydrazides **2a**-**2e** as white solids in 74-81% yield.

4-ethyloxybenzoic acid *N*-4-iodobenzoyl hydrazide **2a**: white solids. Yield: 77%. ^{1}H NMR (400 MHz, DMSO-d_6): 10.53 (s, 1H), 10.37 (s, 1H), 7.92 (d, J = 8.4 Hz, 2H), 7.89 (d, J = 8.4 Hz, 2H), 7.70 (d, J = 8.0 Hz, 2H), 7.04 (d, J = 8.8 Hz, 2H), 4.09 (t, J = 6.4 Hz, 2H), 1.48(t, J = 6.8 Hz, 3H).
4-butyloxybenzoic acid *N*-4-iodobenzoyl hydrazide **2b**: white solids. Yield: 74%. ^{1}H NMR (400 MHz, DMSO-d_6): 10.53 (s, 1H), 10.38 (s, 1H), 7.92 (d, J = 8.4 Hz, 2H), 7.89 (d, J = 8.4 Hz, 2H), 7.70 (d, J = 8.0

Hz, 2H), 7.04 (d, J = 8.8 Hz, 2H), 4.05 (t, J = 6.4 Hz, 2H), 1.85 ~ 1.78 (m, 2H), 1.58 ~ 1.48 (m, 2H), 0.99 (t, J = 7.2 Hz, 3H).

4-hexyloxybenzoic acid *N*-4-iodobenzoyl hydrazide **2c**: white solids. Yield: 79%. ^{1}H NMR (400 MHz, DMSO-d$_6$): 10.53 (s, 1H), 10.37 (s, 1H), 7.92 (d, J = 8.4 Hz, 2H), 7.89 (d, J = 8.4 Hz, 2H), 7.70 (d, J = 8.0 Hz, 2H), 7.04 (d, J = 8.8 Hz, 2H), 4.04 (t, J = 6.4 Hz, 2H), 1.91-1.59 (m, 2H), 1.55-1.11 (m, 6H), 0.88 (t, J = 6.0 Hz, 3H). ESI-MS: m/z: 465.02 $[M-1]^-$.

4-octyloxybenzoic acid *N*-4-iodobenzoyl hydrazide **2d**: white solids. Yield: 81%. ^{1}H NMR (400 MHz, DMSO-d$_6$): 10.53 (s, 1H), 10.38 (s, 1H), 7.92 (d, J = 8.4 Hz, 2H), 7.89 (d, J = 8.4 Hz, 2H), 7.70 (d, J = 8.0 Hz, 2H), 7.04 (d, J = 8.8 Hz, 2H), 4.04 (t, J = 6.4 Hz, 2H), 1.82-1.65 (m, 2H), 1.48-1.18 (m, 10H), 0.87 (t, J = 6.0 Hz, 3H). ESI-MS: m/z: 493.13 $[M-1]^-$.

4-decyloxybenzoic acid *N*-4-iodobenzoyl hydrazide **2e**: white solids. Yield: 78%. ^{1}H NMR (400 MHz, DMSO-d$_6$): 10.53 (s, 1H), 10.38 (s, 1H), 7.92 (d, J = 8.4 Hz, 2H), 7.89 (d, J = 8.4 Hz, 2H), 7.70 (d, J = 8.0 Hz, 2H), 7.04 (d, J = 8.8 Hz, 2H), 4.04 (t, J = 6.4 Hz, 2H), 1.86-1.60 (m, 2H), 1.58-1.05 (m, 14H), 0.88 (t, J = 6.0 Hz, 3H). ESI-MS: m/z: 521.25 $[M-1]^-$.

General Procedure for Synthesis of Compounds 3a-3e

To a 100 mL round bottom was added compound **2a**-**2e** (5 mmol) and $SOCl_2$ (25 mL). The mixture was refluxed for 8 h and cooled to room temperature. Excessive $SOCl_2$ was removed by vacuum distillation and the residue was subsequently poured into iced water (50 mL). The crude solid was collected and washed several times with water and ethanol, and purified by silica gel column chromatography using ethyl acetate/dichloromethane (v/v=1:15) as an eluent.

2-(4-ethyloxybenyl)-5-(4-iodobenyl)-1,3,4-oxadiazole **3a**: white solid. Yield, 87%. ^{1}H NMR (400 MHz, *CDCl$_3$*): 8.05 (d, J = 8.8 Hz, 2H), 7.88 (d, J = 8.4 Hz, 2H), 7.84 (d, J = 8.4 Hz, 2H), 7.02 (d, J = 8.8 Hz, 2H), 4.09 (t, J = 6.4 Hz, 2H), 1.48(t, J = 6.8 Hz, 3H). ESI-MS: m/z: 393.09 $[M+H]^+$.

2-(4-butyloxybenyl)-5-(4-iodobenyl)-1,3,4-oxadiazole **3b**: white solid. Yield, 83%. ^{1}H NMR (400 MHz, *CDCl$_3$*): 8.05 (d, J = 8.8 Hz, 2H), 7.88 (d, J = 8.4 Hz, 2H), 7.84 (d, J = 8.4 Hz, 2H), 7.02 (d, J = 8.8 Hz, 2H), 4.05 (t, J = 6.4 Hz, 2H), 1.85 ~ 1.78 (m, 2H), 1.58 ~ 1.48 (m, 2H), 0.99 (t, J = 7.2 Hz, 3H). ESI-MS: m/z: 421.15 $[M+H]^+$.

2-(4-hexyloxybenyl)-5-(4-iodobenyl)-1,3,4-oxadiazole **3c**: white solids. Yield, 88%. ^{1}H NMR (400 MHz, *$CDCl_3$*): 8.05 (d, *J* = 8.8 Hz, 2H), 7.88 (d, *J* = 8.4 Hz, 2H), 7.84 (d, *J* = 8.4 Hz, 2H), 7.02 (d, *J* = 8.8 Hz, 2H), 4.04 (t, *J* = 6.4 Hz, 2H), 1.87-1.77 (m, 2H), 1.53-1.43 (m, 2H), 1.41-1.31 (m, 4H), 0.92 (t, *J* = 6.8 Hz, 3H). ESI-MS: m/z: 449.20 $[M+H]^+$.

2-(4-octyloxybenyl)-5-(4-iodobenyl)-1,3,4-oxadiazole **3d**: white solids. Yield, 84%. ^{1}H NMR (400 MHz, *$CDCl_3$*): 8.05 (d, *J* = 8.8 Hz, 2H), 7.89 (d, *J* = 8.8 Hz, 2H), 7.85 (d, *J* = 8.4 Hz, 2H), 7.02 (d, *J* = 9.2 Hz, 2H), 4.04 (t, *J* = 6.4 Hz, 2H), 1.87-1.77 (m, 2H), 1.53-1.43 (m, 2H), 1.42-1.25 (m, 8H), 0.92 (t, *J* = 6.8 Hz, 3H). ESI-MS: m/z: 477.28 $[M+1]^+$.

2-(4-decyloxybenyl)-5-(4-iodobenyl)-1,3,4-oxadiazole **3e**: white solids. Yield, 85%. ^{1}H NMR (400 MHz, *$CDCl_3$*): 8.05 (d, *J* = 8.8 Hz, 2H), 7.89 (d, *J* = 8.8 Hz, 2H), 7.84 (d, *J* = 8.4 Hz, 2H), 7.02 (d, *J* = 9.2 Hz, 2H), 4.04 (t, *J* = 6.4 Hz, 2H), 1.87-1.77 (m, 2H), 1.56-1.45 (m, 2H), 1.44-1.03 (m, 12H), 0.92 (t, *J* = 6.8 Hz, 3H). ESI-MS: m/z: 505.35 $[M+1]^+$.

General Procedure for Synthesis of Compounds 4a-4e

To a round bottom flask (100 mL) were added **3a-3e** (1 mmol), *p*-(methoxycarbonyl)phenylbornic acid (1.5 mmol), K_2CO_3 (5 mmol) and $PdCl_2(PPh_3)_2$ (4 μmol) as a catalyst in a solution mixture of dioxane (9 mL) and water (3mL) refluxed for 2 h under nitrogen. The reaction mixture was evaporated in vacuo, the solid residue was extracted by CH_2Cl_2 twice (2 × 50 mL), subsequently dried with anhydrous $MgSO_4$, filtered and concentrated at reduced pressure. The crude product was purified by silica gel column chromatography using ethyl acetate/dichloromethane (v/v = 1:20) as an eluent. The reaction products were obtained as white solids in 74-79% yield.

Methyl 4-(4-(5-(4-(ethyloxy)phenyl)-1,3,4-oxadiazol-2-yl)biphenyl)benzoate **4a**: pale yellow solid, yield, 78%. ^{1}H-NMR (400 MHz, $CDCl_3$): δ (ppm) = 8.22 (d, *J* = 8.4 Hz, 2H), 8.15 (d, *J* = 8.4 Hz, 2H), 8.09 (d, *J* = 8.8 Hz, 2H), 7.79 (d, *J* = 8.4 Hz, 2H), 7.73 (d, *J* = 8.4 Hz, 2H), 7.03 (d, *J* = 8.8 Hz, 2H), 4.13 (q, *J* = 6.8 Hz, 2H), 3.96 (s, 3H), 1.47 (t, *J* = 6.8 Hz, 3H). ^{13}C NMR (400 MHz, CDCl3): δ 166.83, 164.73, 163.80, 161.84, 144.18, 142.92, 130.28, 129.64, 128.76, 127.88, 127.37, 127.12, 123.67, 116.11, 114.97, 63.79, 52.29, 14.74. ESI-MS: m/z: 401.2 $[M+1]^+$.

Methyl 4-(4-(5-(4-(butyloxy)phenyl)-1,3,4-oxadiazol-2-yl)biphenyl)benzoate **4b**: white solid, yield, 77%. ^{1}H-NMR (400 MHz,CDCl$_3$): δ (ppm)= 8.22 (d, *J* = 8.4 Hz, 2H), 8.15 (d, *J* = 8.4 Hz, 2H), 8.08 (d, *J* = 8.8 Hz, 2H), 7.79 (d, *J* = 8.4 Hz, 2H), 7.73 (d, *J* = 8.4 Hz, 2H), 7.03 (d, *J* = 8.8 Hz, 2H), 4.05 (t, *J* = 6.4 Hz, 2H), 3.96 (s, 3H), 1.85 ~ 1.78 (m, 2H), 1.58 ~ 1.48 (m, 2H), 1.00 (t, *J* = 7.2 Hz, 3H). ^{13}C NMR (100 MHz, CDCl$_3$): δ166.80, 164.73, 163.76, 162.04, 144.14, 142.86, 130.27, 129.64, 128.72, 127.84, 127.35, 127.10, 123.67, 116.03, 114.98, 67.97, 52.27, 31.18, 19.23, 13.89. ESI-MS: m/z: 429.1 [M+1]$^+$.

Methyl 4-(4-(5-(4-(hexyloxy)phenyl)-1,3,4-oxadiazol-2-yl)phenyl)benzoate **4c**: white solid, yield, 79%. ^{1}H NMR (400 MHz, CDCl$_3$): δ (ppm) = 8.21 (d, *J* = 8.4 Hz, 2H), 8.14 (d, *J* = 8.4 Hz, 2H), 8.07 (d, *J* = 8.8 Hz, 2H), 7.77 (d, *J* = 8.4 Hz, 2H), 7.71 (d, *J* = 8.4 Hz, 2H), 7.02 (d, *J* = 8.8 Hz, 2H), 4.03 (t, *J* = 6.8 Hz, 2H), 3.96 (s, 3H), 1.86-1.79 (m, 2H), 1.52-1.45 (m, 2H), 1.37-1.35 (m, 4H), 0.92 (t, *J* = 6.8 Hz, 3H). ^{13}C NMR (400 MHz, CDCl$_3$): δ 166.75, 164.68, 163.70, 161.99, 144.09, 142.80, 130.23, 129.60, 128.67, 127.79, 127.30, 127.05, 123.63, 115.99, 114.94, 68.26, 52.23, 31.56, 29.08, 25.67, 22.60, 14.05. ESI-MS: m/z: 457.34 [M+1]$^+$. HRMS [M+H]$^+$ Calcd for $C_{28}H_{28}N_2O_4$: 457.2122; Found: 457.2135.

Methyl 4-(4-(5-(4-(octyloxy)phenyl)-1,3,4-oxadiazol-2-yl)phenyl)benzoate **4d**: White solid, yield: 74%.^{1}H NMR (400 MHz, CDCl$_3$): δ (ppm)= 8.22 (d, J = 8.4 Hz, 2H), 8.15 (d, J = 8.4 Hz, 2H), 8.08 (d, J = 8.8 Hz, 2H), 7.78 (d, J = 8.4 Hz, 2H), 7.72 (d, J = 8.4 Hz, 2H), 7.03 (d, J = 8.8 Hz, 2H), 4.04 (t, J = 6.8 Hz, 2H), 3.96 (s, 3H), 1.86-1.79 (m, 2H), 1.55-1.49 (m, 2H), 1.46-1.25 (m, 8H), 0.90 (t, J = 6.8 Hz, 3H). ^{13}C NMR (100 MHz, CDCl$_3$): δ166.78, 164.71, 163.74, 162.02, 144.14, 142.86, 130.25, 129.63, 128.70, 127.83, 127.33, 127.08, 123.66, 116.01, 114.96, 68.28, 52.24, 31.81, 29.35, 29.24, 29.13, 26.00, 22.67, 14.12. ESI-MS: m/z: 485.30 [M+1]$^+$.

Methyl 4-(4-(5-(4-(decyloxy)phenyl)-1,3,4-oxadiazol-2-yl)phenyl)benzoate **4e**: White solid, yield: 76%.^{1}H NMR (400 MHz, CDCl$_3$): δ (ppm)=8.22 (d, J = 8.4 Hz, 2H), 8.15 (d, J = 8.4 Hz, 2H), 8.08 (d, J = 8.8 Hz, 2H), 7.78 (d, J = 8.4 Hz, 2H), 7.72 (d, J = 8.4 Hz, 2H), 7.03 (d, J = 8.8 Hz, 2H), 4.04 (t, J = 6.4 Hz, 2H), 3.96 (s, 3H), 1.86-1.79 (m, 2H), 1.52-1.43 (m, 2H), 1.41-1.21 (m, 12H), 0.89 (t, J = 6.8 Hz, 3H). ^{13}C NMR (100 MHz, CDCl$_3$): δ166.80, 164.74, 163.77, 162.04, 144.17, 142.90, 130.26, 129.64, 128.72, 127.86, 127.35, 127.10, 123.68, 116.04, 114.98, 68.29, 52.26, 31.90, 29.57, 29.38, 29.33, 29.13, 26.00, 22.69, 14.14. ESI-MS: m/z: 513.36 [M+1]$^+$.

Synthesis of Ia-Ie

The compound **4a-4e** was saponified using an excess of KOH in a THF/methanol (2:1) solution under reflux for 8 h. The solution was cooled, diluted with water and acidified with HCl. The crude solid **5a-5e** was collected and was used in the next step reaction without further purification. Yield: 75-90%. To a round-bottom flask (100 mL) were added **5a-5e** (1 mmol) and thionyl chloride $SOCl_2$ (5 mL).The mixture was refluxed for 6 h and the excessive $SOCl_2$ was removed by vacuum distillation. Dry pyridine (2 mL), dichloromethane (20 mL), DCC (0.3 g) and (*R*)-2-octanol (1.5 mmol) were added to the round-bottom flask. The mixture was refluxed for 10 h. The resulting mixture was cooled, diluted with water and extracted by CH_2Cl_2 twice (2 × 50 mL). The organic solution was washed twice with water (30 mL) and once with 20 mL of brine. The organic phase was dried with anhydrous $MgSO_4$, filtered and concentrated. The crude product was purified by silica gel column chromatography using ethyl acetate/ dichloromethane (v/v=1:40) as an eluent. The reaction products were obtained as white solids in 62-69% yield.

(R)-1-methylheptyl 4-(4-(5-(4-(ethyloxy)phenyl)-1,3,4-oxadiazol-2-yl)phenyl)benzoate **Ia**: pale yellow solid, yield, 65%. ^1H NMR (400 MHz, $CDCl_3$): 8.21 (d, *J* = 8.0 Hz, 2H), 8.15 (d, *J* = 8.0 Hz, 2H), 8.08 (d, *J* = 8.4 Hz, 2H), 7.78 (d, *J* = 8.4 Hz, 2H), 7.71 (d, *J* = 8.0 Hz, 2H), 7.02 (d, *J* = 8.4 Hz, 2H), 5.14-5.23 (m,1H), 4.12 (q, *J* = 6.8 Hz, 2H), 1.72-1.78(m, 1H), 1.61-1.67 (m,1H), 1.45-1.49 (t, *J* = 6.8 Hz, 3H), 1.25-1.39 (m, 7H), 0.89 (t, *J* = 6.8 Hz, 3H).,^{13}C NMR (400 MHz, $CDCl_3$): 165.91, 164.71, 163.80, 161.83, 143.98, 143.06, 130.48, 130.20, 128.74, 127.86, 127.36, 127.04, 123.63, 116.16, 114.98, 71.97, 63.78, 36.09, 31.76, 29.18, 25.44, 2.60, 20.11, 14.71, 14.08. ESI-MS: m/z: 498.8 $[M+1]^+$. HRMS $[M+H]^+$ Calcd for $C_{31}H_{35}N_2O_4$: 499.2591; Found: 499.2599.

(R)-1-methylheptyl 4-(4-(5-(4-(butyloxy)phenyl)-1,3,4-oxadiazol-2-yl)biphenyl)benzoate **Ib**: white solid, yield, 63%. ^1H NMR (400 MHz, $CDCl_3$): 8.22 (d, *J* = 8.4 Hz, 2H), 8.15 (d, *J* = 8.4 Hz, 2H), 8.08 (d, *J* = 8.8 Hz, 2H), 7.78 (d, *J* = 8.4 Hz, 2H), 7.72 (d, *J* = 8.4 Hz, 2H), 7.03 (d, *J* = 8.8 Hz, 2H), 5.14-5.23 (m,1H), 4.06 (t, *J* = 6.4 Hz, 2H), 1.79-1.86 (m, 2H),1.72-1.78 (m, 1H),1.63-1.69 (m, 1H), 1.48-1.56 (m, 2H), 1.23-1.43 (m, 11H), 0.98-1.02 (t, *J* = 7.2 Hz, 3H), 0.86-0.90 (t, *J* = 6.8 Hz, 3H). ^{13}C NMR (100 MHz, $CDCl_3$): 165.93, 164.76, 163.82, 162.06,

144.01, 143.08, 130.49, 130.21, 128.74, 127.88, 127.38, 127.05, 123.66, 116.11, 115.02, 71.99, 67.99, 36.09, 31.76, 31.18, 29.18, 25.44, 22.60, 20.12, 19.22, 14.08, 13.85. ESI-MS: m/z: 527.0 $[M+1]^+$. HRMS $[M+H]^+$ Calcd for $C_{33}H_{39}N_2O_4$: 527.2904; Found: 527.2910.

(R)-1-methylheptyl 4-(4-(5-(4-(hexyloxy)phenyl)-1,3,4-oxadiazol-2-yl)biphenyl)benzoate **Ic**: white solid, yield, 66%. ^{1}H NMR (400 MHz, $CDCl_3$): 8.22 (d, *J* = 8.4 Hz, 2H), 8.15 (d, *J* = 8.4 Hz, 2H), 8.08 (d, J = 8.8 Hz, 2H), 7.78 (d, *J* = 8.4 Hz, 2H), 7.72 (d, *J* = 8.4 Hz, 2H), 7.03 (d, *J* = 8.8 Hz, 2H), 5.14-5.23 (m, 1H), 4.05 (t, *J* = 6.4 Hz, 2H), 1.79-1.86(m, 2H),1.72-1.78 (m, 1H), 1.62-1.68 (m, 1H), 1.45-1.53 (m, 2H), 1.22-1.44 (m, 15H), 0.91-0.94 (t, *J* = 6.8 Hz, 3H), 0.86-0.90 (t, *J* = 6.8 Hz, 3H). ^{13}C NMR (100 MHz, $CDCl_3$): 165.92, 164.74, 163.79, 162.05, 143.98, 143.05, 130.48, 130.21, 128.72, 127.86, 127.36, 127.04, 123.64, 116.08, 115.01, 71.98, 68.31, 36.09, 31.77, 31.58, 29.19, 29.11, 25.70, 25.45, 22.61, 20.12, 14.09, 14.05. ESI-MS: m/z: 555.0 $[M+1]^+$. HRMS $[M+H]^+$ Calcd for $C_{35}H_{43}N_2O_4$: 555.3217; Found: 555.3225.

(R)-1-methylheptyl 4-(4-(5-(4-(octyloxy)phenyl)-1,3,4-oxadiazol-2-yl)phenyl)benzoate **Id**: white solid, yield: 69%.^{1}H NMR (400 MHz, *$CDCl_3$*): 8.22 (d, *J* = 8.4 Hz, 2H), 8.15 (d, *J* = 8.4 Hz, 2H), 8.08 (d, *J* = 8.8 Hz, 2H), 7.78 (d, *J* = 8.4 Hz, 2H), 7.72 (d, *J* = 8.4 Hz, 2H), 7.03 (d, *J* = 8.8 Hz, 2H), 5.14-5.23(m, 1H), 4.05 (t, *J* = 6.4 Hz, 2H), 1.80-1.86 (m, 2H), 1.73-1.78(m, 1H), 1.63-1.68(m, 1H), 1.46-1.53(m, 2H), 1.26-1.43 (m, 19H), 0.85-0.93 (m, 6H). ^{13}C NMR (100 MHz, *$CDCl_3$*): 165.93, 164.76, 163.81, 162.07, 144.00, 143.08, 130.49, 130.21, 128.74, 127.88, 127.37, 127.05, 123.65, 116.08, 115.02, 71.99, 68.32, 36.09, 31.82, 31.77, 29.36, 29.25, 29.19, 29.15, 26.02, 25.45, 22.68, 22.61, 20.12, 14.12, 14.09. ESI-MS: m/z: 583.3 $[M+1]^+$. HRMS $[M+H]^+$ Calcd for $C_{37}H_{47}N_2O_4$: 583.3530; Found: 583.3525.

(R)-1-methylheptyl 4-(4-(5-(4-(decyloxy)phenyl)-1,3,4-oxadiazol-2-yl)phenyl)benzoate **Ie**: white solid, yield: 62%.^{1}H NMR (400 MHz, *$CDCl_3$*): 8.20 (d, *J* = 8.0 Hz, 2H), 8.14 (d, *J* = 8.0 Hz, 2H), 8.06 (d, *J* = 8.8 Hz, 2H), 7.77 (d, *J* = 8.0 Hz, 2H), 7.71 (d, *J* = 8.4 Hz, 2H), 7.02 (d, *J* = 8.8 Hz, 2H), 5.14-5.23 (m, 1H), 4.03 (t, *J* = 6.4 Hz, 2H), 1.80-1.85 (m, 2H), 1.73-1.78 (m, 1H), 1.61-1.68(m, 1H), 1.46-1.52 (m, 2H), 1.25-1.40 (m, 23H), 0.85-0.92 (m, 6H).^{13}C NMR (100 MHz, *$CDCl_3$*): 165.90, 164.74, 163.78, 162.05, 143.97, 143.03, 130.48, 130.20, 128.72, 127.85, 127.35, 127.03, 123.65, 116.08, 115.00, 71.97, 68.31, 36.09, 31.91, 31.77, 29.58, 29.40, 29.34, 29.19, 29.15, 26.02, 25.45, 22.70, 22.61, 20.12, 14.14, 14.09. ESI-MS: m/z: 611.2 $[M+1]^+$. HRMS $[M+H]^+$ Calcd for $C_{39}H_{51}N_2O_4$: 611.3843; Found: 611.3841.

Acknowledgements

This work was financially supported by the Natural Science Foundation of Tianjin (Grant Nos. 17JCTPJC51100 and 18JCYBJC20700), which is gratefully acknowledged.

References

1. Han, J. (2013) 1,3,4-Oxadiazole based liquid crystals. *Journal of Materials Chemistry C*, **1**, 7779-7797.
2. (a) Han, J., Xi, Z., Wang, F., Bu, L., and Wang, Y. (2018) Synthesis, liquid crystalline and photoluminscent properties of 1,3,4-oxadiazole derivatives: From calamitic monomers, H-Shaped dimers to calix[4]arene-based tetramers. *Dyes and Pigments*, **154**, 234-241;
   (b) Saha, S. K., Paul, M. K., Chandran, A., Khanna, P. K., and Biradar, A. M. (2017) Low-temperature nematic phase in asymmetrical 1,3,4-oxadiazole bent-core liquid crystals possessing lateral methoxy group. *Liquid Crystals*, **44**, 1739-1750.
3. (a) Han, J., Wang, Z., and Wu, J. (2018) Room-temperature fluorescent liquid crystalline dimers based on discotic 1,3,4-oxadizole. *Liquid Crystals*, **45**(7), 1047-1054;
   (b) Han, J., Wang, Z.-Z., Wu, J., and Zhu, L.-R. (2015) A room-temperature liquid crystalline polymer based on discotic 1,3,4-oxadizole. *RSC Advances*, **5**, 47579-47583;
   (c) Lin, K.-T., Lee, G.-H., and Lai, C. K. (2015) Mesogenic heterocycles formed by bis-pyrazoles and bis-1,3,4-oxadiazoles. *Tetrahedron*, **71**, 4352-4361;
   (d) Lin, K.-T., Kuo, H.-M., Sheu, H.-S., and Lai, C. K. (2013) Unsymmetric 1,3,4-oxa(thia)diazoles of quinoxaline–naphthalene conjugates. *Tetrahedron*, **69**, 9045-9055;
   (e) Girotto, E., Eccher, J., Vieira, A. A., Betchtold, I. H., and Gallardo, H. (2014) Luminescent columnar liquid crystals based on 1,3,4-oxadiazole. *Tetrahedron*, **70**, 3355-3360;
   (f) Lin, K. T., and Lai, C. K. (2016) Phase crossover in columnar tris-(1,3,4-oxadiazoles) with pendant quinoxalines. *Tetrahedron*, **72**, 7579-7588;
   (g) Westphal, E., Prehem, M., Bechtold, I. H., Tschierske, C., and Gallardo, H. (2013) Room temperature columnar liquid crystalline phases of luminescent non-symmetric star-shaped molecules containing two 1,3,4-oxadiazole units. *Journal of Materials Chemistry C*, **1**, 8011-8022;
   (h) Vieira, A. A., Carero, E., Romero, P., Gallardo, H., Serrano, J. L.,

and Sierra, T. (2014) H-bonded complexes containing 1,3,4-oxadiazole derivatives: mesomorphic behaviour, photophysical properties and chiral photoinduction *Journal of Materials Chemistry C*, **2**, 7029-7038;

(i) Pathak, S. K., Gupta, R. K., Nath, S., Rao, D. S. S., Prasad, S. K., and Achalkumar, A. S. (2015) Columnar self-assembly of star-shaped luminescent oxadiazole and thiadiazole derivatives. *Journal of Materials Chemistry C*, **3**, 2940-2952;

(j) Pradhan, B., Pathak, S. K., Gupta, R. K., Gupta, M., Pal, S. K., and Achalkumar, A. S. (2016) Star-shaped fluorescent liquid crystals derived from s-triazine and 1,3,4-oxadiazole moieties. *Journal of Materials Chemistry C*, **4**, 6117-6130;

(k) Sivadas, A. P., Kumar, N. S., Prabhu, D. D., Varghese, S., Prasad, S. K., Rao, D. S. S., and Das, S. (2014) Supergelation via Purely Aromatic π–π Driven Self-Assembly of Pseudodiscotic Oxadiazole Mesogens. *Journal of American Chemical Society*, **136**, 5416-5423.

4. (a) Parra, M. L., Hidalgo, P. I. Soto-Bustamante, E. A., Barbera, J., Elgueta, E. Y., and Trujillo-Rojo, V. H. (2008) Novel chiral calamitic liquid crystalline oxadiazoles as ferroelectric materials. *Liquid Crystals*, **35**, 1251-1262;

   (b) Parra, M. L., Hidalgo, P. I., and Elgueta, E. Y. (2008) Synthesis and mesomorphic properties of oxadiazole esters derived from (R)-2-octanol, (S)-2-n-octyloxypropanol and (2S,3S)-2-chloro-3-ethylpentanol. *Liquid Crystals*, **35**, 823-832;

   (c) Parra, M. L., Elgueta, E. Y., Jimenez, V., and Hidalgo, P. I. Novel amides and Schiff's bases derived from 1,3,4-oxadiazole derivatives: synthesis and mesomorphic behavior. *Liquid Crystals*, **36**, 301-317.
5. Chiang, I.-H., Long, C.-J., Lin, H.-C., Chuang, W.-T., Lee, J.-J., and Lin, H.-C. (2014) Broad ranges and fast responses of single-component blue-phase liquid crystals containing banana-shaped 1,3,4-oxadiazole cores. *ACS Applied Materials & Interfaces*, **6**, 228-235.
6. Han, J., Wang, F.-L., Zhang, F.-Y., and Zhu, L.-R. (2010) Synthesis, mesomorphic behaviour and photo-luminescent property of new mesogens containing 1,3,4-oxadiazole fluorophore. *Liquid Crystals*, **37**, 1521-1527.
7. (a) Majumadar, K. C., Shyam, P. K., Rao, D. S. S., and Prasad, S. K. (2011) Occurrence of unusually wide thermal range enantiotropic twist grain boundary TGBC* phases in unsymmetrical cholesterol and oxadiazole based liquid crystalline dimers. *Journal of Materials Chemistry*, **21**, 556-561;

   (b) Majumadar, K. C., Shyam, P. K., Rao, D. S. S., and Prasad, S. K. (2012) Oxadiazole-based unsymmetrical chiral liquid crystal dimers: synthesis and mesomorphic properties. *Liquid Crystals*, **39**, 1358-1367.

8. (a) Hirar, Y., Babu, S. S., Praveen, V. K., Yasuda, T., Ajayaghosh, and A., Kato, T. (2009) Anisotropic self-assembly of photoluminescent oligo(p-phenylenevinylene) derivatives in liquid crystals: An effective strategy for the macroscopic alignment of π-gels. *Advanced Materials*, **21**, 4029-4033;
(b) Mizoshita, N., and Kato, T. Liquid-crystal composites composed of photopolymerized self-assembled fibers and aligned smectic molecules. *Advanced Functional Materials*, **16**, 2218-2224.
9. Dieking, I. (2003) *Texture of Liquid Crystals*, Wiley-VCH Verlag, Germany.
10. Demus, D., and Richter, L. (1984) *Textures of Liquid Crystals*, Verlag Chemie, Germany.
11. Kumar, S. (2001) *Liquid Crystals: Experimental Study of Physical Properties and Phase Transition*, Cambridge University Press, UK.
12. Han, J., Chui, S. S.-Y., and Che, C.-M. (2006) Thermotropic liquid crystals based on extended 2,5-disubstituted11,3,4-oxadiazoles: Structure-property relationships, variable-temperature powder X-ray diffraction, and small-angle X-ray scattering studies. *Chemistry: An Asian Journal*, **1**, 814-825.

Chapter 6

A Facile Synthetic Route to Novel Thermotropic Liquid Crystalline Polymers by Bulk Reaction or Reactive Extrusion and Characterization of Their Mesophases

Kyoung Hwan Oh, Hoyeon Kim and Yongsok Seo*

RIAM, School of Materials Science and Engineering, College of Engineering, Seoul National University, Kwanakro-1, Kwanakgu, Seoul, 08826, Republic of Korea
**Corresponding author*: ysseo@snu.ac.kr

6.1 Introduction

Thermotropic liquid crystalline polymers (TLCPs) have outstanding physical properties such as high stiffness, high strength, high heat and chemical resistance, exceptional barrier properties and excellent dimensional stability which make them a super-engineering material with a wide application potential [1-5]. Their unique structures of mesogenic units, which maintain a liquid crystalline order in the melt state, enable them to acquire these excellent physical properties [1,2]. Liquid crystalline polymers can be divided into two groups; lyotropic liquid crystalline polymers which form a liquid crystalline structure in solution and thermotropic liquid crystalline polymers (TLCPs) which form a liquid crystalline structure in the melt state. Since TLCPs maintain ordered structure in the melt state, these can be used in the production of so-called *in-situ* composites by blending with common engineering thermoplastics in conventional processing equipment [6-13]. Most TLCPs show high thermal stability which implies that their processing is not easy due to high melting temperatures. Some TLCPs degrade before melting that exceeds the decomposition temperatures. Thus, in general, it is not easy to compound them with other common thermoplastic polymers to generate *in-situ* composites because thermoplastic matrices should withstand the high transition or processing temperatures of TLCPs in order to convert the TLCP phase into fibril forms that reinforce the thermoplastic matrix [6-15].

Liquid Crystalline Polymers, edited by Vikas Mittal

Since TLCPs are mostly synthesized via condensation polymerizations, they generally have low molecular weights. High molecular weights polymers have not only better mechanical properties than low molecular weight polymers, but also high melting temperatures. For ease of processing and more versatile applications, high molecular weight TLCPs with high mechanical properties, but low melting or processing temperatures are in high demand. Especially, in the last two decades, TLCPs with low transition or processing temperatures have been sought. We reported a new facile process to prepare a new kind of thermotropic liquid crystalline polymers through polyaddition process [16,17]. It consists of a reaction between a diepoxide compound and a primary monoamine (aniline [16] or 1-naphthyl amine[17]) which links two epoxy groups to form linear chains. This approach is noteworthy because it allows for the production of high molar mass TLCPs since no by-products are generated which need to be removed, but produced TLCPs have low processing temperatures. This means that the control on the processing temperatures of the TLCPs is possible by varying the structure of the diepoxy compound. It is also noteworthy that these polymers can be synthesized both via a reactive extrusion process and a bulk polymerization in a reaction chamber. Previously, TLCPs have not been synthesized by a reactive extrusion process in a single step. This is very important for the mass production of TLCPs. Therefore, both objectives of producing TLCPs having high molecular weights but low melting or processing temperatures could be achieved simultaneously. In this chapter, we recapitulate the recently described new TLCPs manufacturing processes [16,17].

6.2 Materials and Methods

6.2.1 Mesogen Synthesis

Two diepoxides, LCE1 (p-phenylene-di 4-(2,3-epoxypropyloxy) benzoate) and LCE3 (p-phenylene-di 4-(4,5-epoxypentyloxy) benzoate) (Scheme 6.1) were prepared following the same procedure in the literature [16,17]. All chemicals were purchased from Aldrich-Sigma, Korea. The synthesis method is summarized here as follows: a mixture of ethyl-4-hydroxybenzoate (0.6 mol, 99 g), 3-bromopropene (0.75 mol, 90 g), K_2CO_3 (0.9 mol, 126 g) and 800 mL of acetone were refluxed for 24 h. After filtering off the solid, the solvent was evaporated. Diethylether (400 mL) and water (200 mL) were poured into the remaining solution. The diethylether layer was separated and washed three times with 200 mL of a 10% NaOH solution. After evaporating away the diethylether, the residue was boiled in 900 mL of an ethanol/water (1:2) solution containing 90 g KOH until the solution was cleared. A 2N HCl solution was added to reach a pH of 2. After filtration, the solids were washed with excess water and recrystallized in ethanol to provide (4-(2-propyloxy) benzoic acid) as a white powder. A mixture of 4-(2-propyloxy) benzoic acid (0.11 mol, 21 g) and $SOCl_2$ (0.4 mol, 29 mL) was refluxed for 2 h. After evaporation of excess $SOCl_2$, 80 mL of pyridine were poured into the solution. Hydroquinone (0.55 mol, 6.3 g) was subsequently added. The mixture was reacted in 1 L of an acidified aqueous solution, filtered and washed with 500 mL of 5% Na_2CO_3 aqueous solution and 500 mL of water. After drying, the solid product (LCE1) was recrystallized in ethyl acetate. LCE1 was purified by column chromatography using CH_2Cl_2 as an eluent. After evaporation of the solvent, the product LCE1 was recrystallized in ethyl acetate. The yield was observed to be 55%. ^{1}H-NMR spectrum ($CDCl_3$): δ (ppm) 8.13 (4H, d), 7.23 (4H, s), 6.99 (4H, d), 4.33 (2H, dd, CH_2 of glycidyl), 4.02 (2H, dd, CH_2 of glycidyl), 3.38 (2H, s, CH of epoxy), 2.92 (2H, dd, CH of epoxy), 2.78 (2H, dd, CH of epoxy).

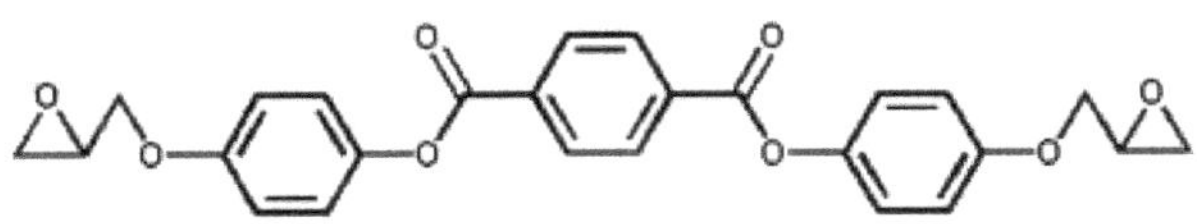

LCE1 (p-phenylene-di[4-(2,3-epoxypropyoxy) benzoate])

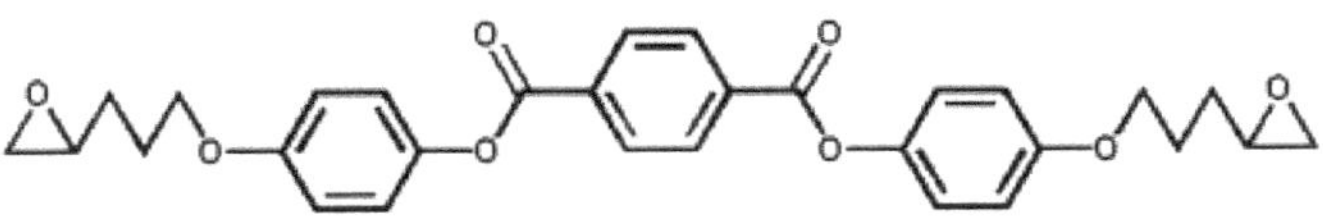

LCE3 (p-phenylene-di[4-(4,5-epoxypentyloxy) benzoate])

Scheme 6.1 Chemical structures of LCE1 and LCE3 [16,17].

The synthetic scheme for LCE3 was the same as LCE1, except using 5-bromopentene instead of 3-bromopropene. Transition temperatures are as follows: crystal to smectic : 110 °C, smectic to nematic :

139 °C, nematic to isotropic : 235 °C. Yield: 69%. ^{1}H-NMR spectrum ($CDCl_3$): δ (ppm) 8.12 (4H, d), 7.24 (4H, s), 6.96 (4H, d), 4.11 (4H, m, CH_2 of glycidyl), 3.01 (2H, m, CH of epoxy), 2.78 (2H, m, CH of epoxy), 2.51 (2H, m, CH of epoxy), 2.01 (4H, m, CH_2 of glycidyl), 1.85 (2H, m, CH_2 of glycidyl), 1.67 (2H, m, CH_2 of glycidyl).

6.2.2 Polymer Synthesis

Melt polymerization of the synthesized monomers (LCE1 and LCE3, having propyl epoxy and pentyl epoxy groups, respectively) with a primary monoamine (aniline or 1-naphthylamine) were first performed in a flask to produce poly(diepoxy aniline) (PDEAs, Scheme 6.2) and poly(diepoxy naphthalene)s (PDENs, Schemes 6.3). The reactive extrusion for PDEN1R or PDEN3R was carried out in a 42 mm Brabender twin screw extruder (AEV651) at a fixed rotation speed of 30 rpm. The extrusion temperatures of the feeding zone/transporting zone/melting zone/die were set as 140/220/220/210 °C, respectively.

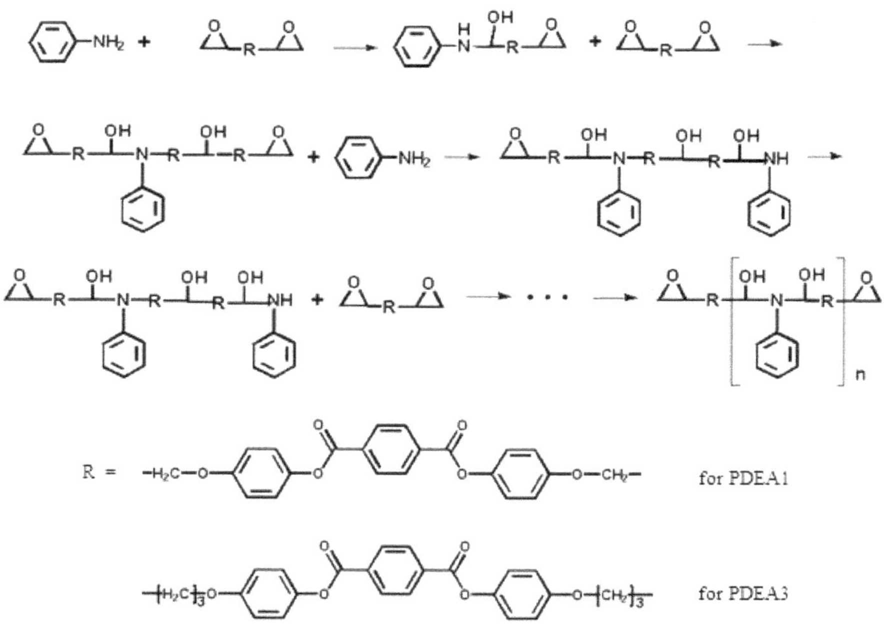

Scheme 6.2 Synthesis of PDEA1 and PDEA3 polymers via polyaddition reaction [16].

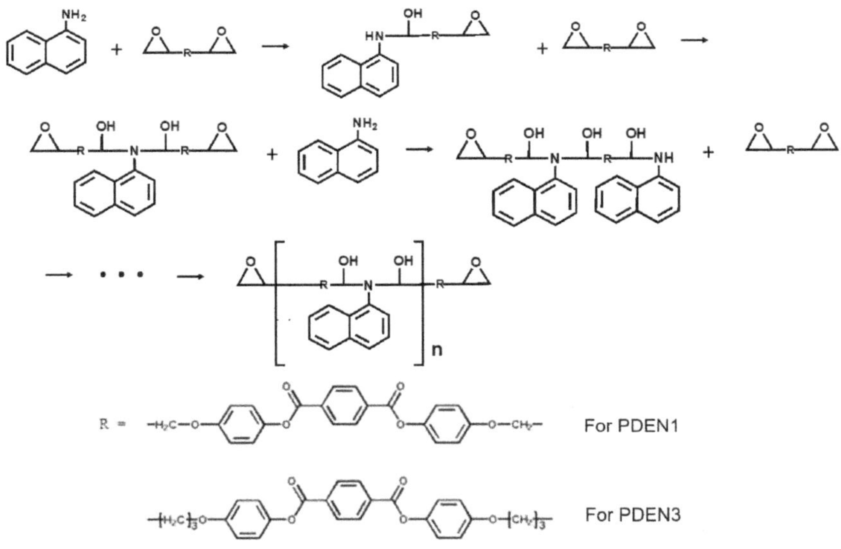

Scheme 6.3 Synthesis of PDEN1 and PDEN3 via a polyaddition reaction [17].

6.2.3 Characterization

The resulting polymers (PDEA1, PDEA3, PDEN1, PDEN1R (PDEN1 synthesized by the reactive extrusion process), PDEN3 and PDEN3R (PDEN3 synthesized by the reactive extrusion process)) were characterized using Fourier transform infrared spectrometry (FT-IR, Nicolet 6700, Thermo Scientific (USA)) with ATR accessory. The spectra were recorded with 32 scans at 8 cm^{-1} resolution. Prior to measurement, TLCP samples were hot pressed at 250 °C and cooled to room temperature to obtain films.

^{1}H-NMR and ^{13}C-NMR spectra were obtained by Bruker spectrometer (300 MHz). Samples were prepared by dissolving in THF-d_8 solvent. The ^{1}H-NMR spectra were referenced to residual TMS (0 ppm) except D_2O (solvent reference, 4.79 ppm) and DMF-d_7 (solvent reference, 8.03 ppm). Chemical shifts of the ^{13}C-NMR spectra were measured relative to DMF-d_7 (67.21 and 25.31 ppm).

Thermal properties such as degradation temperature and transition behaviors of TLCPs were analyzed using thermogravimetric analysis (TGA, TGA/DSC1, Mettler Toledo) and differential scanning calorimetry (DSC, DSC 823e, Mettler Toledo). Under a nitrogen atmosphere, samples were heated from 25 °C to 450 °C at 10 °C/min for

TGA. For DSC measurements, samples were first heated from 25 °C to 320 °C at 10 °C/min, cooled to 25 °C and heated to 320 °C again at the same heating rate under a nitrogen atmosphere.

Gel permeation chromatography (GPC, Viscotek model 250) was performed with an RI750F refractive index detector. Waters Styragel HP 4E and Styragel HR 5E columns were used with THF as eluent at a flow rate of 1 mL/min at 40 °C. The approximate calibration of the column was performed using polystyrene standards.

Thermal transition behaviors of polymer morphologies were observed by optical polarization microscopy (POM, BX51, Olympus) using the polarization mode. Samples were prepared by hot pressing between two glass slides at 250 °C. The measurement temperature range was from 25 °C to 320 °C with Mettler Toledo FP82HT hotstage. The polymers were also extruded to produce fibrils using a mini-extruder (DACA 5000) with a pulling unit. The morphology was observed using scanning electron microscopy (SEM, Hitachi S 2200C). The fractured sample surface in liquid nitrogen was coated with Pt/Pd before SEM observation.

The 2-D high temperature X-ray diffraction (XRD) measurement was performed with D8 Discover (Bruker) using the GADDS (General Area Detector Diffraction System) program. XRD scans were performed with CuKα radiation from 25 °C to 230 °C.

6.3 Results and Discussion

6.3.1 TLCPS Synthesized by the Reactions Between LCE1 or LCE3 with Aniline (PDEA1 and PDEA3)

Figure 6.1 shows the FT-IR spectra of the polymers. The sharp absorption peak at 914 cm^{-1} is due to the vibration of the oxirane ring in LCE1. The intensity of this peak is significantly lowered whereas broad shoulder absorption peaks are evident at 3450 cm^{-1} for the TLCPs which are due to the stretching vibrations of the OH groups generated by the reaction between the oxirane rings and amine group of aniline (Scheme 6.2). These results confirm that the epoxide groups on the LCEs have reacted with the amine group and polymerized to form the TLCPs (PDEA1 and PDEA3).

Analysis of the ^{1}H NMR spectra of a representative monomer and polymer (PDEA1) was performed with Chemsketch software; by assigning the signals to the corresponding protons and carbon atoms, the polymerization was confirmed (Figure 6.2(a); ^{1}H-NMR spectrum

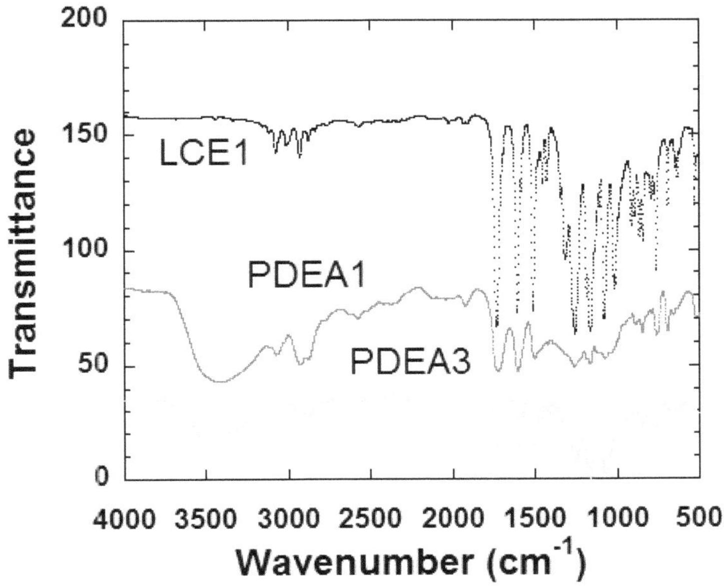

Figure 6.1 FTIR of the LCE1, PDEA1 and PDEA3. Reproduced from Reference 16 with permission from Royal Society of Chemistry.

($CDCl_3$): δ (ppm) 8.13 (4H, d), 7.23 (4H, s), 6.99 (4H, d), 4.33 (2H, dd, CH_2 of glycidyl), 4.02 (2H, dd, CH_2 of glycidyl), 3.38 (2H, s, CH of epoxy), 2.92 (2H, dd, CH of epoxy), 2.78 (2H, dd, CH of epoxy). The mesogenic monomer LCE1 contains five protons on its phenyl rings and its two aliphatic groups, including the methyl group of the spacer unit and the methyl group on the oxirane ring. The aromatic proton signals appeared at 6.5~8.2 ppm, while the TLCP methyl group signals are at 3.2~5.3 ppm and the corresponding protons of the LCEs appeared at 2.5 and 4.1 ppm respectively. The reaction between the primary amine group and epoxide ring (a ring opening reaction via nucleophilic attack on the amine) is accompanied by the appearance of new peaks at 4.7 and 5.3 ppm. The ^{13}C NMR spectrum of PDEA1 is presented in Figure 6.2(b). The characteristic peaks are consistent with those from the polymer structures. The doublet at 3.5 ppm is due to the hydrogen atoms in aniline and disappears as the reaction progresses. The appearance of the aniline aromatic peak at 149.2 ppm confirms the formation of the polymer. The spectral results clearly exhibited the structures of the formed polymers. The epoxy group reacts vigorously with the amine group in order to produce a

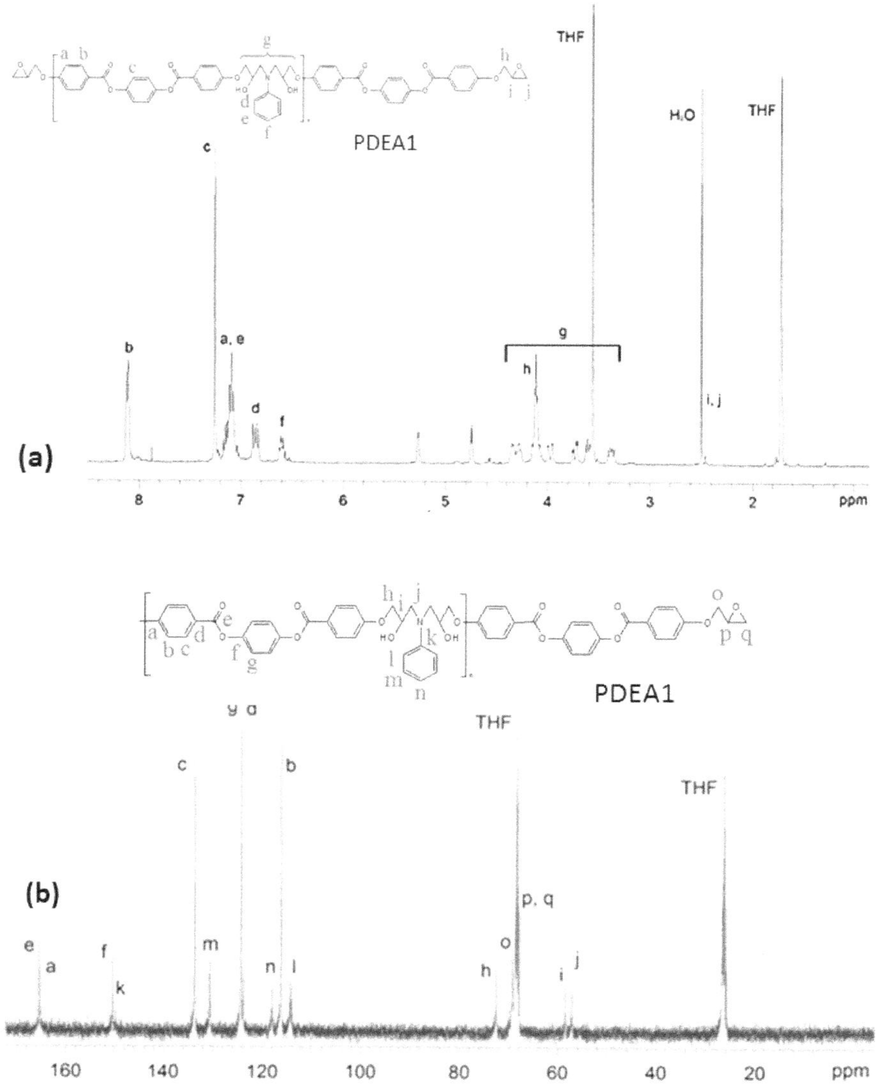

Figure 6.2 (a) ^1H NMR and (b) ^{13}C NMR of PDEA1. Reproduced from Reference 16 with permission from Royal Society of Chemistry.

compound containing a secondary amine that reacted further with other epoxy groups to form linear polymer chains (Scheme 6.2). The reaction between the epoxy groups and secondary amine (formed after the reaction of the first epoxy group with the amine group) is known to be less favorable than that with the primary amine [16-18].

However, this reaction was found to proceed rapidly to yield the polymers.

The 2-D wide angle X-ray diffraction patterns of the synthesized polymers are presented in Figure 6.3, which show that there are inner layer reflections on meridional scanning at a temperature below T_m. Sharp inner peaks are observed at a diffraction angle of $2\theta = 7.6°$, which corresponds to 11.7 Å according to Bragg's law. Broad outer peaks were observed at a diffraction angle of $2\theta = 19{\sim}20°$, which corresponds to 4.4~4.6 Å on the equatorial scan. This diffraction pattern is typical for LC chains with a lateral-packing 3~6 Å [2,19,20]. The broad peak in the wide angle region indicates that the ordering of the TLCP chains is liquid-like. As the temperature increases to 230 °C, the small angle reflections disappear whereas the broad peak at wide angles remains intact, suggesting that the chain regularity collapses and a nematic phase forms at this temperature [16,21].

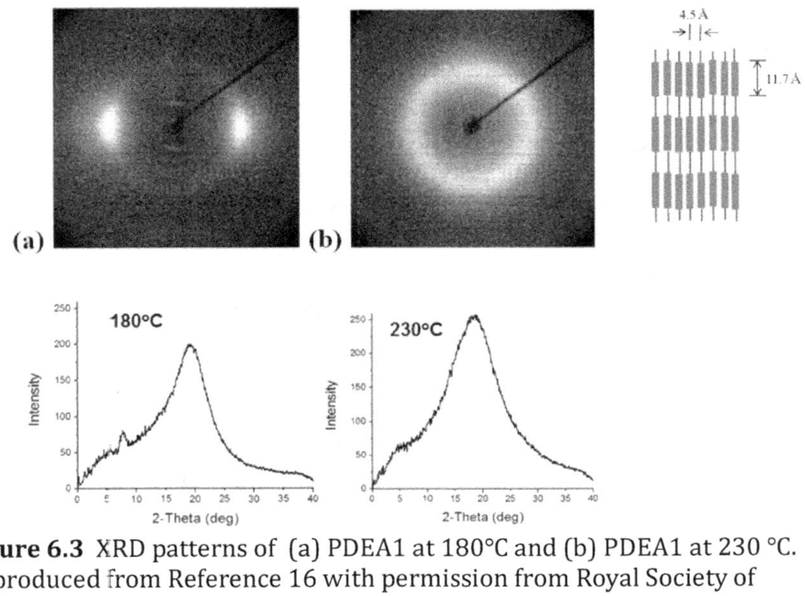

Figure 6.3 XRD patterns of (a) PDEA1 at 180°C and (b) PDEA1 at 230 °C. Reproduced from Reference 16 with permission from Royal Society of Chemistry.

The thermal behaviors of the polymers were investigated by using differential scanning calorimetry (DSC) and thermogravimetric analysis (TGA). The thermograms of both synthesized polymers showed a single endothermic peak during the heating scans attributed to the melting point (Figure 6.4(a)). The complex thermal history of the samples was removed by recording the second heating thermograms to determine their glass transition (T_g) and melting temperatures (T_m). The T_g and T_m values are listed in Table 6.1. T_g is related to the

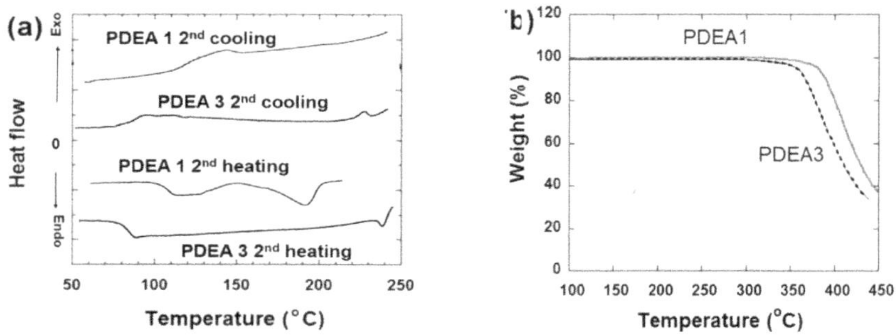

Figure 6.4 (a) DSC thermograms and (b) TGA degradation curves. Reproduced from Reference 16 with permission from Royal Society of Chemistry.

flexibility of the chains and corresponds to the ease of chain relaxation. Increasing the length of the flexible spacer in the backbone decreases the rigidity of the polymer molecules by reducing the mesogene proportion. Thus, PDEA3 has a lower T_g than PDEA1. The polymer with a longer alkyl group, PDEA3, has higher T_m value than the polymer with the short alkyl group (LCE1), PDEA1. Table 6.1 also provides the degradation temperatures (T_{dg} of 5 wt% loss). Both polymers have a high thermal stability (Figure 6.4(b)): degradation starts at 350 °C that proceeds in a single stage. This indicates that the presence of the flexible spacer groups does not affect the high thermal stability of the lengthy mesogen cores with ester link units. In spite of their high thermal stability, both polymers have a low melting

Table 6.1 Molar mass and thermal properties of polymers[a] (Adapted from Reference 16 with permission from Royal Society of Chemistry)

Samples	M_n	M_w	PDI	T_g (°C)	T_m (°C)	T_{dg} (5 wt% °C)
PDEA1	24,100	56,100	2.3	95	189	358
PDEA3	10,100	31,400	3.1	74	239	355

[a] T_g: glass transition temperature, T_m: melt temperature, T_{dg}: 5 wt% degradation temperature.

temperature (189 °C for PDEA1 and 249 °C for PDEA3), so these can be easily processed in a general processing equipment such as an extruder or an injection molder.

The optical textures of the polymers near the transition temperature are presented in Figure 6.5. During heating, both polymers exhibited a Schlieren-texture characteristic of the nematic mesophase near the transition temperature. The images were collected during the second heating process, thus, all samples exhibit the characteristic birefringence texture even at lower temperatures. On cooling the samples to room temperature after the first heating, the colorful textures remained nearly unchanged, implying that the mesophase of the polymers formed at high temperatures is stable. The surface morphologies of the fibers drawn from the melt were examined with SEM (Figure 6.6). All samples exhibit highly oriented TLCP fibrils. The surfaces of the fibers after cryogenic fracturing in liquid nitrogen contain protruding fibrils. This appearance of protruding fibrils on the fractured surfaces is consistent with continuous fibril morphology on the TLCP surface aligned parallel to the fibril direction [10,22].

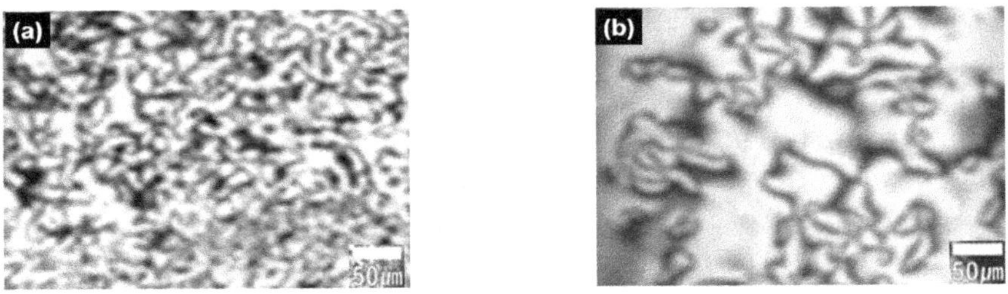

Figure 6.5 POM images of (a) PDEA1 at 190 °C and (b) PDEA 3 at 250 °C. Reproduced from Reference 16 with permission from Royal Society of Chemistry.

6.3.2 TLCPS Synthesized by the Reactions between LCE1 or LCE3 with 1-naphhtyl amine (PDEN1, PDEN3, PDEN1R and PDEN3R)

The FT-IR spectra of the polymers are demonstrated in Figure 6.7. The sharp absorption peak at 914 cm^{-1} is due to the vibration of the oxirane ring in LCE1. The intensity of this peak is significantly reduced for the TLCPs, while broad shoulder absorption peaks at 3450 cm^{-1} for the TLCPs are evident due to the stretching vibrations of the

Figure 6.6 SEM micrographs of polymers (fractured surface normal to fibril direction (left); parallel to fibril direction (middle); slant view of the normal surface (right)). (x2000 magnification). Reproduced from Reference 16 with permission from Royal Society of Chemistry.

OH groups produced by the reaction between the oxirane rings and amine group of 1-naphthylamine (cf. Scheme 6.3) [17]. These results also confirm that the epoxide groups on the LCEs reacted first with the amine group of 1-naphhtylamine and next with the produced secondary amine. The same steps were alternatively repeated to produce the high molar mass polymers (PDEN1 and PDEN3 in Scheme 6.3). Although the size of 1-naphthyl group is larger than that of the phenyl group of aniline (Scheme 6.2), it does not deteriorate the reactivity of the amine group.

The analysis of the ^{1}H NMR spectra of the polymers (PDEN1, PDEN3, PDEN1R and PDEN3R) were repeated with Chemsketch software. The polymer structures were identified by assigning the signals to the corresponding protons and carbon atoms. The mesogenic monomer LCE1 contains five protons on its phenyl rings and its two aliphatic groups, including the methyl group of the spacer unit and the methyl group on the oxirane ring. All protons can be uniquely identified in the spectrum (Figure 6.8(a)). Although it is not shown here, two spectra from PDEN1 and PDEN1R match exactly with each other, thus, confirming that these have the same structure. The signals of the aromatic protons appeared at 6.5~8.2 ppm, whereas the methyl

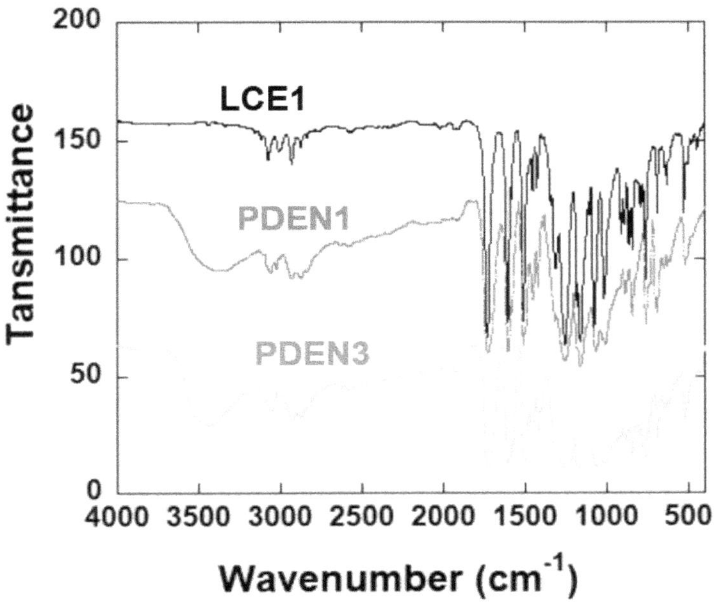

Figure 6.7 FTIR of the LCE1, PDEA1 and PDEA3. Reproduced from Reference 17 with permission from Royal Society of Chemistry.

group signals of the TLCPs are at 3.2~5.3 ppm. Corresponding protons of the LCEs appear at 2.5 and 4.1 ppm, respectively [16]. The reaction between the primary amine group and the epoxy ring (a ring-opening reaction via nucleophilic attack on the amine) is accompanied by the appearance of new peaks at 4.7 and 5.3 ppm. The doublet at 3.5 ppm is due to the hydrogen atoms in 1-naphthylamine and disappears as the reaction progresses. The ^{13}C NMR spectrum of PDEA1, PDEN1 and PDENR is listed in Table 6.2. The characteristic peaks are consistent with those of the polymer structures. The appearance of the aromatic naphthalene peak at 150.1 and 150.5 ppm confirms the formation of the polymer. These spectral results also clearly demonstrate the structures of the polymers. In analogy to the aniline reaction in Scheme 6.2, the first epoxy group reacted vigorously with the amine group to produce a compound containing a secondary amine that further reacted with another epoxy group to form linear polymer chains (Scheme 6.3). We were again able to confirm the reactivity of the secondary amine group which is lower than the primary amine group, but still reactive enough to rapidly form the targeted polymers [19,20].

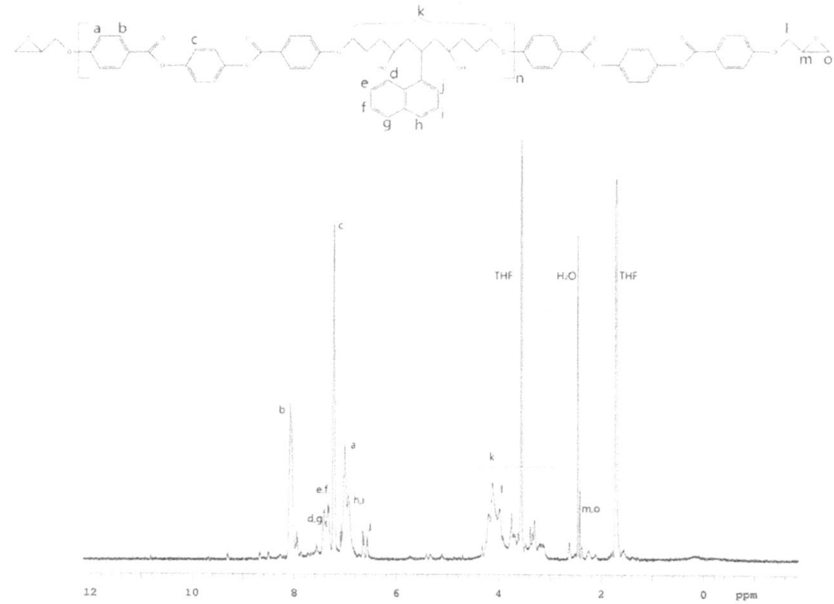

Figure 6.8 ^{1}H NMR of PDEN1. Reproduced from Reference 17 with permission from Royal Society of Chemistry.

The number and weight average molar masses obtained using gel permeation chromatography are listed in Table 6.3. The weight average molar masses of the bulk synthesized polymers (PDEN1 and PDEN3) are relatively high, although the polydispersity index (PDI) values of PDEN1 and PDEN3 are 2.0 and 3.2, respectively. PDEN3 has a lower molar mass than PDEN1 because the diffusivity of the LCE3 monomer is lower than that of the LCE1 monomer. This is the same result as for PDEA1 and PDEA3 [16]. Large PDI values indicate that the molar mass distribution is quite broad. In the case of bulk polymerization, the viscosity increased as the polymerization progressed, preventing the diffusion of LCEs. The polymers synthesized by the reactive extrusion process had better mixing, but a short residence time in the extruder did not allow sufficient reaction time to produce high molecular weight polymers. However, LCE1 could still produce higher molar mass than LCE3. Due to the low boiling point of aniline (184 °C), the reactive extrusion process could not be used for PDEA1 or PDEA3 synthesis while 1-naphthylamine (b.p. 301 °C) could be applied for the reactive extrusion because of its high boiling point. As mentioned above, PDEN1 and PDEN3 have relatively high

Table 6.2 ^{13}C-NMR spectra of LCPs (δ in ppm) (Adapted from Reference 17 with permission from Royal Society of Chemistry)

LCPs	**PDEN1, PDEN1R**	**PDEA3**	**PDEN3, PDEN3R**
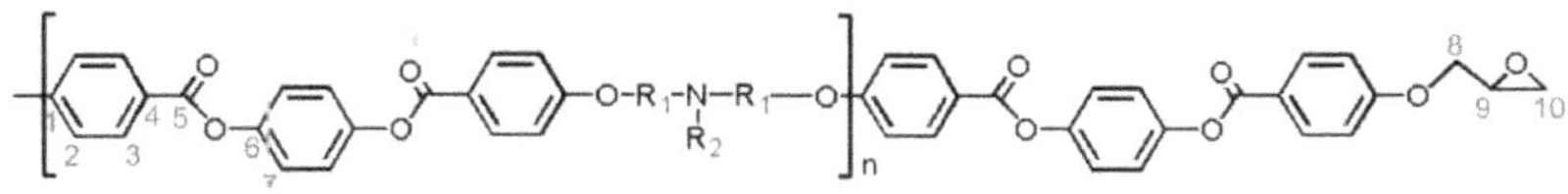			
R1	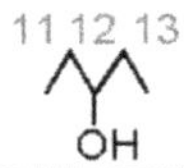	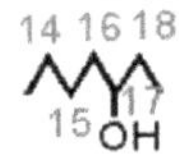	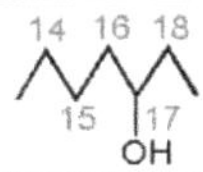
C atom **R2**	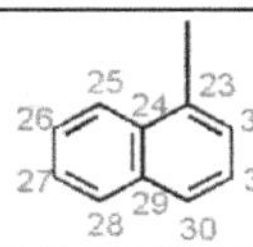	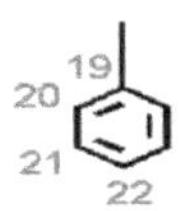	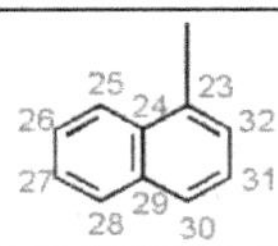
1	164.7	163.5	164.8
2	116.1	114.3	116.1
3	132.7	131.9	133.0
4	123.7	121.7	123.3
5	165.3	164.5	165.8
6	150.1	148.4	148.6
7	123.8	122.4	123.5
8	70.5	68.9	71.1
9	51.2	51.0	52.2
10	45.3	45.8	45.9
11	72.8		
12	68.3		
13	60.3		
14		68.6	69.5
15		24.5	25.8
16		31.2	31.7
17		68.0	68.8
18		66.2	65.0
19		149.2	
20		113.1	
21		128.9	
22		117.0	
23	150.1		150.5
24	132.4		132.8
25	125.7		125.0
26	127.3		127.1
27	127.5		130.0
28	129.3		131.0
29	136.4		136.5
30	124.3		124.3
31	128.1		130.3
32	117.2		117.2

molar masses compared to other engineering polymers as these were produced by the polyaddition reaction without any by-products.

Table 6.3 Molar mass and the thermal properties of the polymers[a)] (Reproduced from Reference 17 with permission from Royal Society of Chemistry)

Samples	M_n	M_w	PDI	T_g (°C)	T_m (°C)	T_{dg} (5wt % °C)
PDEN1	16200	32400	2.0	100	182	357
PDEN1R	10400	18200	1.8	95	182	352
PDEN3	14900	46500	3.2	71	193	358
PDEN3R	8,600	13,600	1.6	73	194	358

[a)]T_g : glass transition temperature, T_m : melt temperature, T_{dg} : 5wt% degradation temperature.

2-D wide angle X-ray diffraction reveals the crystalline structure of the synthesized polymers whose patterns are shown in Figure 6.9. Similar to PDEA1 and PDEA3 polymers, it exhibits inner layer reflections on meridional scan at a temperature below T_m. Again, sharp

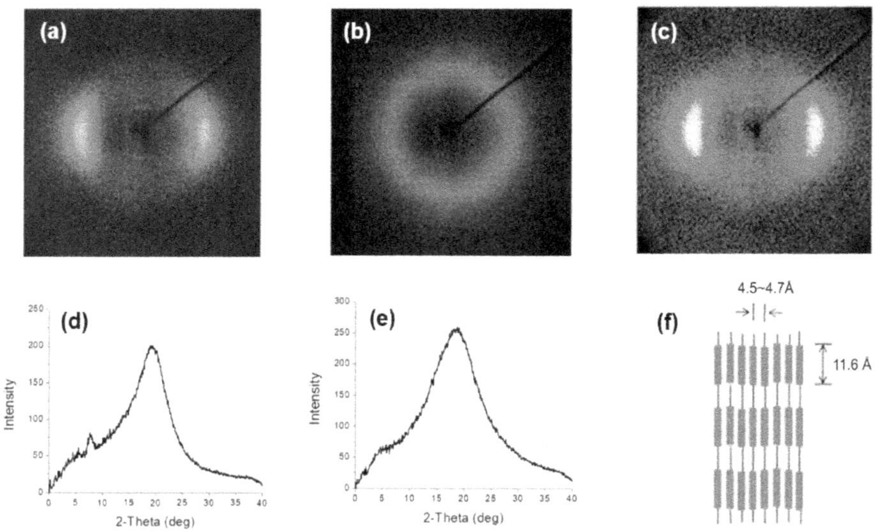

Figure 6.9 XRD patterns of (a) PDEN1 at 170°C, (b) PDEN1 at 230°C, (c) PDEN1R at 170°C, (d) PDEN1 at 170°C, (e) PDEN1 at 230°C, (f) schematic of PDEN1 LC ordering. Reproduced from Reference 17 with permission from Royal Society of Chemistry.

inner peaks are observed at a diffraction angle of 2θ = 7° and wider outer peaks at a diffraction angle of 2θ = 19~20°, corresponding to 4.5~4.7 Å on the equatorial scan. As mentioned above, the diffraction pattern corresponding to the lateral-packing of 3~6 Å is quite close to the PDEAs [1,16,21,22]. The broader peak in the wide angle region compared to PDEA1 or PDEA3 indicates that the ordering of the PDEN1 and PDEN3 chains is more liquid-like. As the temperature rises to 230 °C, the small angle reflections disappear while the broad peak at wide angle remains intact, indicating that there is a break-down of chain regularity and a nematic phase collapse at this temperature [23].

Differential scanning calorimetry (DSC) and thermogravimetric analysis (TGA) were used to investigate the thermal behaviors of the polymers. During the heating scans, single endothermic melting point was observed in the thermograms (Figure 6.10(a)). The T_g and T_m

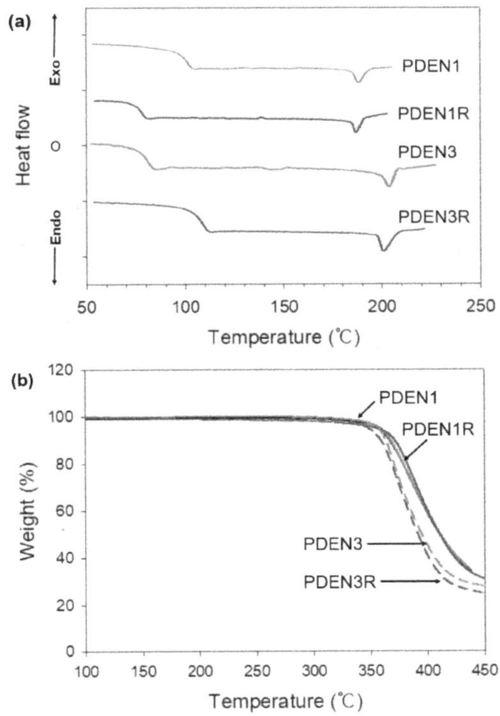

Figure 6.10 (a) DSC thermograms and (b) TGA degradation curves. Reproduced from Reference 17 with permission from Royal Society of Chemistry.

values are listed in Table 6.3. The polymer with a longer alkyl group, PDEN3, has higher T_m values than the polymer, PDEN1, with a short alkyl group (LCE1), which is consistent with the case of PDEA1 and PDEA3. On the other hand, PDEN3 has a lower T_g than PDEN1. Table 6.3 also provides the degradation temperatures (T_{dg} of 5 wt% loss) of the polymers. All polymers have high thermal stability (Figure 6.4(b)) and degradation starts at temperatures above 350 °C. The degradation proceeds in a single stage, confirming that the presence of the flexible spacer groups does not affect the high thermal stability of the lengthy cores with ester linkers. Both polymers have a low melting temperature (175 °C for PDEN1 and 194 °C for PDEN3), so these can be easily processed in an extruder or injection molder.

The optical textures of the polymers near the transition temperature are shown in Figure 6.11. Upon heating to the melt temperature (T_m), both polymers exhibit birefringence characteristics of a nematic liquid crystalline mesophase texture (the marbled nematic mesophase) near the transition temperature [24]. The images were also collected during the second heating process; all samples show the characteristic birefringent texture even at temperatures lower than the transition temperature. Therefore, the colorful textures remained almost unchanged when the samples were cooled to 25 °C after the first heating. This indicates that the mesophase of the polymers formed at high temperatures is quite stable. The surface morphologies of the melt-drawn fibers are also displayed in Figure 6.12.

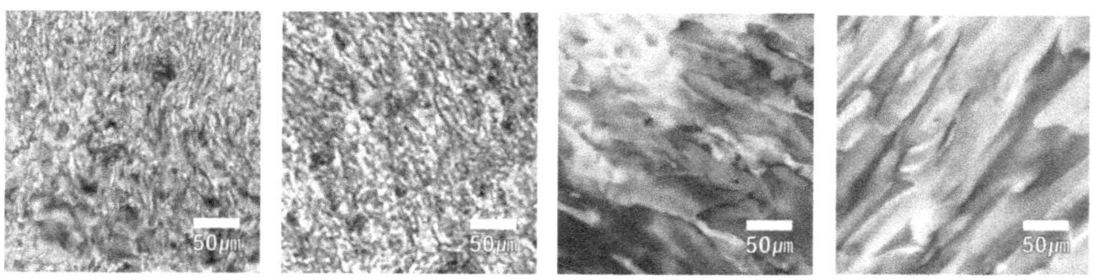

PDEN1 at 170°C **PDEN1R at 170°C** **PDEN3 at 190°C** **PDEN3R at 190°C**

Figure 6.11 POM images of PDEN1, PDEN1R, PDEN3, and PDEN3R respectively. Reproduced from Reference 17 with permission from Royal Society of Chemistry.

Similar to PDEA1 and PDEA3, all polymer samples contain highly oriented TLCP fibrils. The surface of the fibers after cryogenic cracking in liquid nitrogen contain protruding microfibrils. The appearance of protruding fibrils on the fractured surfaces is consistent with that of continuous fibril morphology on the PDEA1 and PDEA3 surfaces, which were observed to be oriented parallel to the fibril direction [10,16,25].

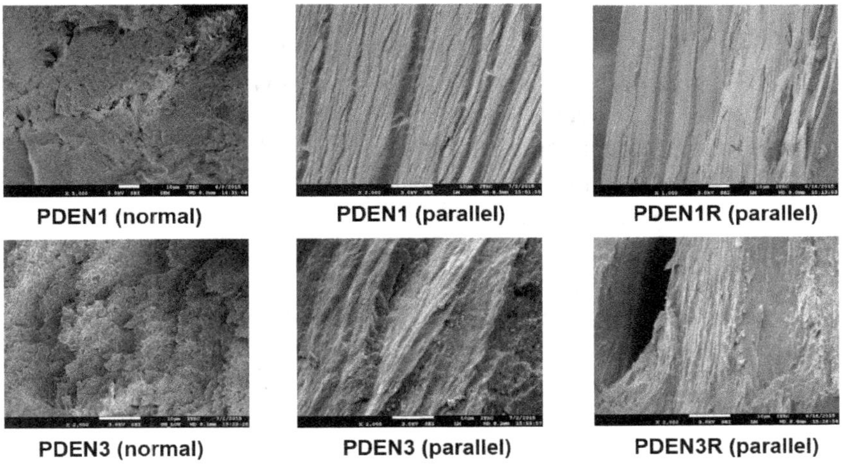

Figure 6.12 SEM micrographs of fractured polymer fibrils (upper row : PDEN1 fractured surface normal to fibril direction (left); parallel to fibril direction (center); PDEN1R fractured surface parallel to fibril direction (right); lower row: PDEN3 fractured surface normal to fibril direction (left); parallel to fibril direction (center); PDEN3R fractured surface parallel to fibril direction (right)), using x2000 magnification. Reproduced from Reference 17 with permission from Royal Society of Chemistry.

6.4 Conclusions and Outlook

A new class of liquid crystalline polymers with low melting temperatures were successfully synthesized by conducting melt reactions between diepoxy molecules and a primary monoamine (aniline or 1-naphthylamine). The synthesis of TLCPs was carried out via a polyaddition process. This approach is of great commercial importance because of its potential for mass production. The synthesized polymers have good thermal stabilities and high decomposition temperatures exceeding 350 °C, as well as low melting temperatures which facilitate their processing. Longer spacer groups in PDEA3 increased the transition temperatures, but the thermal stability of all polymers remained comparable due to the not-so-long spacer groups (methyl and propyl). Both polymers (PDEA1 and PDEA3) exhibit a nematic liquid crystal mesophase in the melt state. The polymers form well-ordered structures. Once stretched, the drawn polymer fibers have a strong orientation on the surface along the flow direction and numerous microfibrillar structures extrude out of the fractured surface perpendicular to the flow direction, indicating that the nematic polymers can be used for *in-situ* composites preparation.

Another type of liquid crystalline polymers with low melting temperatures have also been successfully synthesized by performing polyaddition reactions or reactive extrusion process between diepoxide molecules and 1-naphthylamine. 1-Naphthylamine could be used in the reactive extrusion process due to its high boiling point (301 °C), whereas the low boiling point of aniline (184 °C) prevented its use in the reactive extrusion process. Although synthesized polymers via reactive extrusion process have low molar masses compared to the same polymers prepared in the bulk due to their short reaction time in the extruder, this approach is commercially very significant because of its potential for the mass production, as mentioned earlier. Polymers synthesized by bulk polymerization have high molar masses, which supports the possibility that high molar mass TLCPs can be produced when a commercial extruder is used to provide sufficient residence time for reactants.

All polymers generated in the study (PDEN1, PDEN3, PDEN1R and PDEN3R) have good thermal stability and high decomposition temperatures in excess of 350 °C, indicating that these can withstand high temperature applications. As mentioned above, the polymers exhibited low melting temperatures, which allows easy processing. Like the aniline based polymers (PDEA1 and PDEA3), longer spacer groups in PDEN3 showed slightly higher transition temperatures, but the thermal stabilities of all polymers remained comparable due to the moderate size of the spacer groups used in the study (methyl and propyl). Synthesized polymers by the reactive polymerization process exhibited nearly the same thermal properties as those synthesized by the bulk polymerization. All polymers showed a nematic liquid crystal mesophase morphology in the melt state as well as well-ordered structures upon cooling.

Acknowledgements

This work was supported by NRF (MSE BK21Plus Program of SNU) and MIKE (No. 10067685, Core Original Materials Development Program).

References

1. Tschierske, C. (2013) Development of structural complexity by liquid-crystal self-assembly. *Angewandte Chemie-International Edition*, **52**, 8828-8878.
2. Chung, T. S. (ed.) *Thermotropic Liquid Crystal Polymers : Thin Polymerization, Characterization, Blends, and Applications*, Technomic Publishing, USA (2001).
3. Thakur, V. K., and Kessler, M. R. (eds.) *Liquid Crystalline Polymers, Volume 1: Structure and Chemistry*, Springer, USA (2016).
4. Donald, A. M., Windle, A. H., and Hanna, S. (2006) *Liquid Crystalline Polymers*, 2nd edition, Cambridge University Press, UK.
5. Burrell II, E. M., and Johnson, J. F. (1979) A review of the status of polymerization in thermotropic liquid-crystalline media and liquid-crystalline monomers. *Journal of Macromolecular Science, Part C: Polymer Reviews*, **17**(1), 137-170.
6. Seo, Y., Hong, S. M., and Kim, K. U. (1997) Structure development during flow of ternary blends of a polyamide (nylon 46), a thermotropic liquid crystalline polymer (poly(ester amide)), and a thermoplastic elastomer (EPDM). *Macromolecules*, **30**, 2978-2988.
7. Seo, Y., Ninh, T. H., Hong, S. M., Kim, S., Kang, T. J., Kim, H. S., and Kim, J. (2006) In situ compatibilizer-reinforced interface between a flexible polymer (a functionalized polypropylene) and a rodlike polymer (a thermotropic liquid crystalline polymer). *Langmuir*, **8**, 3062-3067.
8. Fleischmann, E. K., and Zentel, R. (2013) Liquid-crystalline ordering as a concept in materials science : From semi-conductors to stimuli-responsive devices. *Angewandte Chemie-International Edition*, **52**, 8810-8827.
9. Seo, Y., Kim, B., Kwak, S., and Kim, K. U. (1999) Structure development during flow of ternary blends of a polyamide (nylon 66), a thermotropic liquid crystalline polymer (poly(ester amide)) and a functionalized polypropylene. *Polymer*, **40**, 4483-4492.
10. Seo, Y., Kim, B., and Kim, K. U. (1999) Morphology and properties of compatibilized ternary blends (nylon 6/a thermotropic liquid crystalline polymer/a functionalized polypropylene) processed under different conditions. *Polymer*, **40**, 4441-4450.
11. Seo, Y., Hong, S. M., Hwang, S. S., Park, T. S., Kim, K. U., Lee, S., and Lee, J. (1995) Compatibilizing effect of a poly(ester imide) on the properties of the blends of poly(ether imide) and a thermotropic liquid crystalline polymer: 1. Compatibilizer synthesis and thermal and rheological properties of the *in situ* composite system. *Polymer*, **36**, 515-524.
12. Seo, Y., Hong, S. M., Hwang, S. S., Park, T. S., Kim K. U., Lee, S. M., and Lee, J. (1995) Compatibilizing effect of a poly(ester imide) on the

properties of the blends of poly(ether imide) and a thermotropic liquid crystalline polymer: 2. Morphology and mechanical properties of the *in situ* composite system. *Polymer*, **36**, 525-534.

13. Lee, S. M., Hong, S. M., Seo, Y., Park, T. S., Hwang, S. S., Kim, K. U., and Lee, J. (1994) Characterization and processing of blends of poly(ether imide) with thermotropic liquid crystalline polymer. *Polymer*, **35**, 519-531.
14. La Mantia, F. R. (1993) *Thermotropic Liquid Crystal Polymer Blends*, Technomic Publishing, USA.
15. Krishwaswamy, R. K., and Baird, D. G. (1999) Influence of a reactive terpolymer on the properties of in situ composites based on polyamides and thermotropic liquid crystalline polyesters. *Polymer*, **40**, 701-716.
16. Oh, K., Kim, H., and Seo, Y. (2017) A facile synthetic route to novel thermotropic liquid crystalline polymers and characterization of their mesophases. *RSC Advances*, **7**, 29772-29778.
17. Oh, K., and Seo, Y. (2018) Synthesis of novel thermotropic liquid crystalline polymers by bulk reaction or reactive extrusion. *Polymer Chemistry*, in print.
18. Mossey-Lesczak, B., and Locharska, M. W. (2011) Liquid crystallinity in polymers – Liquid crystalline epoxy resins. In: *Liquid Crystalline Compounds and Polymers as Materials of the XXI Century: From Syntheses to Applications*, Iwan, A., and Schab-Balerzak, E. (eds.), Research Signpost, India, pp. 125-152.
19. Enikolopiyan, N. S. (1976) New aspects of the nucleophilic opening of epoxide rings. *Pure & Applied Chemistry*, **48**, 317-328.
20. Blanco, M., Corcuera, M. A., Riccard, C. C., and Mandryar, I. (2005) Mechanistic kinetic model of an epoxy resin cured with a mixture of amines of different functionalities. *Polymer*, **46**, 7989 -8000.
21. Watanabe, J., and Hayashi, M. (1989) Thermotropic liquid crystals of polyesters having a mesogenic p,p'-bibenzoate unit. 2. X-ray study on smectic mesophase structures of BB-5 and BB-6. *Macromolecules*, **22**, 4083-4088.
22. Kantor, S. W., Sung, T. C., and Atkins, E. D. T. (1992) Thermotropic hydrocarbon main-chain liquid crystalline polymers based on a biphenyl mesogen: synthesis and characterization. *Macromolecules*, **25**, 2789-2795.
23. Zhu, Z., Zhi, J., Liu, A., Tang, H., Qiao, W., Wan, X., and Zhou, Q. (2007) Effects of polymerization method on structure and properties of thermoplastic polyurethanes. *Journal of Polymer Science, Part A: Polymer Chemisty*, **45**, 800-847.
24. Ryu, C., Seo, Y., Hwang, S. S., Hong, S. M., Park, T. S., and Kim, K. U. (1994) Miscibility and mechanical properties of poly(ether imide)/liquid crystalline poly(ester imide) blends. *International Polymer Processing*, **9**, 266-272.

25. Chen, B., Tsay, S., and Chen, J. (2005) Synthesis and properties of liquid crystalline polymers with low T_m and broad mesophase temperature ranges. *Polymer*, **46**, 8624-8633.

Chapter 7

Segmented Liquid Crystalline Copolyesters

Katherine V. Heifferon and Timothy E. Long*

Macromolecules Innovation Institute, Department of Chemistry, 1075 Life Science Circle, Virginia Tech, Blacksburg, VA 24061, USA
**Corresponding author*: telong@vt.edu

7.1 Introduction

Over the last 30 years, the field of segmented liquid crystalline (LC) copolyesters expanded significantly. IUPAC defines segmented copolymers as "copolymers containing phase domains of microscopic or smaller size, with the domains constituted principally of single types of structural units" [1]. The understanding of the expanse of polymer architectures and structures that facilitate phase separation into nanodomain morphologies experienced great growth since its inception along with the expansion of knowledge in the role that LC properties play in this phase separation. This chapter explores the synthesis and characterization of segmented liquid crystalline copolyesters, delving into the parameters required for phase separation and the impact of the domain morphology.

7.1.1 History of Polyesters

The first exploration and fundamental understanding of step-growth polymers began with the work of Carothers at DuPont in the 1930's. These studies began with the most well-known step-growth polymers, polyesters [2,3]. This work elucidated the fundamental processes of step-growth polymerization (condensation) through the intentional study of aliphatic polyesters. The authors identified important characteristics, such as functionality, degree of polymerization and conversion among others. Unfortunately, these polymers remained unsuitable for commercialization due to low melting temperatures and other unfavorable properties such as sensitivity to hydrolysis.

Liquid Crystalline Polymers, edited by Vikas Mittal

In the 1940's, the introduction of a rigid aromatic monomer into the polymer backbone revealed stiffening in the polymer chains resulting in high melting temperature fibers. This polyester, synthesized from terephthalic acid and ethylene glycol and known as poly(ethylene terephthalate) (PET), pushed polyesters into commercialization. Further variation of ethylene glycol with a wide variety of aliphatic diols led to an extensive family of polymers [4,5]. Today, PET production encompasses 70% of all PP&A (polyester, polyamide, and acrylic) fiber production and 7% of the world's plastic [6]. The food packaging industry extensively uses PET in blow molded and biaxially oriented bottles, accounting for 30% of its global demand [7]. Aside from the apparent usefulness of PET's semi-crystalline nature and rigidity, one of the major benefits arose from its recyclability back to the monomeric state. This depolymerization uses techniques, such as methanolysis, to break the polymer chains and reform oligomers/monomers. Reuse of these products in repolymerization represents a large field of study for companies, such as Eastman Chemical Company, DuPont and Hoechst-Celanese Company [8]. As research of polyesters grew in the 1970's, exploration into a wide range of unique morphologies and structures provided the scientific community with block copolymers and liquid crystalline polyesters. The relevancy of the polyester field of chemistry continues to this day and demonstrates roots in a deep history of industrial development.

7.1.2 History of Liquid Crystalline Polymers

A liquid crystal refers to molecules that undergo an intermediate phase that lies between a solid and a liquid state. These molecules exhibit more order than an isotropic liquid but less than a 3-dimensional crystal structure and their presence in nature provides many unique and useful materials [9]. Friedrich Reinitzer, an Austrian botanist, first discovered liquid crystals while experimenting with cholesterol in 1888. Otto Lehmann (his collaborator) and Reintzer coined the term mesophase, after the Greek word mesos meaning intermediate, to represent this unique phase [10]. In 1922, Georges Freidel further presented a classification scheme for the three possible morphological phases of this state; nematic, smectic, and cholesteric phases [11,12]. Freidel's work sealed the acceptance of liquid crystals in the scientific community, debunking previous contradictory concepts. Further study of synthetic molecules confirmed that LC properties arose from small molecules comprised of rod- or disk-like shapes, or amphiphilic molecules exhibiting anisotropy in their structure [11,13]. Rod- and disk-shaped rigid molecules encompass the field of thermotropic LC molecules which induce the formation of a mesophase through thermal means. These materials exhibit a LC mesophase at temperatures above the melting point of the crystal structure (T_m). The isotropic temperature (T_i) represents the upper limit to the mesophase at which complete disorder occurs. Amphiphilic molecules transition into a mesophase through the dissolution of the molecule upon introduction of a solvent, termed lyotropic. Though not the focus of this review, however, these materials intersect with many aspects of our daily lives, such as the dissolution of soap into water as well as in our cell membranes (dissolution of phospholipids in water). Small molecule LCs facilitated the development of a wide range of applications including LC displays in television, spatial light modulators and temperature sensors, among others [14,15].

In 1956, Paul Flory hypothesized the ability to incorporate LC molecules into polymers [16,17]. Flory deemed the key parameters impacting LC properties as geometry and molecular structure utilizing lattice theories. This study proposed an aspect ratio (length/diameter) of 6 or greater was necessary for polymers to achieve LC properties. Utilizing this understanding of the aspect ratio, Dupont scientists developed one of the first commercialized polymeric LC, Kevlar, which exhibited lyotropic properties. Further study produced the first thermotropic LCPs from a select group of companies including Eastman-Kodak, Amoco, Dupont and Celanese [14,18]. Thermotropic main-chain polymers excelled in industry due to the useful thermoplastic properties, such as shear thinning, high chemical and thermal resistance, low dielectric constants and low moisture uptake [14]. During this time, main-chain LC structures took the form of fully-aromatic polymers (polyarylates) due to their high rigidity. Vectra, poly(6-hydroxy-2-napthoic acid-co-4-hydroxybenzoic acid), emerged as one of the first LC polyesters, developed by Ticona/Celanese that achieved commercialization despite its high cost [19]. The primary uses of this polymer include electronic parts and high-performance fibers [10]. Eastman-Kodak also explored the reaction of p-acetoxybenzoic acid with PET to form the first aromatic-aliphatic copolyester exhibiting thermotropic properties [10,20-22]. As this field grew, the next logical scientific step became the synthesis of LC block copolymers which could incorporate LC properties into other polymers through phase separated morphologies.

7.1.3 History of Segmented Copolymers

Polyurethanes comprised many of the first reported phase separated copolymers [23-25]. These polymers formed segregated domains considered hard- and soft-segments that impart stiffness from the physical crosslinking of the hard-block and elasticity from the flexibility of the soft-block. Poly(styrene-*b*-isoprene-*b*-styrene) demonstrates another widely studied thermoplastic elastomer synthesized through a more controlled anionic polymerization technique in the 1960's [26,27]. Most of these early works focused on eliminating macroscale phase separation but maintaining microscale phase separated domains utilizing a block copolymer architecture to impart physical properties from both polymeric units into the final polymer structure.

In the field of polyesters, the 1970's represented a boom in block copolymer research though it was DuPont's Hytrel (poly(oxytetramethylene)-*b*-poly(butylene terephthalate)) that eventually found commercial success as a thermoplastic elastomer [28]. Segmented copolymers rose to prominence in LC polymers as blend compatibilizers and mechanical property adaptors [29,30]. The advent of controlled polymerization techniques, such as anionic polymerization, enabled much smaller dispersity in block size opening the door for well-controlled morphologies including lamellae, hexagonal cylinders, bicontinuous gyroid and body-centered cubic arrays of spherical micelles [31]. These morphological structures facilitated the development of a wide variety of applications including, but not limited to, drug delivery, soft lithography and the development of porous materials. This area of research exploded over the last 60 years and continues to develop with the synthesis of new polymers. Liquid crystalline segmented copolymers represent a rapidly growing field in which many interesting orientation properties and corresponding commercial applications lie.

7.2 Synthetic Overview

The synthetic methods required to synthesize segmented LC copolyesters incorporate a range of techniques from polycondensation to controlled radical and other chain-growth methods. The basics of polymer synthesis within the areas of polyesters and LC polymers acts as a platform for understanding the synthetic requirements of segmented LC copolyesters.

7.2.1 Basics of Polyester Synthesis

The family of polymers known as polyesters contains at least one ester functionality in every repeating unit. The formation of an ester occurs through a variety of different carbonyl reactions, most commonly transesterification or esterification, as expanded upon in Figure 7.1. These reactions occur through attack on the carbonyl carbon

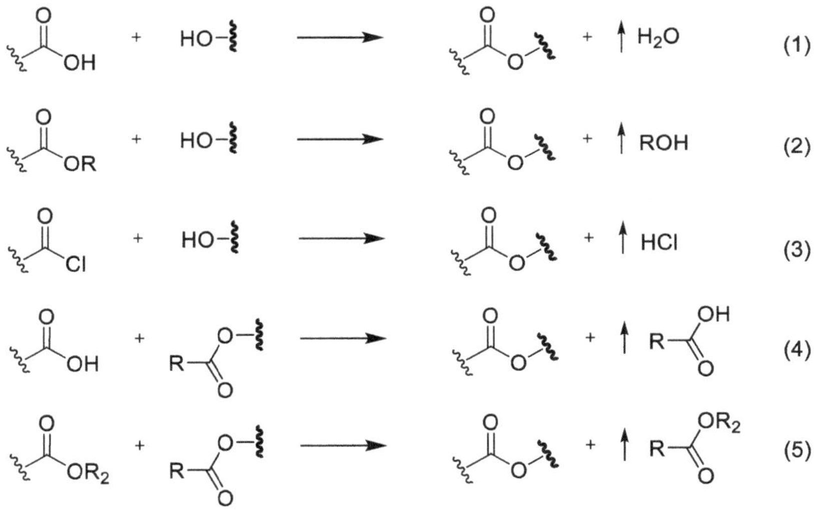

Figure 7.1 Examples of reactions utilized in the synthesis of polyesters. Adapted from Rogers and Long [28].

resulting in the production of a tetrahedral intermediate. Reforming the carbonyl causes elimination of a leaving group and inevitably the formation of the ester functionality [28]. Polymerization to form polyesters, in all cases except ring-opening polymerization, follows step-growth kinetics. As with all step-growth polymerizations, there are six essential criteria that must be achieved in order for the polymerization of a linear polymer to reach high molecular weight (MW) [28,32].

1. Stoichiometry of the functional groups forming the polymer must equal 1.0
2. Monomer functionality (*f*) must equal 2.0 to maintain linearity and prevent branching
3. Absence of side-reactions

4. Efficient removal of condensates formed during polymerization (when applicable)
5. High reaction conversions (>99.9%)
6. Accessibility of reactive end-groups

These essential criteria relate back to the Carothers equation, representing the kinetics of MW growth in a step-growth polymer, as outlined in eq. 1, where X_n represents the degree of polymerization, f_{avg} the average functionality and ρ the extent of conversion.

$$X_n = \frac{2}{2 - \rho f_{avg}} \quad (1)$$

AB or AA/BB monomers build MW through the initial formation of oligomers, such as dimers, trimers, etc. This occurs through coupling of small molecules after complete monomer consumption. The final stage of MW growth occurs only at >90% conversion and results in the formation of linear high MW polymers, as demonstrated in Figure 7.2.

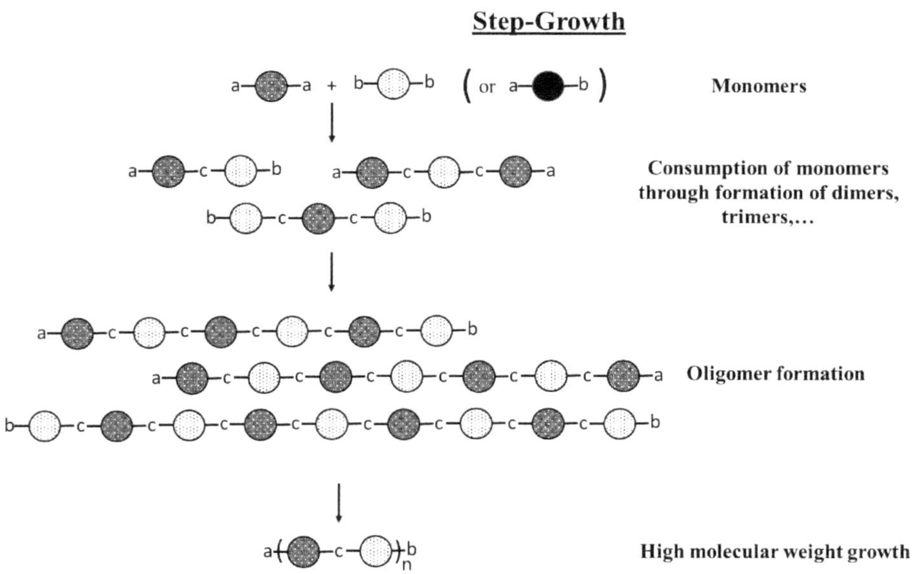

Figure 7.2 Graphical illustration of the mechanism of molecular weight growth for step-growth polymers.

The most common type of polymerization for polyesters utilizes high temperature bulk reactions known as melt polycondensation. As utilized in the synthesis of PET, this occurs through either direct polyesterification or transesterification. These reactions proceed under solvent-less conditions upon heating the monomers above their respective T_m resulting in a homogenous melt. Slow increase of the reaction temperature affords monomer consumption while maintaining a homogenous solution and allowing time for oligomer formation to prevent distillation of the growing chains. The final stage of the reaction includes the application of vacuum to distill off the condensate and form high MW polymers through Le Chatelier's principle resulting in a drastic viscosity increase [8,28,32]. Commonly, excess diol, such as ethylene glycol in the case of PET, is initially added to the polymerization then removed during the final vacuum stage to bring the stoichiometry of the reaction back to the necessary 1:1. This limits error that occurs from early distillation or inaccuracies while measuring monomers. Scheme 7.1 illustrates a common high temperature bulk procedure utilized in the synthesis of PET [33].

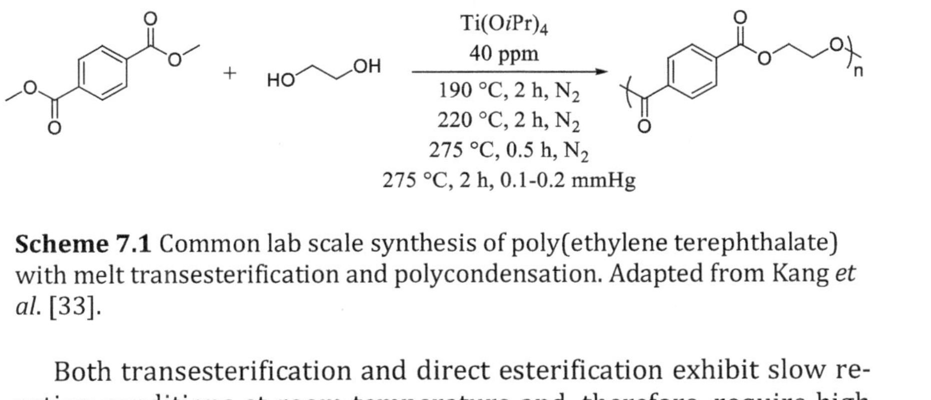

Scheme 7.1 Common lab scale synthesis of poly(ethylene terephthalate) with melt transesterification and polycondensation. Adapted from Kang *et al.* [33].

Both transesterification and direct esterification exhibit slow reaction conditions at room temperature and, therefore, require high temperatures to push the reaction to completion. Catalyst incorporation increases reaction rates and ensures the formation of high MW polymers. Organometallics compounds and metal alkoxides comprise some of the most widely used catalysts for polyesterification reactions. The study of these catalytic mechanisms led to the understanding that Lewis acids, such as zinc and manganese, activate the ester group through complexation with the carbonyl oxygen. This complexation favors nucleophilic attack, lowering the activation energy for these reactions. In contrast, Weingart [34] elucidated that antimony and titanium alkoxide catalysts undergo ligand exchange reactions, as illustrated in Figure 7.3. Titanium alkoxides demonstrate a high efficiency in the catalysis of both esterification and transesterification (polycondensation) reactions resulting in their addition at the start of the reaction [8]. On the contrary, antimony trioxide (Sb_2O_3) exhibits a higher activity toward polycondensation (the final reaction stage) than esterification resulting in its addition just before the vacuum stage to assist in MW growth. The deleterious toxicity of antimony catalysts influences the shift toward titanium based catalysts and novel organocatalysts [35].

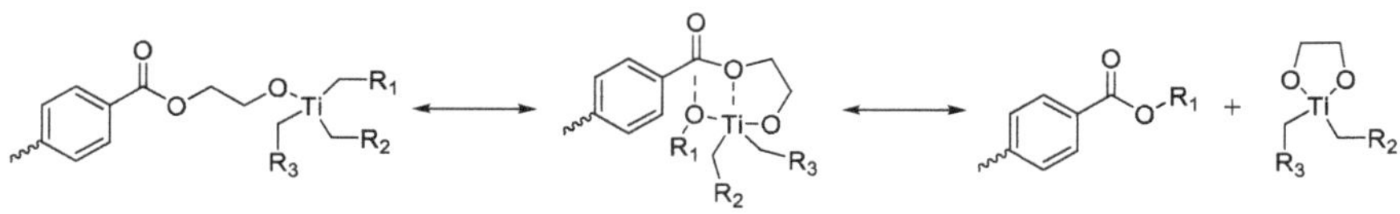

Figure 7.3 Mechanism of ligand exchange during polymerization using titanium catalysts. Proposed by Weingart and adapted from Schiers and Long [8,34].

In the case of less reactive alcohols, like phenols, slow reaction rates limit the use of ester/carboxylic acid functional groups. The conversion of phenolic monomers to the acetoxy functionality promotes the reaction between a carboxylic acid and phenol enabling their polymerization in the bulk. The one-pot method utilizes the addition of anhydrides, like acetic anhydride and pivalic anhydride, to form a mixed anhydride *in-situ* with the carboxylic acid monomer improving the leaving groups of the functional group. The pre-acetylation of phenolic monomers results in the same formation of mixed anhydrides through a two-step route. Many reports have studied the mechanism of this reaction and attributed the improved reactivity to the carboxy-ester interchange reaction between acetoxyaryl groups and carboxylic acid functional groups resulting in the elimination of acetic acid [19,36]. The temperature of these reactions must be at least 20 °C greater than the T_m of the monomers and increased stepwise to maintain a low viscosity during the polymerization. Common polymers synthesized with this technique include fully-aromatic polyesters, such as LC copolymer Vectra and amorphous polyarylate poly(bisphenol-terephthalate).

Melt polycondensation exhibits a very useful and industrially relevant technique towards the synthesis of polyesters. The environmentally friendly nature of this reaction due to the lack of solvent and minimal post-polymerization work up, drove it to prominence. Unfortunately, high temperature reactions are more prone to side-reactions which can impact the final polymer backbone structure and MW. A few examples of this include the formation of diethylene glycol through etherification of ethylene glycol in the synthesis of PET or the formation of THF during synthesis of PBT [8]. These high temperature conditions also inhibit polymerization with any thermally sensitive monomers, such as those utilized in unsaturated polyesters. The conversion of monomers to more activated carbonyls with better leaving groups and the use of solvents facilitate lower temperature reactions [8].

The polyester synthesis between acid chlorides and alcohols typically takes place in solution to reduce side-products resulting from the formation of HCl. Due to the better leaving group, these reactions occur at lower temperatures than those necessary for esters and carboxylic acids. The solvent selection plays a key role in reaction rate; more polar solvents facilitate fast reaction rates [37]. It is important that the final polymer remain soluble in the solvent selected to prevent precipitation from stunting MW growth. HCl evolution enables reaction tracking and end-point determination [38]. Catalysts in this technique often take on the form of tertiary amines, such as pyridine or triethylamine, which also act as HCl scavengers [28,39-43]. The tertiary amine forms a quaternary ammonium salt in the presence of HCl inhibiting reverse-reactions and side-reactions from occurring during the polymerization or post-processing [28]. In many cases, solution polymerization at high temperatures occur to improve solubility of polymers/monomers that do not dissolve well at room temperature or exhibit crystallinity. During these polymerizations, the use of high boiling point solvents prevents distillation.

Solution polymerizations using activating agents stimulate higher monomer reactivity, but their consumption during the reaction limits the label catalysts. Phosphorus derivates represent a large portion of activating agents, triphenyl phosphine and diphenylchlorophosphate representing the most effective. These compounds form salts in the presence of pyridine in turn enabling the activation of carboxylic acids, as demonstrated in Scheme 7.2. Rogers and Long [28] described a wide range of other activating agents that can be utilized in polyester synthesis.

The use of interfacial polymerization with acid chloride functionalized monomers has been demonstrated in a few cases within the segmented LC copolyester literature. The technique restricts alcohol

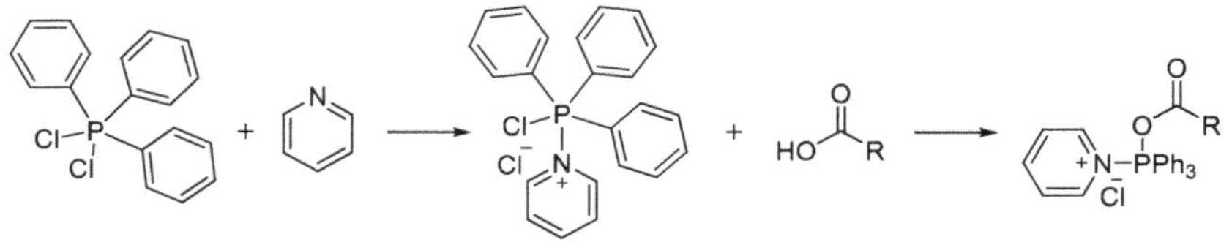

Scheme 7.2 Formation of triphenyl phosphine salts in the presence of pyridine and the resulting carbonyl activation. Adapted from Rogers and Long [28].

to a phenolic monomer due to the required formation of an alcoholate ion in aqueous solutions; impossible for aliphatic diols. The reaction conditions for this technique requires a two-phase solvent system (aqueous/organic), surfactant, water-soluble base and organic-soluble monomers/final polymer. The alcoholate ion of the aromatic phenol forms through a reaction with alkali bases, such as NaOH. Surfactants (phase-transfer agents) increase the interfacial area between the two solvents where the reaction occurs helping to control the reaction rate [28]. Examples of possible surfactants include fatty acid sulfates and quaternary ammonium salts, such as benzyltributylammonium bromide or tetrabutylammonium hydrogensulfate [28,44-46].

These polymerization techniques utilize step-growth kinetics resulting in broad MW dispersity (Đ), equivalent to 2. Ring-opening polymerization exhibits the only polymerization technique for polyesters that achieves narrow Đ as it follows chain-growth kinetics. Examples of this synthesis include polymerization of lactides or glycolides [47,48]. This polymerization occurs in bulk or solution at temperatures typically ranging from room temperature to > 100 °C. Catalysis for this technique ranges from standard metallic systems, like titanium alkoxides and Sb_2O_3, to novel organocatalysts that Kiesewetter *et al.* [48] describe in depth.

7.2.2 Synthesis of Thermotropic LC Polymers

Classification of LC polymers initiates with the differentiation between lyotropic and thermotropic. This chapter focuses on the area of thermotropic LC polyesters which comprise materials that transition into the LC state through application of thermal energy. Three main morphologies of LC mesophases form from thermotropic LCPs; nematic, smectic, and cholesteric [14,49,50]. Nematic exhibits one-dimensional order within the microscopic unit element (domain) in the polymer matrix. The molecules arrange themselves in a specific direction within the domain and the limited intermolecular forces enable the molecules to easily pass each other, allowing mobility in the system. As illustrated in Figure 7.4, other than orientation, the molecules arrange randomly. Smectic LCPs possess "two-dimensional" ordering within the domains. In this case the mesogenic monomers arrange to form layers or planes. The direction or orientation of the

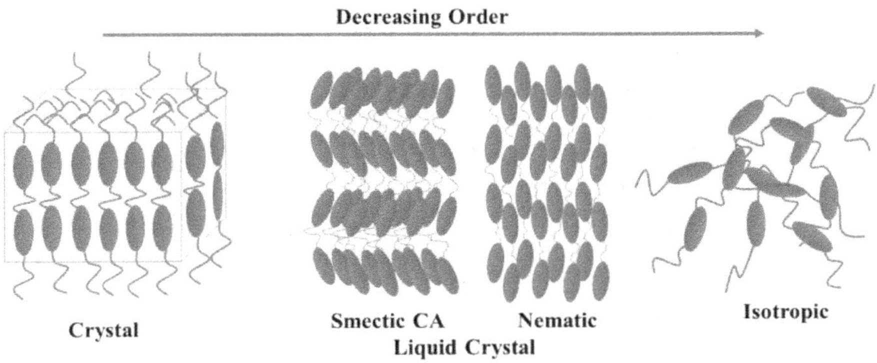

Figure 7.4 Basic graphical illustration of the differences between crystalline, liquid crystalline, and isotropic states for main-chain LC polymers.

mesogens within these layers enable differentiation between a variety of unique smectic LC phases. The smectic A (S_A) mesophase possesses orientation in which the long axis of the rod shaped mesogen lies perpendicular to plane of the layer [14,49,50]. Smectic mesophases that exhibit tilting of the mesogens with respect to the layer encompasses a smectic C (S_C) but if the tilt of the mesogen flips its direction with each new layer, as illustrated in Figure 7.4, the mesophase would identify as smectic CA (S_{CA}) [14,49-51]. Cholesteric LCPs exhibit similarities to nematic phases in that layers do not form, but rather the molecules only possess an orientation director. The difference between cholesteric and nematic lies in the twist of the director resulting in a helical path, like DNA. The origin of this phase results from a chiral bond within the chemical structure of the polymer repeat unit [14,49,50].

Chung [51] and Ciferri [52] describe a wide range of other classifications depending on the packing structure within the mesophase structure.

Main-Chain LC Polymers (MCLCP)

The thermotropic family of LCPs comprises two main polymer architectures; side-chain (SCLCP) and main-chain polymers (MCLCP). Main-chain polymers possess the mesogenic unit (monomer/ structure imparting LC properties) within the backbone of the polymer. These linear polymers, often synthesized through step-growth methods, take on molecular structures, such as polyesters and polyamides. Side-chain differs from main-chain due to the mesogenic unit hanging off the backbone of the polymer, as illustrated in Figure 7.5. Side-chain polymers commonly utilize controlled chain-growth polymerization techniques, such as anionic polymerization and radical addition fragmentation chain transfer (RAFT) polymerization. Their ability to align easily under magnetic or electric fields fostered the use of SCLCPs in lithographic fields. MCLCPs most notably exhibit high mechanical properties, significant shear thinning in the melt and orientation under mechanical deformations promoting their use in high strength fibers and injection molded structural applications [28].

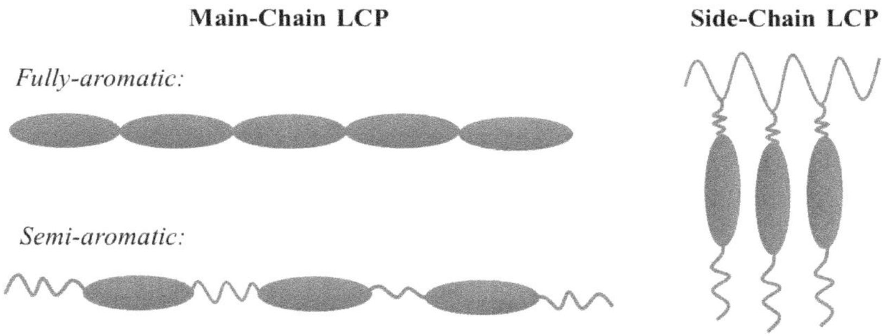

Figure 7.5 Basic architectural differences between main-chain and side-chain LCPs. Ovals represent mesogenic units while the wavy lines represent flexible chains.

LC polyesters fall into the field of main-chain LCPs and take on two different backbone structures, semi-aromatic and fully-aromatic. The first polyesters found with LC characteristics utilize fully-aromatic structures. Flory's original computational hypothesis indicated that the formation of LC properties in a polymer requires a structure with a high degree of rigidity and linearity [16,17]. As a result, scientists initially synthesized very rigid aromatic backbones but quickly discovered that completely linear backbones resulted in T_m's greater than the polymer degradation temperature making it impossible to form a stable LC state. Neither poly(4-hydroxybenzoic acid) (Ekonol) or poly(1,4-terephthalate) melt before their degradation temperature at 600 °C [53-55]. Researchers have used four main approaches to reduce the melting temperature of LCPs in order to achieve a melt-processable polymer that benefits from the significant shear thinning prominent in the LC state:

1) Copolymerization with different linear monomers (mesogens)
2) Copolymerization with non-mesogenic monomers
3) Polymerization of monomers with bulky lateral substituents
4) Introduction of flexible aliphatic spacers [20,28]

Copolymerization of different rigid linear (para) monomers, such as terephthalic acid, 4-hydroxybenzoic acid, 2,6-hydroxynapthoic acid and hydroquinone, comprises common industrially processes [14,28]. This type of LCP maintains the linearity required to produce LC order but disrupts the regularity required for crystalline chain packing through randomization of the structure. Vectra and Xydar, two commercially available fully-aromatic LCPs, utilize this approach to achieve T_m < 300 °C [19,56,57]. The copolymerization of linear mesogenic repeat units with non-mesogenic monomers represents another method to reduce T_m. Typically, these non-mesogenic monomers use kinked aromatic structures due to functional groups placed at meta position on the ring. Examples of this include isophthalic acid, resorcinol and 2,5-substituted thiophene. These structures disrupt the lateral interactions of the chains affording T_m reduction, but unfortunately also exhibit a disruption of the LC properties through interruption of the chain rectilinearity [14,28]. The addition of a bulky lateral group represents a very successful method for disrupting lateral interactions without impacting the rectilinearity of the chain. These monomers take on the structure of various *p*-aromatic monomers with the addition of benzyl rings or halide groups in the ortho or meta positions. Rogers and Long [28] provide an extensive list of various bulky substituent based monomers utilized in LCPs. Fully-aromatic LCPs, such as those described here, form nematic mesophases because of their rigid/linear but non-regular structure impeding the formation of smectic layers [14,28].

The addition of flexible spacers between the mesogenic monomers results in the formation of semi-aromatic LCPs. The structural architecture of semi-aromatic LCPs, illustrated in Figure 7.5, results in significant decreases in transition temperatures. Unfortunately, the polymer degradation temperature also reduces. Watanabe *et al.* and other groups [58-62] extensively studied the structure-property-relationships of semi-aromatic LCPs utilizing the mesogenic monomer, dimethyl 4,4'-bibenzoate (4,4'BB). Considering only linear aliphatic spacers, it was elucidated that a spacer length greater than 3, but less than 9 methylene units exhibits LC properties. These polymers afford $T_m < 250$ °C and $T_i < 300$ °C and decrease in their endothermic transitions as the aliphatic spacer length increases revealing an odd-even oscillation. The chain flexibility and regularity of the 4,4'BB based semi-aromatic polyesters with linear spacers result in the formation of smectic mesophases. Specifically, aliphatic spacers with an odd number of methylene units result in a S_{CA} mesophase morphology while those with even numbered spacers form S_A morphologies [58-61]. This effect results from the extended chain confirmation that occurs in LCPs with odd and even spacers, as demonstrated in Figure 7.6 through the extended chain structures of poly(pentamethylene 4,4'-bibenzoate) and poly(hexamethylene 4,4'-bibenzoate) [63]. Utilizing spacers with additional R-groups off the backbone disrupts the close packing of the polymer chains resulting in a disruption of crystallinity and, under specific circumstances, a transition to a nematic mesophase rather than smectic [64]. In most cases, the addition of R-groups off the chain results in the formation of a LC glass due to the complete inhibition of crystallization but the continued presence of a S_A mesophase. Poly(4,4'BB-2-methyl propylene) (poly(BB-5(3-Me)) represents an example of a LC glass [65-70]. Copolyesters of poly(ethylene terephthalate) with hydroxybenzoic acid represent a non-traditional semi-aromatic LCP which utilize PET as the spacer between the rigid mesogens. Similar to fully-aromatic polymers, the randomized structure due to copolymerization results in a nematic mesophase [71-73]. Other mesogenic monomers utilized in the synthesis of semi-aromatic LCP include 4,4'-biphenol, stilbene, 4,4'-dihydroxy-2,2'-dimethyl-azoxy-benzene and many more outlined by Demus [50]. These monomers form a highly versatile family of structures able to achieve different mesophase morphologies while maintaining processable melting/isotropic temperatures [8,46,50].

All the methods mentioned so far, with the exception of semi-aromatic LCPs, utilize phenolic monomers. As described in the previous section, phenolic monomers require more reactive functional groups

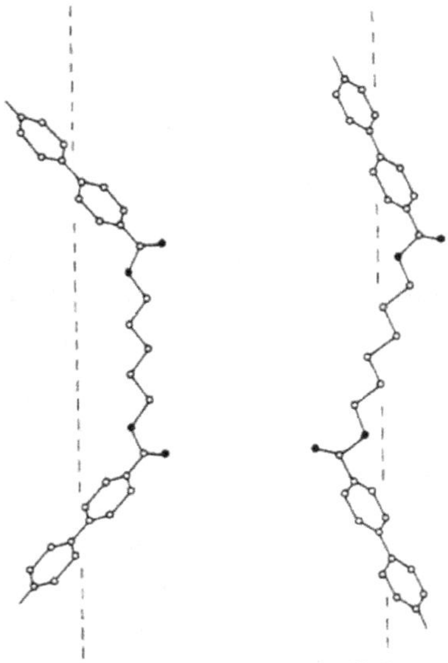

Figure 7.6 Extended chain structure of poly(hexamethylene 4,4'-bibenzoate) (BB-6) and poly(pentamethylene 4,4'-bibenzoate) (BB-5). Reproduced from Reference 63 with permission from American Chemical Society.

than carboxylic acids and alcohols in order to polymerize. Acidolysis polymerization represents a widely utilized polymerization method for the formation of fully-aromatic LCPs [74-76]. Fully-aromatic LCPs often polymerize in a turbid state because of the presence of nematic LC phases at the polymerization temperatures [10,14]. This melt polycondensation technique often utilizes no solvent or catalyst and still achieves high MW polymers. Other groups have also shown that the introduction of catalysts, like dibutyl tin oxide, improves reaction rates between less reactive/nucleophilic monomers [77,78]. Unfortunately, due to the monomer choice, this technique often requires high temperatures (>250 °C) limiting thermally sensitive monomers due to degradation and side reactions. Also, the acetoxy functionality introduces side-reactions through the formation of ketenes or acetolysis which other polymerization methods eliminate [79]. This technique represents an uncommon polymerization method for semi-aromatic polymers but is vastly utilized for fully-aromatic LCPs. In the segmented LCP literature, the report of Jannesari *et al.* [80] represents the only work to utilize acidolysis for semi-aromatic polyesters due to the choice of a phenolic mesogenic monomer and a aliphatic diacid.

The use of acid chloride functionalities instead of carboxylic acid moieties represents a more widely utilized monomer in the polymerization of main-chain LCPs throughout the segmented LCP literature. The greater reactivity of these monomers in esterification reactions occurs from the better leaving group, chloride, on the carbonyl. This functional group enables reactions with phenolic monomers without the need for acetylation. A few studies on LCPs demonstrate the possibility of solvent-less melt polycondensation methods using these monomers, but side-reactions and the need for lower temperatures make the use of solution polymerization a preferred approach [28,81-83]. Solution polymerizations for fully-aromatic polymers often occur at temperatures higher than those utilized for aliphatic polyesters (50-250 °C) due to the limited solubility of these polymers/monomers [29,84-91]. A few segmented LCP studies have also utilized the addition of activating agents, such as triphenylphosphine dichloride (TPPC) and diphenylchlorophosphate (DPCP), as described for polyester synthesis [92]. Angeloni *et al.* [93] represent the first group to utilize interfacial polymerization in the synthesis of segmented LC polyesters. The benefit of this polymerization method came from the use of a lower reaction temperature, to prevent a thermally unstable monomer from degrading during the polymerization. LC polymers that utilize these techniques comprise both semi-aromatic and fully-aromatic LCPs [44,45,94-96].

Melt transesterification represents the most utilized synthesis method for semi-aromatic LC polyesters. Most 4,4'BB based LCPs use this polymerization method with a range of different aliphatic diols [58,67,97-99]. The versatility of this technique and the low processing temperatures of the structures formed using this method resulted in its widespread use in the formation of segmented LCP. The Long group has utilized this polymerization method on several occasions for the synthesis of semi-aromatic LC polyesters and segmented LCP [98,100-102]. For instance, Nelson *et al.* synthesized high-performance segmented LC copolyesters utilizing this technique [98]. The authors heated the monomers and 40 ppm of titanium isopropoxide catalyst from 190-275 °C over the course of 4.5 h under N_2 flow. A final vacuum step (0.15 mmHg) at 275 °C for 2 h enabled the formation of high MW polyesters. The use of titanium alkoxide catalysts has been most prevalent in these studies due to the non-toxic nature and high activity [28,35].

Side-Chain LC Polymers (SCLCP)

A flexible polymeric backbone, mesogenic monomer, spacer and tail comprise the structure of a SCLCP. The polymer backbone in segmented systems range from flexible amorphous polymers, to semi-crystalline or LC polymers. Flexible backbones give rise to easier ordering of the mesogenic units, therefore, the most widely used backbones include polymethacrylates, polyphosphazenes, polysiloxanes, ethylene oxides and different polyacrylate derivates [9,50]. In order to ensure the polymer coiling does not influence the LC properties of the mesogenic unit, it must be decoupled from the polymer backbone through use of a spacer [9]. This allows the mesogen to exhibit long-range order and orientation while promoting entropy in the backbone to balance the flexibility of the chain. The length of the spacer and tail drastically alters the mesophase morphology. For SCLCP with rod-like mesogens, short tails/spacers result in a nematic mesophase, while long tails and long spacers allow enough decoupling from the backbone to order into a smectic mesophase [103]. Altering the structure of the tail from an aliphatic unit to perfluorinated consistently alters the mesophase to S_A due to the greater rigidity of the chain [103]. Cyano and nitro moieties impart positive dielectric anisotropy when used in tails [9]. Emmerling *et al.* [104] demonstrated the synthesis of a novel SCLCP utilizing a nitro tail which afforded multiple smectic and nematic phases.

The mesogenic unit choice depends on whether the small molecule analog exhibits LC properties due to the axial ratio and polarity of the molecule. The molecular geometry usually comprises a rod-shape, disk-shape or amphiphile (lyotropic). In comparison to main-chain LCPs, the mesogenic monomer unit does not change its axial ratio when polymerized. As a result, SCLCP ordering properties resemble the low MW LC much closer than main-chain LCPs. Rigid rod mesogenic units usually comprise combinations of two or three aromatic rings, as Demus [50] comprehensively lists. Disk-shaped (discotic) mesogens usually link aromatic rings to form structures like triphenylene exhibiting a planar rigid disk shape [9,105]. Discotic SCLCP are quite uncommon in the segmented LC copolyester literature, therefore, these will not be focused on in this review. Semi-fluorinated alkyl side chains represent a non-phenyl based mesogen that exhibits high rigidity in comparison to CH_2 units and better molecular organization resulting from their low surface energy [106,107]. In another study, diblock copolymers of polyacrylate derivates with semi-fluorinated side chains and poly(methyl methacrylate) (PMMA) were synthesized utilizing atom transfer radical polymerization (ATRP) [67].

Figure 7.7 illustrates the three key synthetic pathways to the production SCLCP. Pathway (1) uses chain-growth polymerization of a LC monomer to form the SCLCP. As seen for the polymerization of styrene or methacrylate-based monomers, polymerization occurs through the vinyl functionality on the monomer while the mesogenic monomer hangs off the vinyl group. Pathway (2) illustrates the design of a difunctional monomer for polycondensation polymerization. The final scheme (pathway (3)) exemplifies post-functionalization to attach a mesogenic monomer after the polymerization is complete [52].

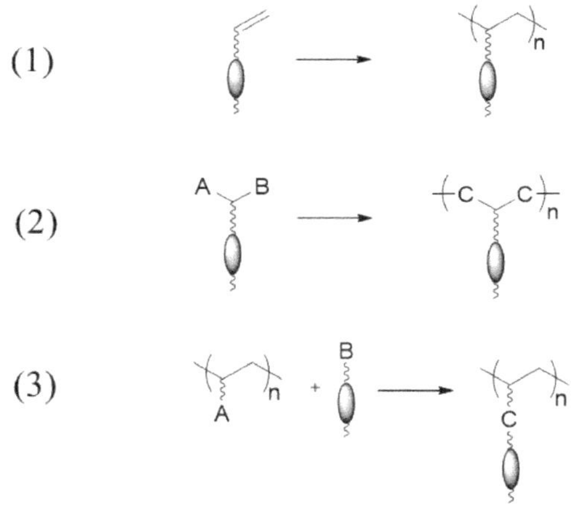

Figure 7.7 Polymerization of SCLCPs using 1) chain-growth polymerization, 2) step-growth polymerization and 3) post-polymerization modification. Adapted from Ciferri [52].

Chain-growth polymerizations use vinyl functionality to achieve high MW through a wide variety of techniques. Though often resulting in polymers with higher dispersity than controlled polymerization methods, free-radical polymerization remains the most commonly used method of polymerization through a vinyl moiety. This method begins the polymerization through thermal or photodecomposition of an initiator to form a radical. This radical propagates through the double bond of the monomer, transferring the activated site. The activated site then goes on to react with another vinyl group and the radical continues transferring and reacting until a termination step neutralizes the radical or two chains combine in a coupling mechanism inhibiting further chain growth [108]. A number of controlled polymerization techniques regulate the kinetics of free-radical polymerizations, such as atom transfer radical polymerization and reversible addition-fragmentation chain transfer polymerization [32]. These methods along with anionic and cationic polymerizations facilitate the polymerization of SCLCP and the selection of polymerization type depends on monomer choice and the desired control over Đ and MW [32].

Ring-opening polymerization also possesses chain-growth kinetics and enables the polymerization of SCLCPs. For instance, the double bond in a cyclic norbornene monomer facilitates polymerization through ring-opening metathesis polymerization. Derivation of these monomers with mesogenic units, like n-alkyloxycyanobiphenyl groups, reveals LC properties upon polymerization [109]. Del Campo *et al.* [110] also demonstrated ring-opening polymerization of epoxide derivates through cationic polymerization also facilitates the formation of SCLCPs [110].

The design of monomers with AA or AB molecular geometries functionalized with SCLCs facilitate their use in step-growth polymerization methods. Mesogenic side-chains located off a rigid aromatic diacid/diester enable their use in polyesters. Pospiech *et al.* [72] explored the use of such monomers in a segmented copolymer with a poly(sulfone) secondary block. This study synthesized a LC polyester segment containing a derivatized isophthalate unit with semi-fluorinated alkyl chain located *meta* to the diacid functionalities. This represents one of the few polyester side-chain LCP examples as this polymer family primarily comprises main-chain LCP structures [72]. Emmerling *et al.* [104] achieved another example of SCLCP through polycondensation polymerization when 4-(4-nitrobenzyloxy)biphenyl-4'-ol underwent etherification with diethyl 6-bromohexylmalonate to form diethyl 6-[4-(4-nitrobenzyloxybiphenyl-4'-oxy]hexylmalonate. This SCLCP, polymerized through melt transesterification with various diols, exhibited multiple smectic and nematic mesophases [104]. Mandal *et al.* [38] further revealed a novel synthetic method toward the formation of SCLCP involving the combination of both polycondensation polymerization and post-processing. This study synthesized aliphatic polyesters through solution polymerization between acyl chloride moieties and derivatized diols, as illustrated in Scheme 7.3. The diols contained pendant propargyl groups affording a clickable functionality on the polymer enabling the

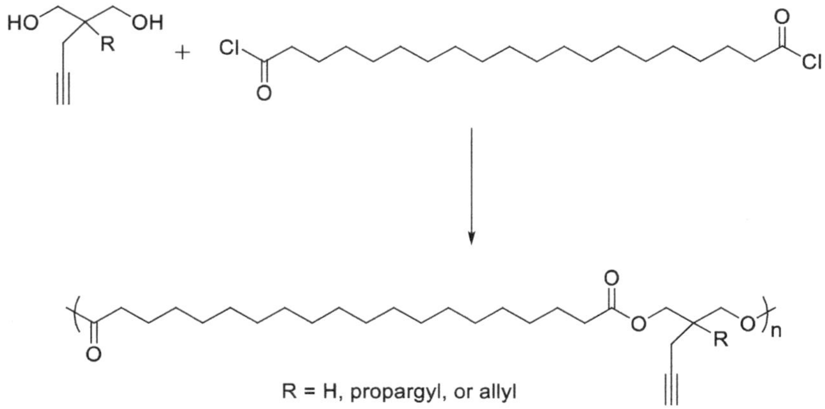

Scheme 7.3 Polymerization of a periodically clickable polyester utilized in the synthesis of a SCLCP. Adapted from Mandal *et al.* [38].

attachment of a semi-fluorinated alkyl mesogens [38]. Post-processing ensured that the mesogenic unit did not inhibit polymerization through impacting solubility or introduction of side-reactions. Anionic polymerization is notoriously sensitive to monomer choice, therefore, post-polymerization affords a wider range of final achievable polymer structures. Manfred et al. [111] utilized the protected monomer, 2-(trimethylsiloxy)ethyl methacrylate (SHEMA), to polymerize SCLCP using anionic polymerization. Deprotection of SHEMA after polymerization afforded 2-hydroxyethyl methacrylate (HEMA) which, when reacted with cholesteryl chloroformate, afforded the SCLCP illustrated in Scheme 7.4 [111].

7.2.3 Synthesis of Segmented LC Copolyesters

Segmented LC copolyesters must consist of three main features. First, the polymer must phase separate through synthesis of block or multiblock architectures. Secondly, one of the blocks must contain a LC polymer. Finally, one block must contain a polyester though the second and third requirements may be combined through use of a LC polyester. Design of these materials occurs using a wide range of block copolymer structures and architectures that result in discrete blocks of monomer units along the polymer chain. This includes linear diblocks

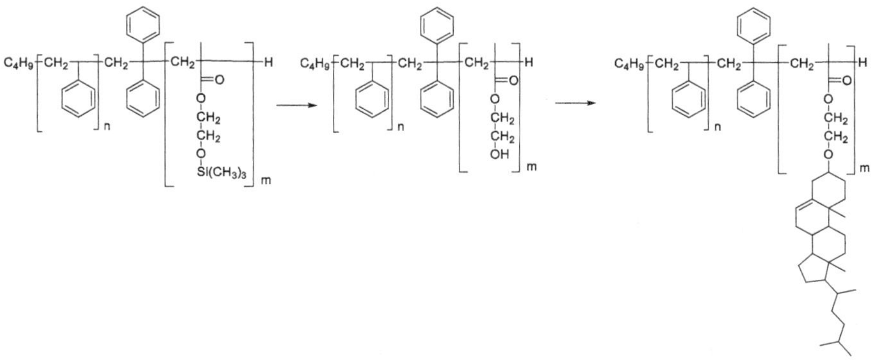

Scheme 7.4 Post-polymerization deprotection of 2-(trimethylsiloxy)ethyl methacrylate and subsequent reaction with cholesteryl chloroformate to form a SCLCP. Adapted from Manfred *et al.* [111].

(AB), triblock (ABA), pentablock (ABABA) and multiblock (segmented, $(AB)_n$) or non-linear graft and star block copolymers. Synthesis of these structures occurs through both step-growth and chain-growth polymerization methods. Due to the presence of polyesters in these (co)polymers, synthesis occurs through either entirely step-growth polymerization or a hybridization of step-growth and chain-growth polymerization. Only one account of a segmented LC (co)polyester through entirely chain-growth polymerization has been located. In this polymerization, Deshmukh *et al.* [109] functionalized norbornene with either one or two cyanobiphenyl mesogens, illustrated in Figure 7.8. A secondary norbornene monomer was

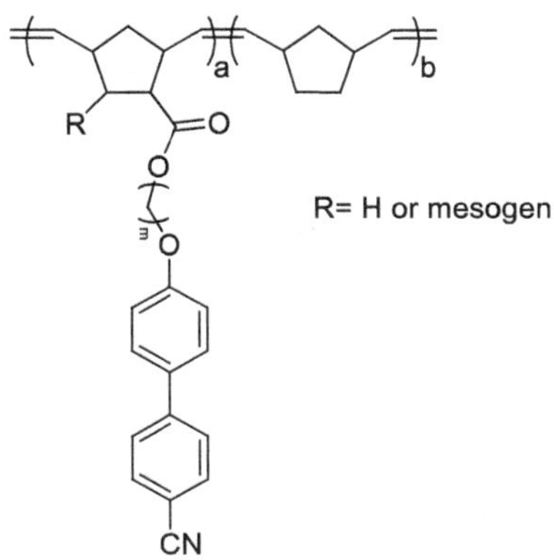

Figure 7.8 Structure of a segmented LC copolyester utilizing only chain-growth polymerization. Adapted from Deshmukh *et al.* [109].

functionalized with poly(lactide) (PLA) through ring-opening polymerization. Ring-opening metathesis polymerization afforded the grafted block copolymer with 75% liquid crystalline content in order to achieve a hexagonally packed cylindrical phase separated cylindrical morphology [109].

Segmented LC Copolyesters Synthesized via Step-Growth Polymerization

Polymerization methods utilizing entirely step-growth reactions comprise half of all the literature on segmented LC copolyesters. Considering the history of multiblock copolymers in the thermoplastic elastomer field, it is not surprising that this architecture possesses a large presence in literature. While uncommon in more recent literature, early studies into LC block copolyesters analyzed block-copolymers containing only polyesters. Mathew *et al.* [112] studied the copolymerization of PET with 4-acetoxybenzoic acid (ABA) through melt esterification. Kinetics studies indicated that the DP limit for the ABA monomers reached 5 and DSC studies supported a random block structure. Although it is difficult to determine whether phase separation occurred in these early studies, an understanding was formed that in order to create block copolymers, transesterification between the segments must be limited [39,112]. In 1991, Tsai *et al.* [113] attempted to control the transesterification between two polyester oligomers to afford block copolyesters. This study utilized the immiscible poly(butylene terephthalate) and poly(butylene adipate) and copolymerized these oligomers through melt mixing with the addition of 1,4-butanediol (BD). The amount of added BD and reaction time controlled the degree of randomization that occurred [113]. Novack *et al.* [89] and Yan and He [41] utilized the natural randomization that occurs through transesterification to introduce LC blocks into commercial PET. Yan and He [41] coupled two HBA molecules with terephthaloyl chloride (TPC) to form a LC triad acid chloride which was incorporated into commercial PET chains through solution polymerization with a triethylene amine/pyridine catalyst. Novack *et al.* [89] performed a similar analysis in which the monomers, 4,4'-dihydroxy-α,ω-diphenoxy decamethylene and TPC, reacted with PET using solution polymerization. In 1994, Ignatious *et al.* [42] further explored this subject, but utilized PBT instead of PET and different diad and triad LC acid chloride monomers. In all of these reported studies, high temperature solution polymerization was performed instead of melt transesterification in an effort to control sequence randomization. Unfortunately, the disperse nature of the blocks made the microphase separated morphology difficult to discern in all studies. Only a blocky structure was truly confirmed due to the presence of endothermic transitions for both the semi-crystalline and LC blocks and optical microscopy analysis [41,42,89].

These analyses enabled the observation that randomization control demonstrates a less practical method in the formation of multiblock copolymers. In order to introduce more control into the polymerization, scientists began utilizing reaction methods and secondary copolymers that inhibited transesterification. Tsai *et al.* [114] utilized methylene-4,4'-diphenylene diisocyanate to link oligomers of a semi-aromatic LCP, poly(pentamethylene 4,4'BB), with poly(tetramethylene adipate) through coupling in DMF in the presence of dibutyltin dilaurate. While this successfully produced high MW phase separated multiblock copolymers, the presence of hydrogen bonding resulted in conformation restrictions which limited the LC properties [114].

Over the years, segmented LC copolyesters with multiblock architectures incorporated a large variety of high performance polymers in the form of oligomers, such as polysulfones (PSU), poly(ether ether ketone) (PEEK), poly(phenylene sulfide) (PPS) and poly(phenylene oxide) (PPO). These high-performance materials utilize step-growth polymerization methods, such as nucleophilic aromatic substitution [32]. Due to the structural differences between these polymers and polyesters, randomization through transesterification becomes impossible. The extensive use of these types of phase separated copolymers in the area of blend compatibilization arises from the incompatibility between these polymers and commercial LC polyesters [85].

PSU represent the most widely studied of these high-performance polymers for segmented LC copolymers. Copolymerization of PSU oligomers with LC polyesters occurs through coupling with LC polyester oligomers or polymerization of mesogenic monomers in the presence of PSU oligomers. Auman and Percec [115] first synthesized PSU-LCP (co)polyesters in 1988 utilizing an acetoxy terminated PSU. Melt acidolysis polymerization with chlorohydroquinone diacetate and trans-1,4-cyclohexanedicarboxylic acid afforded high MW block copolymers with phase separated properties. The immiscibility of the amorphous PSU (MW > 2300 g/mol) with the LC phase of the monomers at the reaction temperatures resulted in chain extended structures whereas the lower MW PSU (M_n = 1400 g/mol) facilitated the formation of multiblock copolymers [115]. Chain-Shu and Ling-Bar [78] followed this study and synthesized PSU-LCP copolymers utilizing the LC monomers, p-acetoxybenzoic acid (ABA), terephthalic acid and hydroquinone diacetate. The copolymer utilized melt acidolysis polymerization from the acetoxy terminated poly(ether sulfone) at different weight % relative to LC monomers. This method revealed a significant amount of unreacted PSU oligomers remaining at the end of reaction that required removal so as not to affect the final properties.

In 1994, Brenda et al. [84] moved to a high temperature (180-250 °C) solution polycondensation utilizing diphenyl ether as the solvent in order to improve the reaction with the oligomers. This study analyzed poly(ether ether sulfone) oligomers (10,000, 7000, and 2500 g/mol) copolymerized with either a semi-crystalline LC polyester, an amorphous LC polyester (LC glass), or a non-LC amorphous polyester, as illustrated in Figure 7.9. This polymerization technique enabled the formation of phase separated block copolymers utilizing

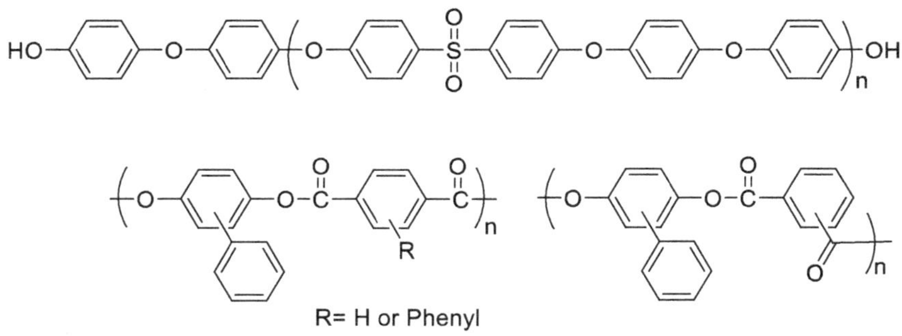

Figure 7.9 Structure of poly(ether ether) sulfone multiblock copolymers with different LC polyester segments. Adapted from Brenda *et al.* [84].

much higher MW PSU oligomers than previously reported because it no longer relied on solubility in the melt [84]. Chang *et al.* [96] utilized solution polymerization with an activating agent, triphenylphosphine/pyridine, in hexachloroethane to polymerize block copolymers from hydroxy terminated PSU oligomers and HBA/HNA LC monomers. Unfortunately, fractionating the polymer with different solvent systems revealed that the use of an activating agent still left unreacted PSU oligomers in the final polymer. A later study by Wang et al. [29] using a wide MW range of hydroxy terminated PSU (30,000–10,000 g/mol) reconfirmed that high temperature solution polymerization utilizing diacid chloride and diphenol functional groups consistently produced multiblock copolymers. This study utilized 2-(3'-trifluoromethylphenyl) hydroquinone (TFMPH) and TPC to produce the LC segment [29]. Pospiech et al. [72,82] in 1996 and 2001 reported the synthesis of PSU-LC block copolymers through the coupling of both a PSU oligomer and LC polyester oligomers rather than polymerization from monomers. These polymerization studies utilized melt acidolysis to couple acetoxy terminated PSU with carboxylic acid terminated PET/HBA copolymers. This method produced homogenous multiblock copolymers eliminating the presence of unreacted oligomers [82]. Other LC systems, such as carboxylic acid terminated semi-fluorinated side-chain polyesters were polymerized utilizing this same method [72].

Yang *et al.* [85] and Zeng *et al.* [30] both produced segmented LC copolymers utilizing poly(ether ether ketone) segments of different backbone structures. Yang *et al.* [85] synthesized BPA based PEEK with hydroxy terminal groups through nucleophilic aromatic substitution achieving an M_n = 7797 g/mol and Đ = 1.2. This oligomer facilitated the formation of block copolymers through high temperature (180-250 °C) solution polymerization using TFMPH and TPC to form the LC segments. Alternatively, Zeng *et al.* [30] polymerized multiblock copolymers using carboxylic acid terminated Vectra oligomers and acetoxy terminated PEEK, exhibited in Scheme 7.5, at an equimolar ratio through melt acidolysis using a zinc acetate/diantimony trioxide (co)catalyst. Heyde *et al.* [90] utilized high temperature solution polymerization to produce block copolymers of PPO and LC polyester segments. Telechelic functionality was added to poly(oxy-1,4-phenylene) post-polymerization using potassium 4-bromophenolate

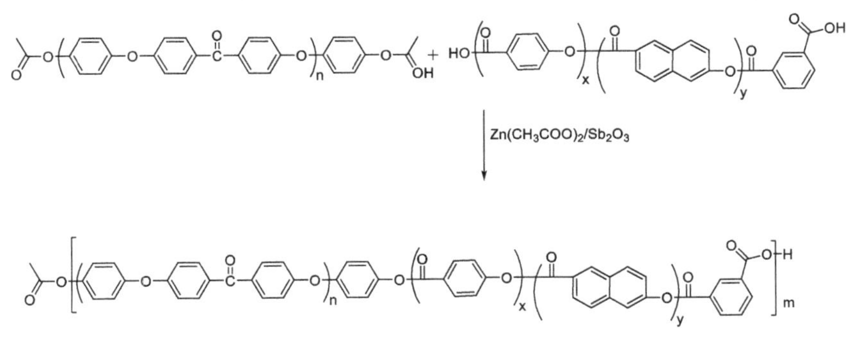

Scheme 7.5 Synthetic route to multiblock copolymers containing PEEK and Vectra segments. Adapted from Zeng *et al.* [30].

and 4,4'-isopropylidenediphenol (BPA). Following similar methods demonstrated for PSU and PEEK, LC monomers (TPC or bromoterephthaloyl dichloride and biphenyl-2,5-diol) polymerized off poly(oxy-1,4-phenylene) to produce multiblock copolymers [90]. Heitz demonstrated the synthesis of PPO-LC block copolymer as well as PPS-LC block copolymers in 1989 [116]. Synthesis of poly(oxy-2,6-dimethyl-1,4-phenylene) and poly(oxy-1,4-phenylene) through Cu catalyzed oxidation of either 2,6-dimethylphenol or bromophenol with a diphenol monomer produced telechelic phenol groups. Reacting 1,4-dichlorobenzene and sodium sulfate produced poly(phenylene sulfide) which encountered functionalization during the second stage of the reaction converting the thiolate end groups to carboxylic acids moieties with chlorobenzoic acid. These oligomers were reacted with the monomers of semi-aromatic LC polyesters (TA and 4,4'-alkylenediphenols) using different carbon spacers through solution polymerization in pyridine at 120 °C [116]. In a more recent study, Gopakumar *et al.* [73] copolymerized carboxylic acid terminate PPS with PET-co-oxybenzoate (PET/HBA) using uncatalyzed melt transesterification at 300 °C.

Two different groups synthesized LC copolymers with non-traditional architectures producing phase separated morphologies using entirely step-growth polymerization methods. Novel PPO macromonomers, α-(2,5-dimethoxyphenyl)-ω-bromooligo(oxy-l,4-phenylene), were synthesized through Ullmann reactions with potassium 4-bromophenolate and 1-bromo-2,5-dimethoxybenzene proceeding cleavage of the ether bond with boron tribromide. This macromonomer polymerized with trifluoromethylterephthaloyl dichloride and t-butylhydroquinone through solution polycondensation in diphenyl ether to form a graft copolymer, illustrated in Figure 7.10 [91]. Mandal *et al.* [38] produced aliphatic polyesters with propargyl side groups enabling a click reaction with semi-flourinated SCLCP.

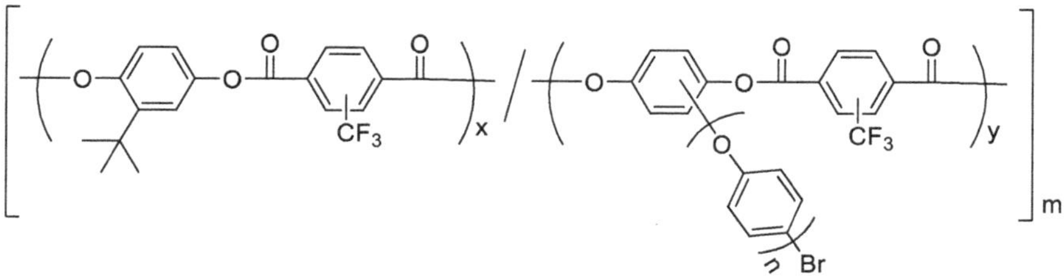

Figure 7.10 Structure of LC copolyester with PPO grafts. Adapted from Norbert *et al.* [91].

This graft copolymer utilized solution polymerization, as mentioned in a previous section, and the foldable aliphatic section promoted phase separation and ordering of the LC side chain groups [38].

Segmented LC Copolyesters Synthesized via Combination of Step-Growth/Chain-Growth Polymerization

The combination of step-growth and chain-growth polymerization to form segmented LC copolyesters enables the synthesis of some of the most well-known thermoplastic elastomers, multiblock copoly(ether-ester)s. The incompatibility between polyethers and polyesters results in macro phase separation for blended systems but forms beneficial nanoscopic phase separation for block copolymers. Thermoplastic elastomers form from multiblock copoly(ether-esters) due to the ability to combine properties from the flexible ether soft-segment and the tough physical crosslinks of the LC polyester hard-segment. The choice of polymerization technique used to synthesize these polymer systems depends heavily on the type of LC polyester. In all cases, oligomers of poly(ether)s, such as polytetramethlyene oxide (PTMO or PTHF) and poly(ethylene oxide) (PEO or PEG), were synthesized or bought ahead of time at a variety of MWs. The synthesis of these poly(ethers) occurs through ring-opening polymerization using a variety of reaction conditions, as Herzberger *et al.* [117] has described. Hydroxy terminated poly(ethers) enables the oligomer to participate in transesterification reactions.

Pospiech *et al.* [77] converted the hydroxy terminal group to a trimellitimide functionality to prevent the characteristic degradation that had been reported for PTMO with hydroxy, acetoxy or acetamido end-groups. Similar to the PSU block copolymers, incorporation of these oligomers into different polymerizations occurred through introduction as a reactive starting material, to become a repeat unit segmented within the final polymer backbone. Due to the lower reactivity of phenol monomers, fully-aromatic LCP used melt acidolysis or reactions between acyl chloride and alcohols either in the melt or in solution. One of the first studies of these polymers in 1991, utilized solution polymerization of TPC and methylhydroquinone with pyridine in 1,1,2,2-tetrachloromethane to polymerize phase separated copoly(ether-esters) with the fully-aromatic LC segments [43]. Sonpatki *et al.* [81,83] also reported the utilization of melt transesterification with acyl chloride moieties to synthesize multiblock copolymers with PTMO segments and fully-aromatic LC segments from TPC and different hydroquinone derivatives (hydroquinone, methylhydroquinone, or chlorohydroquinone). The trimillitimide terminated PTMO (750, 1000, and 2000 g/mol) mentioned previously utilized melt-acidolysis polymerization with acetoxy terminated HBA and an acetoxy terminated trimer, illustrated in Scheme 7.6 [77].

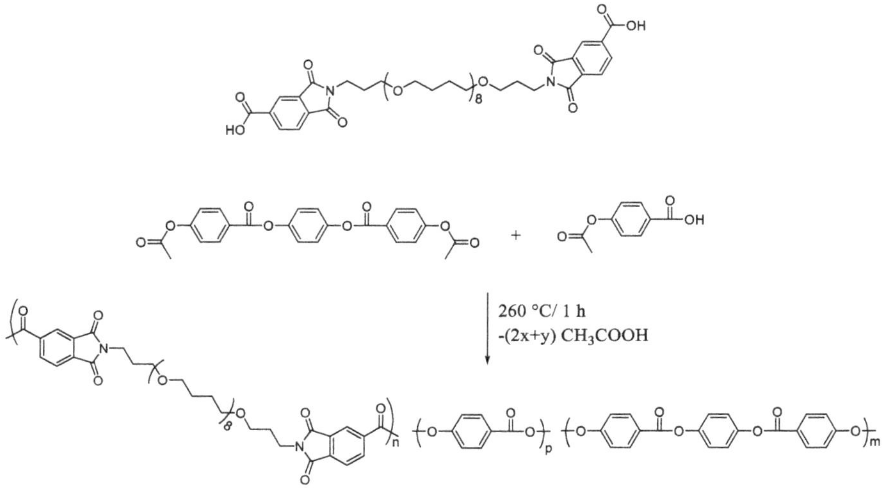

Scheme 7.6 Melt acidolysis polymerization for the synthesis of segmented LC polyester with PTMO soft-segment. Adapted from Pospiech *et al.* [77].

Semi-aromatic LCPs exhibit successful polymerization through melt transesterification of diester-diol monomers. Successful application of this polymerization method to multiblock copoly(ether-esters) resulted in an array of polymer structures. Tsai *et al.* [118-120] extensively studied the incorporation of PTMO oligomers into multiblock copolymers with a series of 4,4'BB based LCP featuring different alkyl spacers through melt-transesterification. Within the same group, Hsuie *et al.* [121,122] synthesized multiblock copolymers with PEO soft-blocks and either poly(hexamethylene-4,4'BB) or poly(pentamethylene-4,4'BB) utilizing melt-transesterification. These studies focused on the change in hydrophilicity of the final polymers through the incorporation of a hydrophilic soft-segment. Nelson *et al.* [98] synthesized novel multiblock copolymers using pluronic oligoethers (poly(propylene glycol)-*block*-poly(ethylene glycol)-*block*-poly(propylene glycol). The polymer synthesis used one-pot melt-transesterification with a titanium tetraisopropoxide catalyst, as illustrated in Scheme 7.7.

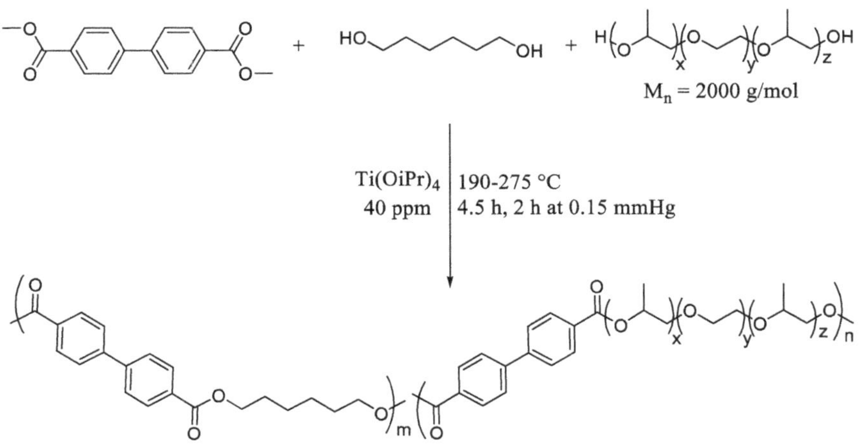

Scheme 7.7 Synthesis of segmented LC polyester with pluronic soft-segment using melt transesterification polymerization. Adapted from Nelson *et al.* [98].

Internal macroinitiators represent an innovative method to synthesize multiblock copolymers of LC polyesters and radically polymerizable segments. Angeloni *et al.* [93] first proposed this technique in 1993. This work synthesized a semi-aromatic LC polyester through interfacial polymerization of 4,4'-octamethylene dioxydibenzoyl chloride, pentamethylene di(4-hydroxybenzoate) and 4,4'-azobis(4-cyanopentanoyl chloride). Upon thermal decomposition of the azo moiety in the polymer backbone, a radical initiation of styrene occurred. This resulted in a multiblock structure due to the different termination mechanisms for PS [93].

Figure 7.11 demonstrates a schematic of the polymerization process. Chiellini *et al.* [45] and Galli *et al.* [95] further analyzed the block formation of this polymerization in 1994. The authors demonstrated the use of this approach to synthesize multiblock copolymers with a combination of side-chain and main-chain LCPs. In these studies, the same semi-aromatic LCP macroinitiator initiated free radical polymerization with two different PMMA based side-chain LCPs [44,94,123]. While this method represents a creative technique to combine step-growth and chain-growth polymerization, Lin *et al.* [124] recently demonstrated a much simpler approach to achieving multiblock copolymers with PS and LC polyesters. This approach functionalized PS oligomers with carboxylic acid end groups, which underwent high temperature solution polymerization with HBA and

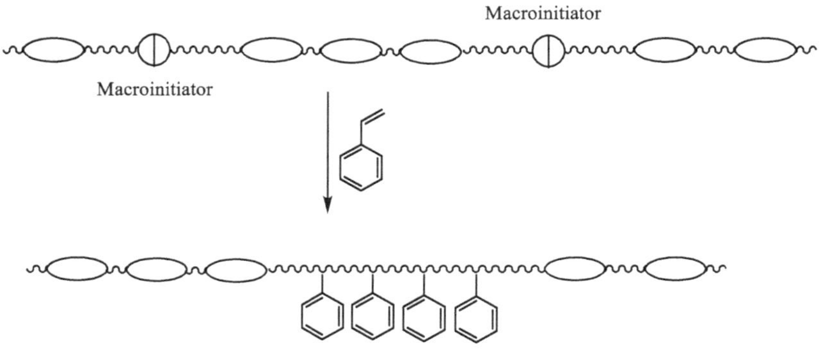

Figure 7.11 Graphical schematic of multiblock copolymers utilizing an internal macroinitiator. Adapted from Chiellini *et al.* [45].

HNA using the activating agent, triphenylphosphine dichloride/pyridine. Acidolysis polymerization of these PS segments would be impossible due to the ceiling temperature residing at 280 °C. The limitation of this approach resulted from the mixture of final products (PS oligomer, LC oligomer and segmented copolymer) requiring fractionation to remove unreacted oligomers [124]. Rather than polymerizing PS off of LC oligomers, Reichelt *et al.* [88] coupled hydroxy terminated oligomers of PS in solution with acid chloride terminated poly(2,2'-dimethyl-4,4'-biphenylene phenylterephthalate) at a 2:1 ratio to produce triblock copolymers. The same group replaced PS with PEG to further study this coupling reaction and form hydrophilic triblock copolymers [86].

Functionalizing the terminal group of LC polyester into macroinitiators, rather than dispersion through the backbone, enables the polymerization of copolymers with controlled architectures such as triblock copolymers. A grouping of studies performed using a semi-aromatic LC polyester (poly(4,4'BB-5(3-Me))) functionalized through post-polymerization modification with a 2-bromo-2-methylpropionyl end-group formed a terminal macroinitiator [65,67,69,70,125]. This macroinitiator facilitated ATRP off the ends of the LCP chains forming triblock copolymers with a variety of outer blocks. Some of these outer blocks include PMMA, PEMA and derivatized PMMA functionalized with either a semi-fluorinated or phenylbenzoate based SCLCP. This method was further utilized with fully-aromatic LC polyester comprised of diphenyl-4,4'-(decane-1,10-diylbis(oxy)dibenzoate) and tert-butylhydroquinone monomers with PS outer blocks [126].

Two examples of graft copolymers utilizing a hybrid chain/step-growth polymerization method have been studied. Heitz *et al.* [127] reported the synthesis of PS graft copolymers in which macromonomers of PS bearing either a 4,4'-methylenediphenol or hydroquinone functional group were synthesized using free-radical polymerization. A two-stage solution condensation first end-capped these macromonomers with derivatives of TPC and then enabled the polymerization with more TPC and tert-butylhydroquinone to form graft copolymers [127]. In another study, a diphenolic OH-terminated PS macromonomer was synthesized to expand upon this work [92]. The Higashi method of DPCP in pyridine with lithium chloride polymerized this macromonomer into a LC polyester. Overall, the studies into the synthesis of segmented LC copolyesters span a large range of polymer architectures and structural characteristics. These unique structures impart many interesting physical properties from magnetic orientation to blend compatibilization.

7.3 Properties of Segmented LC Copolyesters

The unique properties of segmented LC copolyesters arise from nanoscopic scale phase separations enabling the bulk properties to encompass a combination of properties from the two types of polymer segments. Unlike block copolymers, random copolymers with non-mesogenic units eventually result in a complete disruption of LC properties, such as is the case with many copolyesters. The phase separated domains of block copolymers maintain the order of the segments within those domains enabling LC or crystalline packing.

$$\chi_{AB} = \frac{V_{AB}}{RT}(\delta_A - \delta_B)^2 \quad \text{with } V_{AB} = (V_A - V_B)/2 \qquad (2)$$

In order to achieve phase separated morphologies once copolymerized, the solubility of each polymer segment within the other along with MW must be considered. The χ-parameter represents the interaction parameter between two different materials and was originally applied to studies of polymers in solution to determine the interaction between the solvent and the polymer. Eq. 2 expresses the parameter in mathematical form illustrating that the interaction parameter comprises the Hildebrand solubility parameters (δ) of each polymer as well as the volume (V_{AB}) of each polymer segment related to MW [128]. This basic parameter helps to determine the MW necessary to cause phase separation during the polymerization of block copolymers. Pospiech *et al.* [72] demonstrated the use of this parameter in the synthesis of multiblock copolymers of polysulfones (PSU) with a range of secondary blocks. The group calculated the χ-parameter between PSU and each composition, as illustrated in Figure 7.12, ranging from 0.89 to 13.68. This study fostered the understanding that for multiblock copolymers with lower χ, higher MW PSU oligomers were necessary to induce phase separation in comparison to copolymers with higher χ, such as the fluorinated SCLCP block. A non-phase separated (co)polymer commonly forms with low MW PSU oligomers.

Polymer	Non-PSU structure in block copolymer	χ
PSU-poly(tetramethylene oxide)	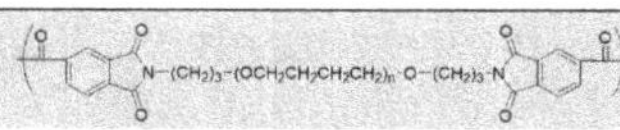	0.89
PSU-poly(ethylene terephthalate-co-oxybenzoate)	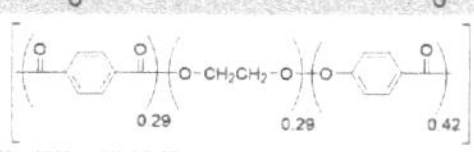	0.07
PSU-poly(ethylene terephthalate-co-polyester imide)	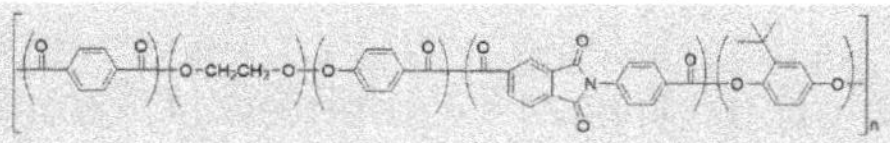	1.61
PSU-poly(ester imide) R= t-butyl or phenyl substituent	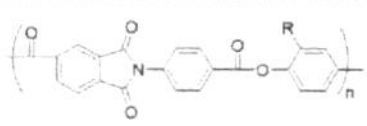	0.75 or 2.22
PSU-poly(phenyl-p-phenylene terephthalate)	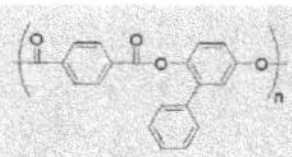	0.45
PSU-semifluorinated polyester multiblock copolymer	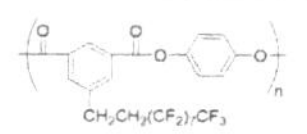	13.68

Figure 7.12 Segmented copolysulfones utilizing a range of secondary block structures exhibiting different χ parameters. Adapted from Pospiech *et al.* [72].

The dispersity of the polymer segments greatly affects the types of observed phase separated morphologies. Unlike multiblock copolymers, which exhibit large PDI between the segments, the narrow dispersity of controlled architectures, such as diblock/triblock copolymers, promote phase separation into ordered long range morphologies. The phase morphology of block copolymers relies heavily on the fractional composition of each block, as illustrated in Figure 7.13, and take 3D structures, such as lamellar, cylindrical and spherical [129-131]. The development of X-Ray scattering to analyze well-ordered morphologies led to in depth understanding of molecular orientation within layers. These structures exhibit significant use in nanotemplating for lithographic applications. Many applications deem controlled morphologies unnecessary and a multiblock copolymer produces proper phase separation to achieve the desired properties.

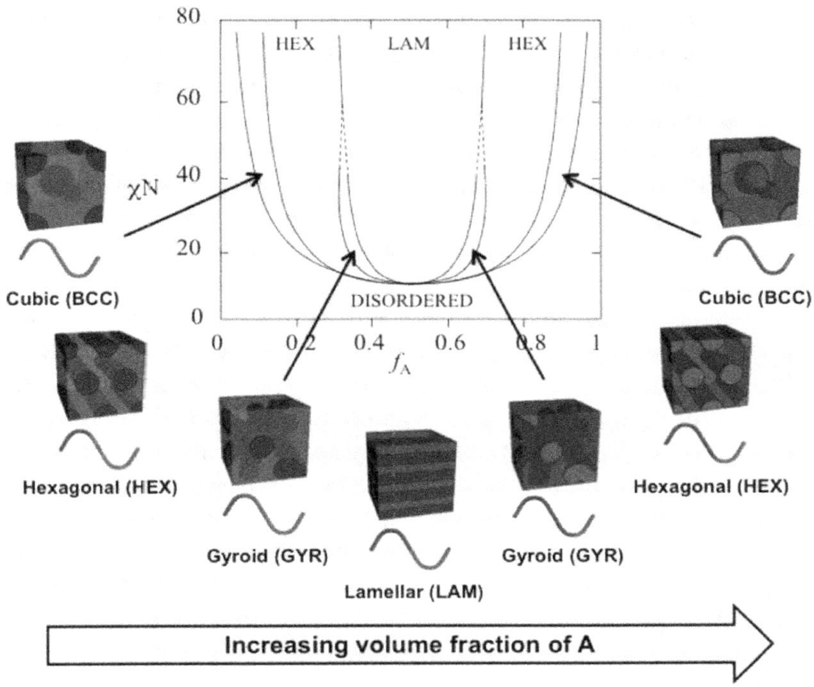

Figure 7.13 Phase diagram of morphological changes within the compositional range of a block copolymer. Reproduced from Reference 131 with permission from MDPI.

Microphase separated block copolymers maintain the properties of each polymer segment within the domain of that material. As a result, the thermal properties of block copolymers generally exhibit the same transitions observed for each separate homopolymer. For example, two glass transition temperatures (T_g) corresponding to the T_g for each separate segment occur in both multiblock copolymers and well-defined morphologies. Ishige *et al.* [67] observed two T_g's for the well-ordered tri-block copolymer. If the homopolymer of the segment exhibits a LC mesophase, ordering occurs within the domains of the LC segment. This results in the observation of a T_i in the block copolymers with the ΔH_i relying on the volume of the segment. The ability to form a LC mesophase within the phase separated domain boundary exhibits a length scale limit. An early study reported that the confinement of a nematic material within a droplet of 350 nm exhibited a well-defined T_i, while reduction of the droplet to less than 35 nm resulted in loss of LC order [132]. Polymer segments also exhibit crystallization with the phase separated domain and a T_m associated with segregated domain melting occurs. In all cases, unless intermixing happens between the segments, the crystalline packing structure and mesophase morphology remains the same between the homopolymers and the block copolymers. The difference in ΔH and slight shifts in the T_m/T_i are attributed to MW differences between the segment lengths and the high MW homopolymer as well as the overall volume change of the ordered segments within the copolymer throughout the series.

The mechanical properties of the bulk materials also comprise a combination of each block's properties. This represents the basis for the development of thermoplastic elastomers. The hard-segment, usually a semi-crystalline or LC polymer, acts as a physical crosslink while the low T_g amorphous segment imparts flexibility. Scientists studying TPEs exhibited interest in LC polyesters because of their high tensile properties. Pospiech *et al.* [77] demonstrated the formation of a TPE with LC hard-segments through the combination of a acetoxy terminated LC trimer with a p-acetoxybenzoic acid chain extender and trimellitimide terminated PTMO. The group noted that when utilizing PTMO, a degree of polymerization between 8-13 afforded phase separation resulting in elastic properties while higher MW soft-segments overwhelmed the HS resulting in LC domain disruption [77]. Sonpatki *et al.* [81] also noted elastomeric properties in multiblock copoly(ether ester). These systems impart inherent tunability onto the bulk mechanical properties during the synthesis through a balance between number of crosslinks and amount of flexibility. In dynamic mechanical analysis, these types of polymers exhibit an extended plateau after the soft-segment T_g and a flow temperature consistent with the disruption of the hard-segment.

For multiblock copolymers utilizing high-performance polymers, thermoplastic elastomers were not the primary goal of copolymerizing. Instead, other benefits arose from blend compatibilization and shear flow modulation. Blend compatibilizing occurs through lowering the surface tension at the interface between two polymers. Block copolymers act as a bridge between the two polymers resulting in a lower degree of phase separation in the blend [29]. Yang *et al.* [85] synthesized multiblock copolymers from a thermotropic LC polyester and PEEK segments to act as a compatibilizer between blends of the homopolymers. The non-compatibilized blends exhibited a strong interface at the surface between the two blends during SEM of the fracture surface. Incorporating different phase separated block copolymers compatibilized the blend resulting in a undiscernible interface [85].

PEEK and PSU high performance polymers exhibit high melt viscosities causing difficulties when processing these materials. Uniquely, LC alignment facilitates significant shear thinning in the melt which resulted in their success/commercialization as composites in injection molded parts. The phase separation of LC segments enables the bulk properties to benefit from the reduced viscosity of the LC polymers resulting in an overall decrease in melt viscosity. Studies demonstrated that the bulk melt viscosity ranges between the values of the two homopolymers in a compositional range of block copolymers [72].

For tri/di block copolymers exhibiting well-ordered morphologies, extensive studies determined the orientation of the LC mesophase within the phase separated domains. These studies utilized a non-crystalline smectic center block and various outer blocks ranging from amorphous to SCLCPs. Koga *et al.* [69,97] determined that for a triblock copolymer series using PEMA outer blocks, as the fraction of amorphous polymer increased, the lamellar thickness remained stable until transitioning into cubes and finally disordered spheres at $\delta > 68\%$. The smectic mesophases form chain folding lamella with extended chain conformations in which the folding surface is oriented parallel to the phase separated lamellae boundary. As the amorphous domain in the lamellae increased, the LC phase counterbalanced by increasing the number of folds that occurred and extending the phase boundary between the LC and PEMA maintaining the lamellar thickness, as observed in Figure 7.14 [69,97]. Studies utilizing another amorphous outer block, PMMA, corroborated the reported ordering [125]. WAXS studies proved in both cases that a S_{CA} mesophase remained within the phase boundaries [69,97,125]. SAXS studies at temperatures greater than the center block T_i indicated that lamellar phase separation relies on the volume fraction of the outer block. Polymers with $< \sim 40\%$ PMMA/PEMA resulted in disruption of microphase separated structure above the T_i of the LC block [97].

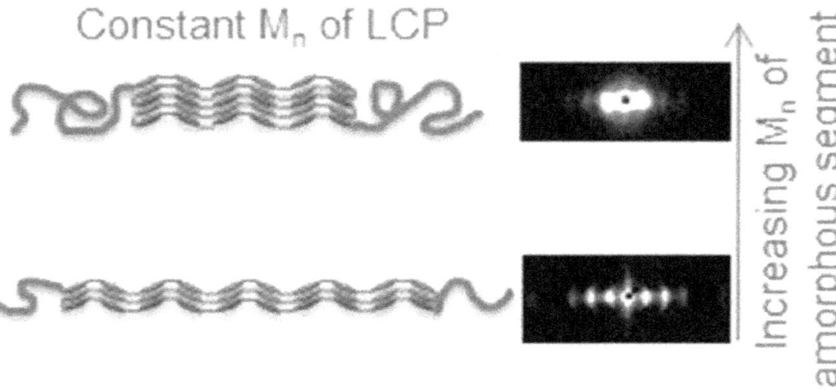

Figure 7.14 Change in lamellar thickness of smectic phase as amorphous domain increases in a triblock copolymer. Reproduced from Reference 97 with permission from American Chemical Society.

The exchange of the outer block to a side-chain LCP did not impact the ordering of the internal S_{CA} polyester. Both a semi-fluorinated and a phenyl benzoate based SCLCP resulted in primarily lamellar phase separated morphologies. The phenyl benzoate based SCLCP exhibited lamellar morphology with volume fractions ranging from 22-56% outer block, then transitioned into a hexagonally packed cylinder at 65%, finally achieving spherical domains at 81% of the SCLCP. The mesophases of each block stayed the same regardless of the phase separated morphology. Unlike the internal block, the SCLCP exhibited a S_A mesophase and in both domains the smectic layers lay parallel to the lamellar interface while the LC director lay perpendicular, as seen in Figure 7.15. This homeotropic LC anchoring at the interface demonstrates the fact that the LC morphology of the internal block influences the outer block as this ordering contradicts previous studies in which non-LC internal blocks exhibited planar anchoring

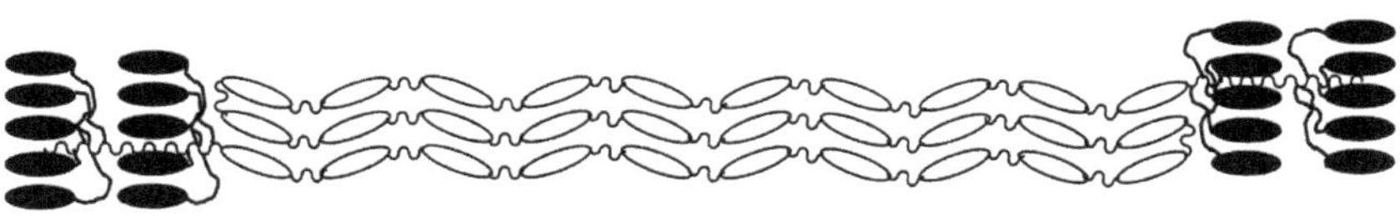

Figure 7.15 Simplified two-phase model of microdomain phase separated structure within a triblock copolymer made up of a MCLCP inner block and SCLCP outer block. Open oval indicative of main-chain LCP structure while filled oval indicates side-chain LCP structure. Adapted from Koga *et al.* [65].

of the SCLCP at the interface [65]. The exchange of the phenyl benzoate SCLCP for a semi-fluorinated block results in a change in the outer block morphology to tilted hexatic with a bilayer structure, likely S_F or S_I. Higher immiscibility between the fluorinated block and the polyester resulted in a well-controlled lamellar morphology from 11-70 % outer block, as well as the maintenance of the lamellar structure at temperatures above the T_i of the SCLCP mesophase. Unlike the aromatic SCLCP, the DP of the outer block impacts the direction of the LC director (long-axis of mesogen). In depth SAXS/WAXS studies elucidated that short outer block chains result in the LC director of the SCLCP laying parallel to the MCLCP and perpendicular to the microphase interface. Increasing the chain length resulted in the opposite orientation to the interface, as seen in Figure 7.16. This suggests a counterplay between the need to match the interfacial area and the energetic favoring to match smectic layer orientation [67]. Due to the low surface free energy of the semi-fluorinated chains, block copolymers with high outer block content (70%) resulted in a perpendicular lamellar structure relative to the silicone surface without extra processing steps [67].

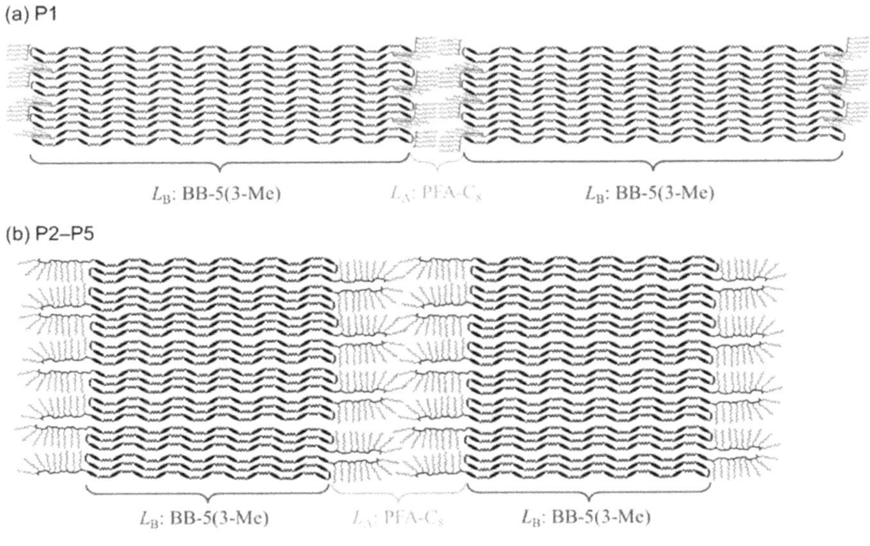

Figure 7.16 Two-phase model of microdomain phase separated structure within a triblock copolymer made up of a MCLCP inner block and fluorinated SCLCP outer block. Reproduced from Reference 67 with permission from American Chemical Society.

Sato *et al.* [126] studied the ordering of an ABA triblock copolyester with a nematic internal block utilizing a fully-aromatic copolyester and an amorphous PS outer block. Unlike smectic mesophases, the nematic ordering results in the LC director laying parallel to the microphase interface. The lamellar spacing in this polymer increases reversibly with increasing temperature, associated with an increase in chain folding, then disrupts completely upon isotropization of the LC segment, as illustrated in Figure 7.17 [126].

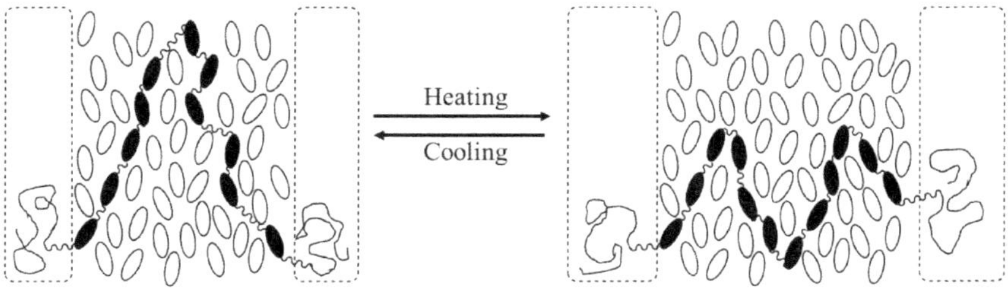

Figure 7.17 Two-phase model of microdomain phase separated structure within a triblock copolymer made up of a nematic LCP inner block and amorphous outer block. Adapted from Sato *et al.* [126].

7.3.1 Orientation of Segmented LC Copolyesters

Orientation of block copolymers enables the formation of nanostructured morphologies over large length scales. Direction of the LC phases within the block copolymer significantly impacts this orientation. Orientation of segmented LC polyesters occurs both through application of mechanical (shear/tensile) strain or in the presence of a magnetic field. Electrical orientation is yet to be examined for these systems.

Mechanical orientation of LC polyesters in commercial processes usually occurs through fiber drawing. Nematic polymers, such as fully-aromatic polyesters, orient the LC director with the pull direction, therefore, in line with the fiber length resulting in the high tensile properties associated with these materials [20]. Smectic polymer orientation relies on the speed of mechanical deformation. Osada *et al.* [133] and Tokita *et al.* [134] showed through WAXS analysis of drawn fibers that fast drawing speeds result in the orientation of the smectic layers perpendicular to the fiber axis, while slow speeds result in parallel layer orientation. This is attributed to flow within the smectic layer when rate of strain is slow enough to prevent domains disruption [133,134]. Koga *et al.* [70] studied the orientation of fibers comprised of a PEMA-[BB-5(3-Me)]-PEMA triblock. The orientation of the phase separated structure and mesophase packing changed depending on the thermal history of the fiber. Fibers that were spun at temperatures above the T_i and quickly quenched resulted in the LC director of the smectic polymer parallel with the microdomain interface due to the polymer chains aligning in the direction of the orientation. Annealing the spun fiber below T_i maintained the chain alignment, but the phase separate morphology migrated so that the lamellae interface returned to the energetically favored position perpendicular to LC director (parallel with smectic layers). As a result, the phase separated lamella structure lay perpendicular to the fiber axis. Heating the spun fiber above the T_i resulted in disruption of the internal block but maintained the phase separated morphology. Cooling back down below T_i resulted in homeotropic anchoring of the mesogens at the interface to the energetically favored orientation and final lamellar microstructure orienting parallel with the fiber axis [70]. Figure 7.18 illustrates this unique orientation behavior. On the contrary, orientation of a triblock copolymer with an internal nematic block resulted in the lamellar layers and the LC director of the nematic polymer parallel to the fiber direction as expected [126].

Magnetic orientation of segmented LC copolyesters has only been demonstrated in one study. Deshmukh *et al.* [109] synthesized novel

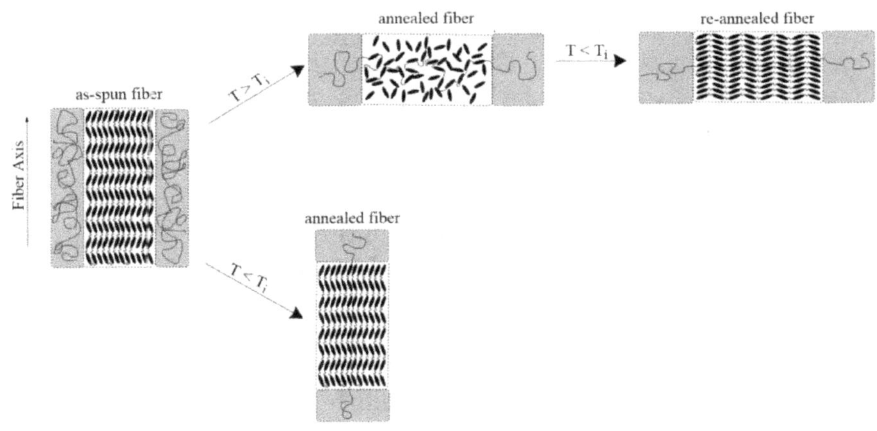

Figure 7.18 Orientation of lamellar and smectic layer structures in fibers with different thermal histories. Fiber axis is vertical. Adapted from Koga *et al.* [70].

brush-like block copolymers with SCLCP and PLA grafted blocks, which exhibited a hexagonally packed cylindrical morphology. The chosen SCLCP exhibited the required magnetic anisotropy. Previous studies indicated that orientation of a LC block copolymer under a magnetic field relies on orientation of the mesogen relative to the microphase interface, while the kinetics of the alignment rely on the viscosity of the material. In the case of the cylindrical morphologies, the mesogens anchored planar to the interface of the cylinder resulting in parallel alignment relative to the magnetic field [135-137]. Orientation occurs upon cooling the sample through its mesophase, while in the presence of the magnetic field. The driving force for alignment results from the ordering of the mesophase below the T_i. In a material, which is weakly segregated, the LC order drives phase separation resulting in a disorder temperature of the phases separated morphology equivalent to the T_i. This enables better alignment due to the low viscosity above the T_i and a drastic decrease in mobility below the T_i freezing the orientation into place. The increased mobility of the material imparted through the addition of plasticizers resulted in effective orientation of cylinders and smectic layer normal parallel to the magnetic field. X-ray scattering elucidated the order parameter revealing a value of P= 0.98, where 0 indicates complete disorder and 1 completely ordered systems [109].

7.4 Conclusions

Segmented LC copolyesters represent a field of block copolymers encompassing intricately controlled architectures to multiblock copolymers utilized in commercial applications. These materials exhibit phase separated morphologies that incorporate the beneficial properties of each polymer and combine them into all-inclusive materials. The choice of segments stems from the desired final properties, as drastically different materials can be synthesized. The combination of flexible ether soft-segments with rigid main-chain LC polyesters form thermoplastic elastomers while exchanging these soft-blocks with high-performance polymers results in high tensile materials that exhibit low melt viscosities and act as compatibilizers to blends. Controlling the synthetic methods to reduce Đ of the blocks enables the formation of well-ordered phase separated structures that form long range nanostructures in fibers and films through mechanical and magnetic orientation. This area of long range nanostructures will benefit greatly from increased research and could be significantly beneficial to the field of lithography bringing scientists one step closer to the computers and transmission devices of the future.

References

1. Work, W. J., Horie, K., Hess, M., and Stepto, R. F. T. (2004) Definition of terms related to polymer blends, composites, and multiphase polymeric materials (IUPAC Recommendations 2004). *Pure and Applied Chemistry*, **76**(11), 1985-2007.
2. Carothers, W. H., and Arvin, J. A. (1929) Studies on polymerization and ring formation. II. Poly-esters. *Journal of American Chemical Society*, **51**, 2560-2570.
3. Carothers, W. H. (1929) Studies on polymerization and ring formation. I. An introduction to the general theory of condensation polymers. *Journal of American Chemical Society*, **51**, 2548-2559.
4. Whinfield, J. R., and Dickson, J. T. (1946) Improvements relating to the formation of highly polymeric substances. Great Britain patent 578,079.
5. Whinfield, J. R. (1946) Chemistry of 'terylene'. *Nature*, **158**, 930-931.
6. Geyer, R., Jambeck, J. R., and Law, K. L. (2017) Production, use, and fate of all plastics ever made. *Science Advances*, **3**, doi: 10.1126/sciadv.1700782.
7. Ji, L. N. (2013) Study on preparation process and properties of polyethylene terephthalate (PET). *Applied Mechanics and Materials*, **312**, 406-410.
8. Scheirs, J., and Long, T. E. (2003) *Modern Polyesters: Chemistry and Technology of Polyesters and Copolyesters*. John Wiley, USA.
9. Collings, P. J., Hird, M., and Huang, C. C. (1998) Introduction to liquid crystals: Chemistry and physics. *American Journal of Physics*, **66**, 551.
10. Tai-Shung, C. (1986) The recent developments of thermotropic liquid crystalline polymers. *Polymer Engineering & Science*, **26**(13), 901-919.
11. Hussein, M. A., Abdel-Rahman, M. A., Asiri, A. M., Alamry, K. A., and Aly, K. I. (2012) Review on: liquid crystalline polyazomethines polymers. Basics, syntheses and characterization. *Designed Monomers and Polymers*, **15**, 431-463.
12. Friedel, G. (1922) Les états mésomorphes de la matière. *Annales de Physique*, **9**, doi: 10.1051/anphys/192209180273.
13. Collings, P. J. (1947) *Liquid Crystals: Nature's Delicate Phase of Matter*, 2nd edition, Princeton University Press, USA.
14. Chung, T.-S. (2001) *Thermotropic Liquid Crystal Polymers: Thin-Film*

Polymerization, Characterization, Blends, and Applications, CRC Press, USA.

15. *Handbook of Liquid Crystals*, Wiley, USA (2014).
16. Flory, P. J. (1956) Phase equilibria in solutions of rod-like particles. *Proceedings of the Royal Society of London. Series A, Mathematical and Physical Sciences*, **234**, 73.
17. Flory, P. J. (1985) Theoretical basis for liquid crystallinity in polymers. In: *Recent Advances in Liquid Crystalline Polymers*, Chapoy, L. L. (ed.), Springer, Netherlands, pp. 99-103.
18. Cowie, J. M. G. (1994) Book Review: *Liquid Crystal Polymers: From Structures to Applications*, Collyer, A. A. (ed.), Elsevier Science Publishers, UK. *Polymer International*, **34**, 122.
19. Calundann, G. W. (1977) Polyester of 6-Hydroxy-2-naphthoic Acid and Para-hydroxy Benzoic Acid Capable of Readily Undergoing Melt Processing, US patent 4,161,470.
20. Jackson, Jr., W. J. (1980) Liquid crystal polymers. IV. Liquid crystalline aromatic polyesters. *Polymer International*, **12**, 154-162.
21. Jackson, W. J., and Kuhfuss, H. F. (1976) Liquid crystal polymers. I. Preparation and properties of p-hydroxybenzoic acid copolyesters. *Journal of Polymer Science: Polymer Chemistry Edition*, **14**, 2043-2058.
22. Jackson, W. J., and Kuhfuss, H. F. (1980) Liquid crystal polymers. III. Preparation and properties of poly(ester amides) from p-aminobenzoic acid and poly(ethylene terephthalate). *Journal of Applied Polymer Science*, **25**, 1685-1694.
23. Speckhard, T. A., Miller, J. A., and Cooper, S. L. (1986) Monte Carlo simulation study of the polymerization of polyurethane block copolymers. 1. Natural compositional heterogeneity under ideal polymerization conditions. *Macromolecules*, **19**, 1558-1567.
24. Peebles, L. H. (1974) Sequence length distribution in segmented block copolymers. *Macromolecules*, **7**, 872-882.
25. Miller, J. A., Speckhard, T. A., and Cooper, S. L. (1986) Monte Carlo simulation study of the polymerization of polyurethane block copolymers. 2. Modeling of premature phase separation during reaction using the two-phase ideal reaction model. *Macromolecules*, **19**, 1568-1574.
26. Holden, G., Bishop, E. T., and Legge, N. R. (1969) Thermoplastic elastomers. *Journal of Polymer Science Part C: Polymer Symposia*, **26**, 37-57.
27. Szwarc, M., Levy, M., and Milkovich, R. (1956) Polymerization initiated by electron transfer to monomer. A new method of formation of block polymers. *Journal of American Chemical Society*, **78**, 2656-2657.
28. Rogers, M. E., and Long, T. E. (2003) *Synthetic Methods in Step-Growth Polymers*, John Wiley & Sons, USA.

29. Wang, T. T., Wang, H. B., Liu, Z. Z., Liu, T., Zhang, H., and Yang, Y. H. (2012) Synthesis and characterization of liquid crystalline copolymers containing poly (ether sulfone) segments. *High Performance Polymers*, **24**, 289-294.
30. Zeng, L. X., Li, R. S., Chen, P., Xu, J. J., and Liu, P. Q. (2016) Synthesis and characterization of thermotropic liquid crystalline polyarylate with ether ether ketone segments in the main chain. *Journal of Applied Polymer Science*, **133**, doi:10.1002/app.43800.
31. Lodge, T. P. (2003) Block copolymers: Past successes and future challenges. *Macromolecular Chemistry and Physics*, **204**, 265-273.
32. Odian, G. G. (1991) *Principles of Polymerization*, John Wiley & Sons, USA.
33. Kang, H., Lin, Q., Armentrout, R. S., and Long, T. E. (2002) Synthesis and characterization of telechelic poly(ethylene terephthalate) sodiosulfonate ionomers. *Macromolecules*, **35**, 8738-8744.
34. Weingart, F. (1994) *Titankatalysatoren bei der Herstellung von Polyethylen-terephthalat - Untersuchungen zu Wirkung und Katalysemechanismus*, PhD thesis, University of Stuttgart, Germany.
35. Nachtergael, A., Coulembier, O., Dubois, P., Helvenstein, M., Duez, P., Blankert, B., and Mespouille, L. (2015) Organocatalysis paradigm revisited: Are metal-free catalysts really harmless? *Biomacromolecules*, **16**, 507-514.
36. Choi, W. S., Padias, A. B., and Hall, H. K. (2000) LCP aromatic polyesters by esterolysis melt polymerization. *Journal of Polymer Science, Part A: Polymer Chemistry*, **38**, 3586-3595.
37. Valery, A. V., and Svetlana, V. V. (1979) Advances in acceptor-catalytic polyesterification. *Russian Chemical Reviews*, **48**, 16.
38. Mandal, J., Prasad, S. K., Rao, D. S. S., and Ramakrishnan, S. (2014) Periodically clickable polyesters: Study of intrachain self-segregation induced folding, crystallization, and mesophase formation. *Journal of American Chemical Society*, **136**, 2538-2545.
39. Higashi, F., Ong, C. H., and Kim, J. H. (1999) Attempt to control monomer sequences in copolycondensations of IPA, TPA, BPA, and PHB. *Journal of Polymer Science, Part A: Polymer Chemistry*, **37**, 2371-2377.
40. Meng, F. B., Zhang, B. Y., Xu, Y., and Liu, J. Y. (2005) Main-chain liquid-crystalline ionomers bearing potassium sulfonate groups. *Journal of Applied Polymer Science*, **96**, 2021-2026.
41. Yan, Q., and He, J. (1995) Segmented liquid crystalline copolyesters and blending with poly(ethylene terephthalate). *Polymer Bulletin*, **34**, 441-447.
42. Ignatious, F., Lenz, R. W., and Kantor, S. W. (1994) Synthesis and characterization of block-copolymers containing rigid liquid-crystalline and flexible butylene terephthalate segments. *Macromolecules*, **27**, 5248-5257.

43. Wang, J., and Lenz, R. W. (1991) Preparation and properties of a liquid crystalline segmented block copolyester. *Polymer Engineering & Science*, **31**, 739-742.
44. Galli, G., Chiellini, E., Laus, M., Bignozzi, M. C., Angeloni, A. S., and Francescangeli, O. (1994) Synthesis and thermal behavior of liquid-crystalline block copolymers containing both main-chain and side-chain mesomorphic blocks. *Macromolecular Chemistry and Physics*, **195**, 2247-2260.
45. Chiellini, E., Galli, G., Angeloni, A. S., Laus, M., Bignozzi, M. C., Yagci, Y., and Serhatli, E. I. (1994) Hybrid thermotropic liquid-crystalline block copolymers. *Macromolecular Symposia*, **77**, 349-358.
46. Huang, J. M., Kuo, J. F., and Chen, C. Y. (1995) Synthesis and characterization of segmented liquid-crystal poly(azoxy polyester-co-polyoxypropylene). *Journal of Polymer Science, Part A: Polymer Chemistry*, **33**, 165-173.
47. Dove, A. P. (2012) Organic catalysis for ring-opening polymerization. *ACS Macro Letters*, **1**, 1409-1412.
48. Kiesewetter, M. K., Shin, E. J., Hedrick, J. L., and Waymouth, R. M. (2010) Organocatalysis: Opportunities and challenges for polymer synthesis. *Macromolecules*, **43**, 2093-2107.
49. *Liquid Crystals*, Stegemeyer, H. (ed.), 3rd volume, Topics in Physical Chemistry, Springer, Germany (1994).
50. Demus, D. (1998) *Handbook of Liquid Crystals*, Wiley-VCH, Germany.
51. Tokita, M., Funaoka, S.-i., and Watanabe, J. (2004) Study on smectic liquid crystal glass and isotropic liquid glass formed by thermotropic main-chain liquid crystal polyester. *Macromolecules*, **37**, 9916-9921.
52. *Liquid Crystallinity in Polymers: Principles and Fundamental Properties*, Ciferri, A. (ed.), VCH Publishers, USA (1991).
53. Economy, J., Storm, R. S., Matkovich, V. I., Cottis, S. G., and Nowak, B. E. (1976), Synthesis and structure of the p-hydroxybenzoic acid polymer. *Journal of Polymer Science, Part A: Polymer Chemistry*, **14**, 2207-2224.
54. Jin, J.-I. and Kang, C.-S. (1997) Thermotropic main chain polyesters. *Progress in Polymer Science*, **22**, 937-973.
55. Economy, J., and Goranov, K. (1994) Thermotropic liquid crystalline polymers for high performance applications. *High Performance Polymers*, **117**, 221-256.
56. Cottis, S. G., Economy, J., and Nowak, B. E. (1972) p-Oxybenzoyl Copolyesters, US Patent 3,637,595.
57. Cottis, S. G., Economy, J., and Wohrer, L. C. (1976) Process for Spinning High Modulus Oxybenzoyl Copolyester Fibers, US Patent 3,975,487.
58. Tokita, M., and Watanabe, J. (2006) Several interesting fields exploited through understanding of polymeric effects on liquid crystals of

main-chain polyesters. *Polymer Journal*, **38**, 611-638.

59. Watanabe, J., Hayashi, M., Nakata, Y., Niori, T., and Tokita, M. (1997) Smectic liquid crystals in main-chain polymers. *Progress in Polymer Science*, **22**, 1053-1087.
60. Watanabe, J., Hayashi, M., Morita, A., and Tokita, M. (1995) Thermotropic liquid crystals of main-chain polyesters having a mesogenic 4,4'-biphenyldicarboxylate unit. 6. Chiral mesophases of polyesters with a (S)-2-methylbutylene spacer. *Macromolecules*, **28**, 8073-8079.
61. Watanabe, J., and Hayashi, M. (1988) Thermotropic liquid crystals of polyesters having a mesogenic p,p'-bibenzoate unit. 1. Smectic A mesophase properties of polyesters composed of p,p'-bibenzoic acid and alkylene glycols. *Macromolecules*, **21**, 278-280.
62. Krigbaum, W. R., Asrar, J., Toriumi, H., Ciferri, A., and Preston, J. (1982) Aromatic polyesters forming thermotropic smectic mesophases. *Journal of Polymer Science: Polymer Letters Edition*, **20**, 109-115.
63. Watanabe, J., and Hayashi, M. (1989) Thermotropic liquid crystals of polyesters having a mesogenic p,p'-bibenzoate unit. 2. X-ray study on smectic mesophase structures of BB-5 and BB-6. *Macromolecules*, **22**, 4083-4088.
64. Perez, E., Fernandez-Blazquez, J. P., Martinez-Gomez, A., Bello, A., Cerrada, M. L., Benavente, R., and Perena, J. M. (2009) Applications of synchrotron X-ray diffraction to the study of the phase behavior in liquid crystalline polymers. In: *Applications of Synchrotron Light to Scattering and Diffraction in Materials and Life Sciences*, Ezquerra, T. A., GarciaGutierrez, M. C., Nogales, A., and Gomez, M. (eds.), Springer, Germany, pp. 157-182.
65. Koga, M., Sato, K., Kang, S., and Tokita, M. (2018) Microphase-separated morphology and liquid crystal orientation in block copolymers comprising a main-chain liquid crystalline central segment connected to side-chain liquid crystalline segments at both ends. *Macromolecular Chemistry and Physics*, **219**, 1700332.
66. Tokita, M., Sugimoto, A., Takahashi, C., Yoshihara, S., van de Watering, R., and Kang, S. (2016) Extended chain lamella formation characteristics of main-chain smectic liquid crystalline copolyesters comprising different length units. *Macromolecules*, **49**(7), 2718-2723.
67. Ishige, R., Ohta, N., Ogawa, H., Tokita, M., and Takahara, A. (2016) Fully liquid-crystalline ABA triblock copolymer of fluorinated side-chain liquid-crystalline A block and main-chain liquid-crystalline B block: Higher order structure in bulk and thin film states. *Macromolecules*, **49**, 6061-6074.
68. Koga, M., and Tokita, M. (2014) Micro-phase separated structures of block copolymers having a semiflexible liquid crystalline main-

chain polyester block. *Kobunshi Ronbunshu*, **71**, 501-507.

69. Koga, M., Abe, K., Sato, K., Koki, J., Kang, S., Sakajiri, K., Watanabe, J., and Tokita, M. (2014) Self-assembly of flexible–semiflexible-flexible triblock copolymers. *Macromolecules*, **47**, 4438-4444.
70. Koga, M., Sato, K., Kang, S. M., Sakajiri, K., Watanabe, J., and Tokita, M. (2013) Influence of smectic liquid crystallinity on lamellar microdomain structure in a main-chain liquid crystal block copolymer fiber. *Macromolecular Chemistry and Physics*, **214**, 2295-2300.
71. Amendola, E., Carfagna, C., Netti, P., Nicolais, L., and Saiello, S. (1993) The Influence of a copolymer as adhesion promoter for polycarbonate and liquid-crystalline polymer blends. *Journal of Applied Polymer Science*, **50**, 83-94.
72. Pospiech, D., Haussler, L., Eckstein, K., Komber, H., Voigt, D., Jehnichen, D., Friedel, P., Gottwald, A., Kollig, W., and Kricheldorf, H. R. (2001) Synthesis and phase separation behaviour of high performance multiblock copolymers. *High Performance Polymers*, **13**, S275-S292.
73. Gopakumar, T. G., Ponrathnam, S., Rajan, C. R., and Fradet, A. (1998) Block copolymers of telechelic poly(phenylene sulfide) and semiaromatic thermotropic liquid crystalline polyester segments. *Journal of Polymer Science, Part A: Polymer Chemistry*, **36**, 2707-2713.
74. Suenaga, J.-i., Owaki, T., and Egawa, M. (2003) Wholly Aromatic Heat-Stable Liquid Crystalline Polyester Resin Composition with Improved Melt Flowability, US patent 6,656,386 B2.
75. Cottis, S. G., Layton, R., and Field, N. D. (1986) Injection Moldable Aromatic Polyesters Compositions and Method of Preparation, US patent 4,563,508.
76. Jackson, Jr., W. J., and Morris, J. C. (1979) Liquid Crystal Copolyesters Containing Terephthalic Acid and 2,6-Napthalenedicarboxylic Acid, US patent 4,169,933.
77. Pospiech, D., Komber, H., Voigt, D., Haußler, L., Meyer, E., Schauer, G., Jehnichen, D., and Bohme, F. (1994) Thermotropic elastomers with poly(oxy-1,4-butanediyl) units in the main chain. *Macromolecular Chemistry and Physics*, **195**, 2633-2651.
78. Chain-Shu, H., and Ling-Bar, C. (1993) Synthesis and characterization of block copolymers of polyether sulfone with liquid crystalline polyesters. *Materials Chemistry and Physics*, **34**, 28-34.
79. Stinson, S. (1997) Polyesterification mechanism study. *Chemical & Engineering News Archive*, **75**, 29.
80. Jannesari, A., Ghaffarian, S. R., and Molaei, A. (2006) The effect of curing reaction on mesophase-rich islands of segmented main chain liquid crystalline oligoesters. *Reactive & Functional Polymers*, **66**, 1250-1262.
81. Sonpatki, M. M., Ravindranath, K., and Ponrathnam, S. (1995) Random thermotropic elastomers: effect of hard/soft-segment lengths

on the properties of liquid-crystalline copoly(ether ester)s. *Polymer*, **36**, 3127-3134.

82. Pospiech, D., Haussler, L., Komber, H., Voigt, D., Jehnichen, D., Janke, A., Baier, A., Eckstein, K., and Bohme, F. (1996) LC multiblock copolymers containing polysulfone segments .1. Synthesis and morphology. *Journal of Applied Polymer Science*, **62**, 1819-1833.
83. Sonpatki, M. M., Ravindranath, K., and Ponrathnam, S. (1994) Segmented copoly(ether-ester) elastomers - influence of hard segment length and substitutions on mesophase formation. *Polymer Journal*, **26**, 804-815.
84. Brenda, S., Heitz, W., Karbach, A., and Wehrmann, R. (1994) Block-copolymers of poly(ether-ether-sulfones) with liquid crystalline non-liquid-crystalline polyesters. *Macromolecular Chemistry and Physics*, **195**, 1327-1339.
85. Yang, Y. H., Zhu, W. C., Jiang, D., Ma, R. T., and; Jiang, Z. H. (2007) Synthesis of multi-block copolymer and its compatibilization to the blends of poly(ether ether ketone) with thermotropic liquid crystalline polymer. *Journal of Applied Polymer Science*, **104**, 35-43.
86. Schulze, U., and Schmidt, H. W. (1998) Synthesis and characterization of ABA-triblock copolymers with poly(ethylene glycol) segments and LC-segments. *Polymer Bulletin*, **40**(2-3), 159-166.
87. Lin, L. L., and Hong, J. L. (2000) Multiple melting behavior of a thermotropic copolyester containing spirobicromane moiety. *Polymer*, **41**, 7471-7481.
88. Reichelt, N., Schulze, U., and Schmidt, H. W. (1997) Synthesis and characterization of ABA-triblock copolymers with polystyrene and main-chain liquid crystalline polyester segments. *Macromolecular Chemistry and Physics*, **198**, 3907-3930.
89. Novack, K. M., Martins, J. C. A., and Gomes, A. S. (1996) Block copolymers with poly(ethylene terephthalate) and liquid crystalline polyester segments: Effect of reaction time. *Polymer Bulletin*, **37**(5), 603-608.
90. Heyde, G., Heitz, W., Karbach, A., and Wehramann, R. (1993) Block copolymers with poly(oxy-1,4-phenylene) and liquid crystalline polyester segments. *Die Makromolekulare Chemie*, **194**, 2741-2758.
91. Norbert, N., and Walter, H. (1990) Poly(oxy-1,4-phenylene) macromonomers for polycondensation. *Die Makromolekulare Chemie*, **191**, 1463-1475.
92. Moriyuki, S., Takashi, M., Kazue, N., Ken-ichi, M., Fumihiko, Y., and Yasushi, O. (1994) Thermotropic liquid-crystalline polyester-graft-polystyrene copolymers from a polystyrene macromonomer with amido linkage. *Macromolecular Rapid Communications*, **15**, 243-250.
93. Angeloni, A. S., Bignozzi, M. C., Laus, M., Chiellini, E., and Galli, G. (1993) Hybrid liquid-crystalline block copolymers based on polyst-

yrene and polyester blocks. *Polymer Bulletin*, **31**(4), 387-392.

94. Laus, M., Bignozzi, M. C., Angeloni, A. S., Francescangeli, O., Galli, G., and Chiellini, E. (1995) Hybrid liquid-crystalline block copolymers I. Synthesis and mesomorphic behavior of polyester–polymethacrylate block copolymers. *Polymer Journal*, **27**, 993-1001.
95. Galli, G., Chiellini, E., Laus, M., Angeloni, A. S., Bignozzi, M. C., and Francescangeli, O. (1994) Hybrid liquid-crystalline block copolymers with polystyrene and polyester blocks. *Molecular Crystals and Liquid Crystals Science and Technology. Section A. Molecular Crystals and Liquid Crystals*, **254**, 429-443.
96. Chang, Y. M., Hsu, C. S., and Liu, M. C. (1996) Synthesis and characterization of segmented copolymers of aromatic polyether sulfone and thermotropic liquid crystalline poly(oxy-1,4-phenylenecarbonyl-co-oxy-2,6-naphthaloyl). *Materials Chemistry and Physics*, **43**, 250-255.
97. Koga, M., Ishige, R., Sato, K., Ishii, T., Kong, S., Sakajiri, K., Watanabe, J., and Tokita, M. (2012) Well-ordered lamellar microphase-separated morphology of an ABA triblock copolymer containing a main-chain liquid crystalline polyester as the middle segment 2: Influence of amorphous segment molecular weight. *Macromolecules*, **45**, 9383-9390.
98. Nelson, A. M., Fahs, G. B., Moore, R. B., and Long, T. E. (2015) High-performance segmented liquid crystalline copolyesters. *Macromolecular Chemistry and Physics*, **216**, 1754-1763.
99. Tokita, M., Takahashi, T., Hayashi, M., Inomata, K., and Watanabe, J. (1996) Thermotropic liquid crystals of polyesters having a mesogenic p,p'-bibenzoate unit. 7. Chain folding in the smectic phase of BB-6. *Macromolecules*, **29**, 1345-1348.
100. Herzberger, J., Meenakshisundaram, V., Williams, C. B., and Long, T. E. (2018) 3D printing all-aromatic polyimides using stereolithographic 3D printing of polyamic acid salts. *ACS Macro Letters*, **7**(4), 493-497.
101. Lin, Q., Pasatta, J., and Long, T. E. (2003) Synthesis and characterization of chiral liquid-crystalline polyesters containing sugar-based diols via melt polymerization. *Journal of Polymer Science, Part A: Polymer Chemistry*, **41**, 2512-2520.
102. Lin, Q., Pasatta, J., Wang, Z.-H., Ratta, V., Wilkes, G. L., and Long, T. E. (2002) Synthesis and characterization of sulfonated liquid crystalline polyesters. *Polymer International*, **51**(6), 540-546.
103. *Liquid Crystal Elastomers: Materials and Applications*, de Jeu, W. H. (ed.), Springer, Germany (2012).
104. Emmerling, U., Diele, S., Schmalfuss, H., Werner, J., Kresse, H., and Lindau, J. (1998) Thermal phase behavior of homologous liquid crystalline side group polyesters in dependence on the polymethylene main chain spacer length. *Macromolecular Chemistry and*

Physics, **199**(8), 1529-1538.

105. Mu, B., Wu, B., Pan, S., Fang, J., and Chen, D. (2015) Hierarchical self-organization and uniaxial alignment of well synthesized side-chain discotic liquid crystalline polymers. *Macromolecules*, **48**, 2388-2398.
106. Katano, Y., Tomono, H., and Nakajima, T. (1994) Surface property of polymer films with fluoroalkyl side chains. *Macromolecules*, **27**, 2342-2344.
107. Hare, E. F., Shafrin, E. G., and Zisman, W. A. (1954) Properties of films of adsorbed fluorinated acids. *The Journal of Physical Chemistry*, **58**, 236-239.
108. Matroniani, R., and Wang, S. H. (2017) Thermal degradation of polymer systems having liquid crystalline oligoester segment. *Polimeros-Ciencia E Tecnologia*, **27**, doi:10.1590/0104-1428.2423.
109. Deshmukh, P., Gopinadhan, M., Choo, Y., Ahn, S.-k., Majewski, P. W., Yoon, S. Y., Bakajin, O., Elimelech, M., Osuji, C. O., and Kasi, R. M. (2014) Molecular design of liquid crystalline brush-like block copolymers for magnetic field directed self-assembly: A platform for functional materials. *ACS Macro Letters*, **3**, 462-466.
110. del Campo, A., Bello, A., and Perez, E. (2003) Synchrotron X-ray study of liquid crystalline polyoxetanes containing two mesogenic groups connected by a flexible spacer in the side chain. *Macromolecular Chemistry and Physics*, **204**, 682-691.
111. Manfred, A., Sven, P., Hartmut, F., Willy, F., and Hermann, U. (1994) Liquid-crystalline side chain block copolymers - synthesis, morphology and LC behaviour. *Macromolecular Rapid Communications*, **15**, 487-496.
112. Mathew, J., Ghadage, R. S., Ponrathnam, S., and Prasad, S. D. (1994) Catalytic synthesis of poly(ethylene terephthalate-co-oxybenzoate) systems via the melt polyesterification route-copolyesterification kinetics. *Macromolecules*, **27**, 4021-4035.
113. Tsai, H.-B., Li, H.-C., Chang, S.-J., and Yu, H.-H. (1991) Copolyesters. *Polymer Bulletin*, **27**, 141-146.
114. Tsai, R. S., Lee, Y. D., and Tsai, H. B. (2002) Block copolyesters of poly(pentamethylene p,p '- bibenzoate) and poly(tetramethylene adipate). *Journal of Polymer Science, Part A: Polymer Chemistry*, **40**, 2626-2636.
115. Auman, B. C., and Percec, V. (1988) Synthesis and characterization of segmented copolymers of aromatic polyether sulfone and a thermotropic liquid-crystalline polyester. *Polymer*, **29**, 938-949.
116. Heitz, W. (1989) Block copolymers with LC-segments. *Makromolekulare Chemie. Macromolecular Symposia*, **26**, 1-8.
117. Herzberger, J., Niederer, K., Pohlit, H., Seiwert, J., Worm, M., Wurm, F. R., and Frey, H. (2016) Polymerization of ethylene oxide, propylene oxide, and other alkylene oxides: Synthesis, novel polymer arc-

hitectures, and bioconjugation. *Chemical Reviews*, **116**, 2170-2243.

118. Tsai, H. B. (1997) Characterization of thermotropic block copolyetheresters by x-ray diffraction and DSC. *Journal of Macromolecular Science-Physics*, **B36**, 175-185.
119. Tsai, H.-B., Lee, C., and Chang, N.-S. (1992) Synthesis and thermal properties of thermotropic block copolyetheresters. *Polymer Journal*, **24**, 157-164.
120. Tsai, H.-B., Lee, D.-K., Liu, J.-L., Tsao, Y.-S., Tsai, R.-S., and You, J.-W. (1995) Block copolyetheresters. *Polymer Bulletin*, **35**, 743-749.
121. Hsiue, L. T., Ma, C. C. M., and Tsai, H. B. (1995) Hydrophilic thermotropic block copolyetheresters with poly(pentamethylene p,p'-bibenzoate) segments. *Macromolecular Chemistry and Physics*, **196**, 3459-3468.
122. Hsiue, L. T., Ma, C. C. M., and Tsai, H. B. (1995) Synthesis and characterization of hydrophilic thermotropic block copolyetheresters. *Journal of Polymer Science, Part A: Polymer Chemistry*, **33**, 1153-1159.
123. Ferri, D., Wolff, D., Springer, J., Francescangeli, O., Laus, M., Angeloni, A. S., Galli, G., and Chiellini, E. (1998) Phase and orientational behaviors in liquid crystalline main-chain/side-group block copolymers. *Journal of Polymer Science, Part B: Polymer Physics*, **36**, 21-29.
124. Lin, C. H. (2013) Synthesis and characterization of segmented copolymers of polystyrene and thermotropic liquid crystalline poly(4-oxybenzoate-co-2,6-oxynaphthoate). *Designed Monomers and Polymers*, **16**, 537-542.
125. Ishige, R., Ishii, T., Tokita, M., Koga, M., Kang, S. M., and Watanabe, J. (2011) Well-ordered lamellar microphase-separated morphology of an ABA triblock copolymer containing a main-chain liquid crystalline polyester as the middle segment. *Macromolecules*, **44**, 4586-4588.
126. Sato, K., Koga, M., Kang, S., Sakajiri, K., Watanabe, J., and Tokita, M. (2013) Lamellar morphology of an ABA triblock copolymer with a main-chain nematic polyester central block. *Macromolecular Chemistry and Physics*, **214**, 1089-1093.
127. Heitz, T., Rohrbach, P., and Hocker, H. (1989) Rigid rods with flexible side chains - A route to molecular reinforcement? *Die Makromolekulare Chemie*, **190**, 3295-3316.
128. Burchard, W. (1983) Solution thermodynamics of non-ionic water-soluble polymers. In: *Chemistry and Technology of Water-Soluble Polymers*, Finch, C. A. (ed.), Springer, USA, pp. 125-142.
129. Matsen, M. W., Schick, M. (1996) Self-assembly of block copolymers. *Current Opinion in Colloid & Interface Science*, **1**, 329-336.
130. Ciach, A., Pekalski, J., and Gozdz, W. T. (2013) Origin of similarity of phase diagrams in amphiphilic and colloidal systems with competing interactions. *Soft Matter*, **9**, 6301-6308.

131. Swann, J. M. G., and Topham, P. D. (2010) Design and application of nanoscale actuators using block-copolymers. *Polymers*, **2**, 454-469.
132. Fischer, H., and Poser, S. (1996) Liquid crystalline block and graft copolymers. *Acta Polymerica*, **47**, 413-428.
133. Osada, K., Koike, M., Tagawa, H., Hunaoka, S.-i., Tokita, M., and Watanabe, J. (2005) Two distinct types of orientation process observed in uniaxially elongated smectic LC melt. *Macromolecules*, **38**, 7337-7342.
134. Tokita, M., Tokunaga, K., Funaoka, S.-i., Osada, K., and Watanabe, J. (2004) Parallel and perpendicular orientations observed in shear aligned SCA liquid crystal of main-chain polyester. *Macromolecules*, **37**, 2527-2531.
135. Osuji, C., Ferreira, P. J., Mao, G., Ober, C. K., Vander Sande, J. B., and Thomas, E. L. (2004) Alignment of self-assembled hierarchical microstructure in liquid crystalline diblock copolymers using high magnetic fields. *Macromolecules*, **37**, 9903-9908.
136. Gopinadhan, M., Majewski, P. W., Choo, Y., and Osuji, C. O. (2013) Order-disorder transition and alignment dynamics of a block copolymer under high magnetic fields by in situ X-ray scattering. *Physical Review Letters*, **110**, 078301.
137. Gopinadhan, M., Majewski, P. W., Beach, E. S., and Osuji, C. O. (2012) Magnetic field alignment of a diblock copolymer using a supramolecular route. *ACS Macro Letters*, **1**, 184-189.

Chapter 8

Smectic Phases of Rod-Like Helical Polymers

Kento Okoshi

Department of Applied Chemistry and Bioscience, Chitose Institute of Science and Technology, 758-65 Bibi, Chitose, Hokkaido 066-8655, Japan
k-okoshi@photon.chitose.ac.jp

8.1 Introduction

The origin of the liquid crystal is largely in the anisotropy of the molecule. Many kinds of materials, e.g., DNA [1,2], viruses [3-7] and colloidal suspensions of crystal particles [8,9], have been observed to form liquid crystals in addition to conventional liquid crystals with rigid mesogenic cores and aliphatic tails, in spite of their different chemical properties. The only thing these materials have in common is that these are rod-shaped. The long axes of the rod-shaped molecules spontaneously line up in parallel to gain the packing entropy that exceeds the loss of orientational entropy as their number density increases. In other words, the rod-like molecules align side by side to increase the overlapping of the depletion zone which the center of gravity of the molecules cannot enter. This is referred to as excluded volume interaction, which forms the basis of liquid crystal formation; in addition, attractive intermolecular interactions play a certain role in liquid crystalline formation. Therefore, it is appropriate to assume a simple physical model that treats liquid crystals as rod-like rigid particles.

It has been theoretically shown that systems of monodisperse-length hard rod-like particles give rise to not only nematic-isotropic transition by Onsager [10] and Flory [11]; moreover, the most common liquid crystalline columnar-smectic-nematic phase transition (Figure 8.1A) has also been studied, based on numerical calculations and computer simulations, only with an excluded volume interaction taken into consideration, to prove that the complexity of liquid crystal varies uniformly with shape anisotropy alone [12-16]. Until now, much of such theoretical treatment has been performed

Liquid Crystalline Polymers, edited by Vikas Mittal

mainly with athermal systems and developed beyond the range that seems to be experimentally verifiable, such as using disc-shaped [17], ellipsoidal [18] and biaxial particles [19] and their mixtures.

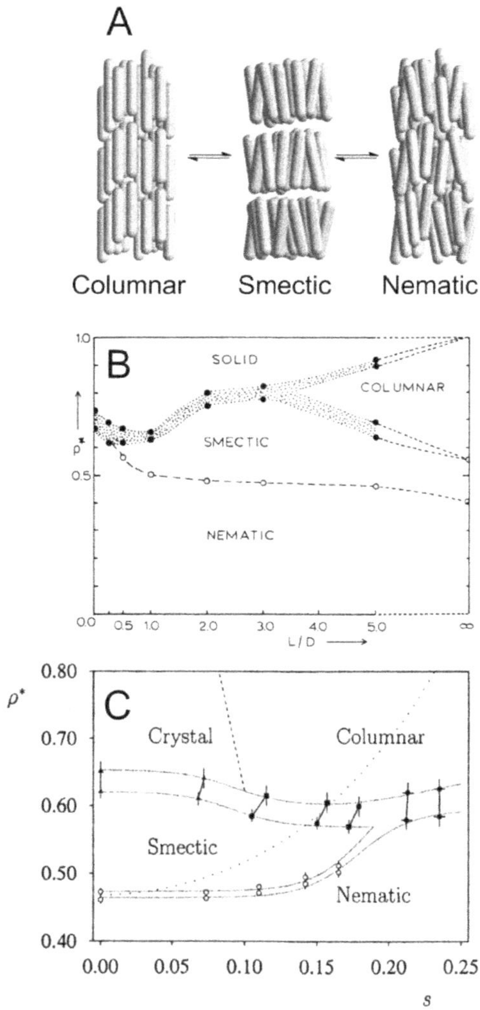

Figure 8.1 (A) Illustration of columnar-smectic-nematic phase transition, (B) phase diagram of monodisperse spherocylinder as a function of the axial ratio of the cylinder and reduced density (reproduced from Reference 14 with permission from American Physical Society) and (C) phase diagram of polydisperse length spherocylinder as a function of its length dispersity and reduced density (reproduced from Reference 27 with permission from AIP publishing).

The experimental verifications for these theoretical results were initially conducted mainly using rod-like biological samples. In the 1970s, detailed verification of Flory's theoretical prediction concerning the nematic-isotropic phase transition was reported using a lyotropic solution of hard-rod-shaped poly(γ-benzyl-L-glutamate) (PBLG) [20]. The smectic phase was discovered for the first time by Wen *et al.* [3] in 1989 in an aqueous solution of tobacco mosaic virus, which is fairly stiff and rod-shaped, and it was also confirmed that a smectic-nematic phase transition takes place as its concentration decreases. A number of studies about the smectic phases of similar rod-like viruses have been reported since then [4-7]. In 1997, Yu *et al.* reported the smectic phase of fully monodisperse PBLG in its swollen film [21], and the formation of a twist grain boundary (TGB) phase [22]. However, this PBLG, with a distinct molecular weight, can be prepared by a very laborious and costly bacterial synthetic method, in which the vectors encoding glutamic acid chains are introduced into Escherichia coli and the corresponding gene product is then benzylated [23]. Therefore, experimental verification of the theoretical predictions made for the liquid crystalline phase behavior of rod-like particles has not attained significant progress due to the lack of easy-to-handle experimental systems, as can be seen from the fact that these have been made entirely with viruses and biopolymers. Thus, it is necessary to develop an experimental system of synthetic rod-like polymers which is less cumbersome.

8.2 Smectic Phases of Rod-Like Polymers

8.2.1 Theoretical Predictions

Although it has long been known that a rod-like particle system shows a nematic phase, widespread acceptance of the existence of the smectic phase has only been achieved recently. The first theoretical indication of the smectic-nematic transition of rigid rod-like particles was suggested by Wadati and Isihara [24] and Hosino *et al.* [12,13], based on the mean field approximation, wherein the free energy, expressed as a function of positional and orientational order of the particles, is minimized. It is interesting to note that a similar result has been derived from a model considering the attractive potential alone around the same time [25,26]. Subsequently, a computer simulation using a cylinder capped with half-spheres called a spherocylinder as a physical model revealed the phase diagram as a function of the axial ratio (L/D) of the cylinder and reduced density (defined as ρ/ρ_{cp}, where ρ_{cp} is the density of closest packing) of the particles (Figure 8.1B) [14]. It showed no isotropic phase due to the approximation of parallel orientation, but clearly showed a columnar-smectic-nematic phase transition as its density decreased. Although the result seemed convincing because it posted a plausible snapshot of the simulation, the columnar phase has been proven to be unstable at least for mono-disperse-length rod-like particle systems in following studies [15,16].

Viruses and biopolymers were initially used to prove the theoretical predictions for mono-disperse-length particle systems because such biological samples scarcely show any variations due to the structural information required to be encoded by genes. However, a computer simulation employing rod-like particles with length dispersity was performed: it showed that a smectic phase formation tolerates dispersity in the length [27]. Figure 8.1C shows the phase diagram for the rod-like particles with an infinite axial ratio drawn as a function of the dispersity of the rod length (s represents the standard deviation of the length distribution of the rod-like particles when the average length is unity) and reduced density. Though the phase behavior was almost the same as in the monodisperse system when the length distribution is small ($s < 0.08$), the smectic phase becomes increasingly unstable with respect to the nematic phase as the length distribution becomes larger, and depletes at $s = 0.18$ which corresponds to $M_w/M_n = 1.03$, assuming a normal distribution. This would allow us to perform more detailed experimental verifications of theoretical predictions with easy-to-handle synthetic polymers, which inevitably possess molecular weight distributions in varying degrees.

8.2.2 Experimental Verifications

The most desirable systems for the experimental verifications of the numerous preceding theoretical studies are synthetic helical polymers due to their main chain stiffness and controllable length. Among these systems, helical polysilanes provide the best results (Figure 8.2A), which can be regarded as rod-shaped particles due to their exceptional stiffness generated by the large steric hindrance between the long and β-branched alkyl side chains on the neighboring repeating units. Since these polymers simply consist of silicon, carbon and hydrogen, these are nonpolar, which makes them ideal experimental systems for the verification of the theoretical studies on the athermal systems, as there is no electrostatic intermolecular interaction but a predominant excluded volume interaction and a weak van der Waals force working between the polymers.

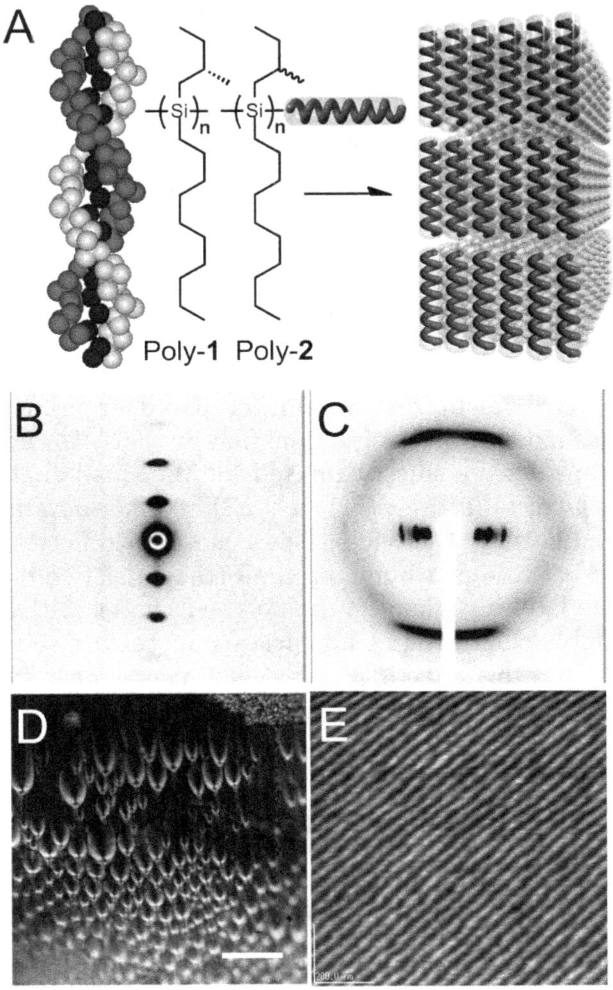

Figure 8.2 (A) Structure of polysilanes and illustration of their smectic liquid crystal formation, (B) side-angle and (C) wide-angle X-ray diffraction patterns of poly-2 (reproduced from Reference 32 with permission from Taylor and Francis) and (D) polarizing optical microscopic image of poly-1 (scale bar: 200 nm). (E) AFM phase image observed on the cast film of poly-1 (scale: 1x1 μm).

The first conclusive report of the smectic phase for synthetic polymers was obtained for poly[n-decyl-(S)-2-methylbutylsilane] (poly-1, Figure 8.2A) with a very narrow molecular weight distribution ($M_w/M_n < 1.2$) [28], which possesses a right-handed rigid helical main chain with a persistence length of up to 85 nm [29,30]. Before this, only ambiguous polarized microscopic observations of the smectic phase had been reported for a synthetic polypeptide with long alkyl side chains [31]. The experimental system should be a lyotropic liquid crystal to verify the theoretical predictions of rod-like particle system with only rigid body repulsion. However, the long alkyl side chains of poly-1 act as a solvent in its thermotropic liquid crystalline state, which makes it possible to avoid experimental barriers such as concentration changes due to volatilization of the solvent, phase separation and sedimentation.

Figures 8.2B-D show small- and wide-angle X-ray scattering patterns of poly-2 (Figure 8.2B, 8.2C) [32] and the polarizing microscopic image of poly-1 (Figure 8.2D), which demonstrate a very narrow molecular weight distribution prepared from polydisperse samples by a simple solvent fractionation method. It is evident that the polysilanes form a smectic phase from the small-angle meridional smectic layer reflections (Figure 8.2B) in the orientational direction perpendicular to the wide-angle equatorial reflections (Figure 8.2C), which can be attributed to the orthogonal two-dimensional lattice of the lateral packing of the polymer chains [28], along with its indisputably clear focal conic microscopic texture (Figure 8.2D), characteristic of the smectic phase, which was almost the same as those observed for low-molecular-weight smectic liquid crystals. The smectic layer of polysilanes can also be directly observed by atomic force microscopy (AFM). Figure 8.2E shows the phase image taken on the surface of a 1-μm-thick cast film of poly-1 after annealing in saturated chloroform vapor for 12 h. The sharply defined smectic layers can be observed to spread beyond the field of view of the microscope, seemingly differently from textbook descriptions of liquid crystal, which have likened the molecular distribution within a smectic layer as sinusoidal. As predicted theoretically, this smectic phase is highly susceptible to its molecular weight distribution and understandably disfavored when the polydispersity is large since the rods with different lengths could not be packed within the same smectic layer [28].

As the temperature increased, the orthogonal lattice of the lateral polymer packing of poly-1 melted away, as demonstrated by disappearance of the wide-angle diffraction peaks (Figures 8.3A), followed by the conversion of the fan-like optical microscopic texture

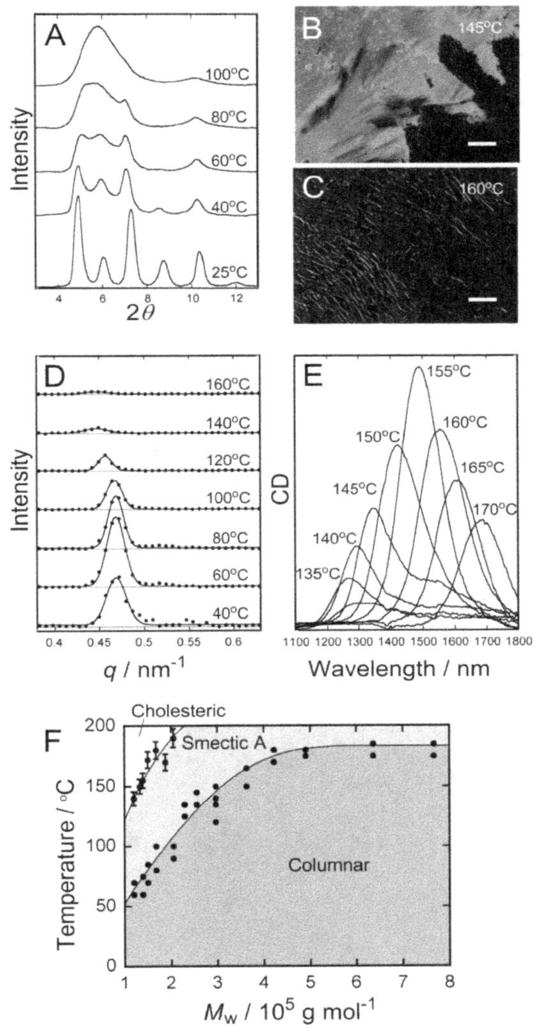

Figure 8.3 (A) Wide-angle X-ray diffraction profiles, (B, C) polarizing optical microscopic images (scale bars: 100 nm) (reproduced from Reference 33 with permission from Taylor and Francis), (D) small-angle X-ray diffraction profiles and (E) CD spectra of poly-1 taken upon increasing the temperature across the smectic-cholesteric liquid crystal phase transition temperature (reproduced from Reference 33 with permission from Taylor and Francis). (F) Molecular weight-temperature phase diagram of poly-1.

of the smectic phase into a Grandjean texture with oily streaks characteristic of the cholesteric phase (Figures 8.3B,C), disappearance of the small-angle smectic layer reflection (Figures 8.3D) and appearance of a cholesteric selective reflection in the circular dichroism (CD) spectrum (Figures 8.3E) [33]. These results clearly show that a columnar-smectic A-cholesteric (chiral nematic) phase transition takes place although the thermal decomposition precedes the smectic A-cholesteric phase transition at around 200 °C, when the molecular weight is 20,000 g/mol or more. The phase transition behavior of poly-1 can be summarized in the phase diagram as a function of the molecular weight and temperature in Figure 8.3F. It should be noted that the theoretically predicted phase transition behavior of athermal lyotropic systems could be experimentally reproduced with polysilane by replacing the increase in temperature with a decrease in concentration since it is a thermotropic system.

A similar smectic phase formation has been reported for polyisocyanide [34] with an extremely stiff backbone due to its tight hydrogen bonding between amide groups on its neighboring side chains, which makes it possible to observe the arrangement of the polymers within a smectic layer by AFM. One can clearly see how the rod-like polymers with a length distribution are packed within a smectic layer in Figure 8.4. It seems that the molecules are arranged

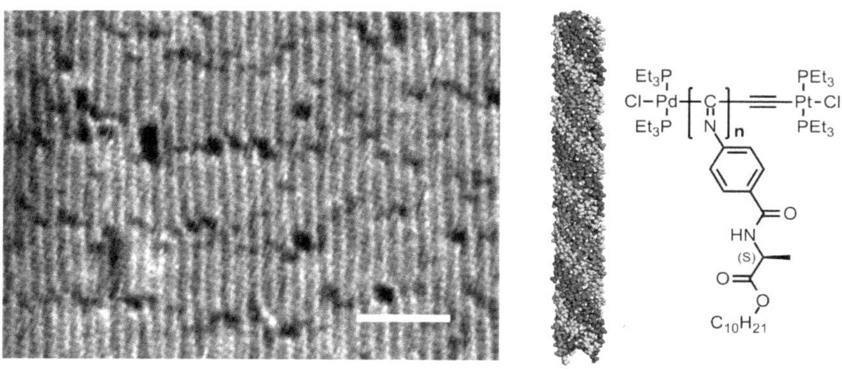

Figure 8.4 AFM phase image observed on the polyisocyanide film on the glass substrate (scale bar: 15 nm) with 3D depiction and chemical structure of the polyisocyanide. Reproduced from Reference 34 with permission from American Chemical Society.

within the smectic layers to enable stacking of the layers by combining the concave and convex curvature of the layers to extend the correlation length of the layer stack even with the M_w/M_n ratio well over the simulated limit of 1.03. This result was also supported by X-ray analysis of the smectic layer reflection obtained during the smectic-cholesteric phase transition using the one-dimensional electron density model [35].

8.3 Smectic Phases of Binary Mixtures of Rod-like Polymers

8.3.1 Theoretical Predictions

It should be interesting to note that smectic layers could form an incommensurate smectic phase to accommodate layers with different layer spacings when the rod-like particles with different lengths are mixed because the smectic layer spacing is largely defined by its component length. Even though it might be more intuitively understandable that the length polydispersity of the component disrupts the smectic layers because the components of different lengths cannot be accommodated in the same smectic layer, theoretical studies using the same technique have been performed for binary mixtures of rod-like particles with different lengths and have shown large variety of liquid crystal phases.

Koda and Kimura [36] have performed a theoretical study on the smectic phases formed in the binary mixtures of parallel rod-like particles (spherocylinders) of different lengths by stability analyses with Onsager approximation. The authors revealed that the smectic phase in which layers of long and short rod-like particles are alternatively stacked was formed by adding a small amount of short rods to the long rods when the length ratio was greater than 2.5. In the abundance of the short rods, the long rods randomly straddled the multiple smectic layers of short rods. If the length ratio was greater than 4.99, the two components were no longer miscible and segregated from each other to form nematic phases in the pure form.

Varga *et al.* [37,38] have investigated the effect of length bidispersity on liquid crystals formed in binary mixtures of long and short rod-like particles and concluded that four different types of smectic phases were formed depending on the mixing ratio and length ratio: 1) smectic A1 phase (named for the sake of convenience), in which short and long rod-like particles are mixed within the same smectic layer regardless of the mixing ratio when the length ratio (long rod length/short rod length) is less than 1.72 (Figure 8.5A); 2) smectic A2 phase, in which two smectic layers of short rods are nested within a smectic layer of abundant long rods when the length ratio is greater than 1.72 and less than 3.00 (Figure 8.5B); 3) smectic A3 phase, in which the long rods randomly straddle a pair of smectic layers of abundant short rods when the length ratio is greater than 1.79 and less than 3.13 (Figure 8.5C), even though the smectic layer ordering is destabilized and leads to the nematic phase for mixing ratios between smectic A2 and A3 phases, and 4) smectic A4 phase, in which a smectic layer of short rods and a smectic layer of long rods are alternately stacked when the long rod is in abundance and the length ratio is greater than 2.56 (Figure 8.5D), which also forms the nematic phase when the short rods are abundant [36,37]. However, systematic experimental verifications of these theoretical studies have not been reported sufficiently because there has been almost no appropriate rod-like particle system with an adjustable length.

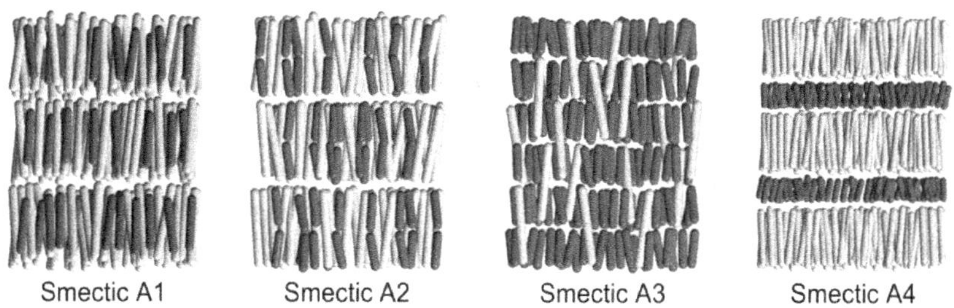

Figure 8.5 Four predicted types of smectic phases in a binary mixture of long and short rod-like particles [37,38].

8.3.2 Experimental Verifications

Structural studies of binary mixtures of poly-1 with different molecular weights (binary mixture of long polymer and short polymer) have been performed to demonstrate the validity of these theoretical predictions. Figure 8.6 shows the smectic layer spacing ratios of the binary mixtures with the mixing ratio of 0.75 (long polymer/binary mixture) over the corresponding short polymers observed by X-ray analyses plotted against the molecular weight ratios (long polymer/short polymer), which are equal to the length ratios [39]. The increase of the smectic layer spacings are almost proportional to the mixing ratios when the molecular length ratio is smaller than 1.7, clearly indicating that the two components are mixed within a smectic layer to form a layer with an average layer spacing, in other words, the smectic A1 phase. However, two reflections that correspond to the first and second order reflections were observed at the same time when the molecular weight ratio is greater than 1.7 and its first order reflection strangely disappeared on decreasing the mixing ratio, whereas the second order reflection remained.

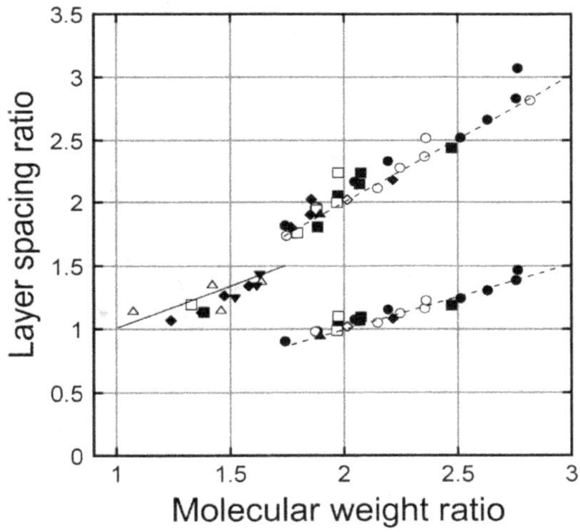

Figure 8.6 Observed layer spacing ratios (binary mixture/short polymer) versus molecular weight ratio (long polymer/short polymer) of binary mixtures of poly-1 at the mixing ratio of 0.75 (long polymer/binary mixture). The circles, squares, triangles, and diamonds represent a series of binary mixtures with short polymers with different molecular weights. Reproduced from Reference 39 with permission from American Chemical Society.

Figure 8.7 shows AFM images of the binary mixtures of poly-1s with the molecular weight ratio of 2.49 with various mixing ratios [39]. As short polymers were added to long polymers, lines along the smectic layers started to appear at the center of the layers of long polymers to divide them into two (Figure 8.7B) and the smectic layers further began to take up a frustrated structure where the smectic layers deviated by a half pitch in some places (Figure 8.7C), which ended up leading to the formation of smectic layers of the short polymer in the whole area (Figure 8.7D). The mechanism underlying this observation is considered to be as follows. As the number of short polymers increases, two smectic layers of short polymers get nested in a smectic layer of long polymers (smectic A2 phase). Upon further increase in the amount of short polymers, the long polymers cannot maintain their smectic layer and start to randomly straddle a pair of smectic layers of the short polymers (smectic A3 phase). The nematic phase that has been predicted at an intermediate mixing ratio was not observed here, indicating that the smectic phase is considerably more stable than the theoretical prediction, probably due to the dispersion interaction acting between the polymers, which is not taken into consideration in the theoretical treatment.

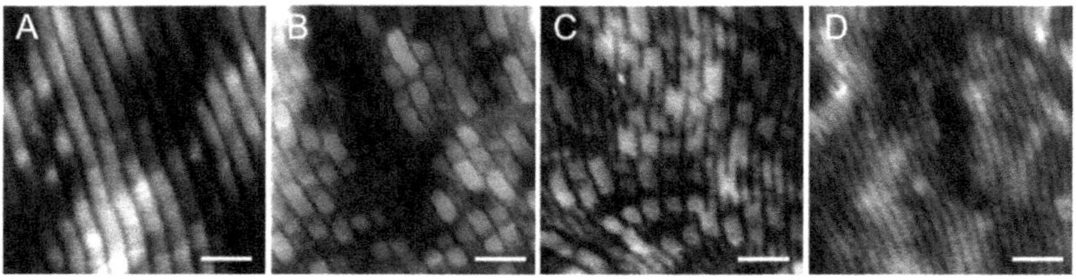

Figure 8.7 AFM images of binary mixtures of poly-1 with the molecular weight ratio of 2.49. Mixing ratios of long polymer: (A)1.00, (B) 0.75, (C) 0.5, and (D) 0.33. Scale bars: 200 nm. Reproduced from Reference 39 with permission from American Chemical Society.

Then, what kind of structure will appear in a binary mixture having an even larger molecular weight ratio? Figure 8.8 shows AFM images of binary mixtures of poly-2 having a molecular weight ratio of 3.14 with various mixing ratios [40]. As the number of short polymers increases, a smectic layer of short polymers gets inserted between the smectic layers of long polymers one by one instead of forming the nested smectic A2 phase (Figure 8.8B), leading to the

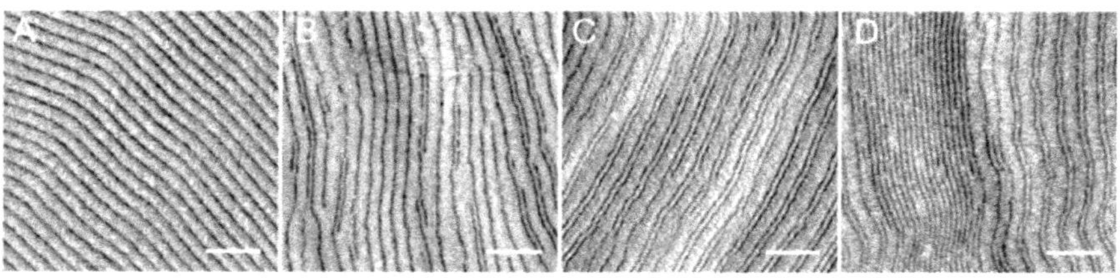

Figure 8.8 AFM images of binary mixtures of poly-2 with the molecular weight ratio of 3.14. Mixing ratios of (A)1.00, (B) 0.83, (C) 0.75, and (D) 0.50. Scale bars: 400 nm. Reproduced from Reference 40 with permission from American Chemical Society.

alternate stacking of thick smectic layers of long polymers and thin smectic layers of short polymers at the mixing ratio of 3/1 (Figure 8.8C), which perfectly reproduces the predicted smectic A4 phase. Unlike the theoretical prediction, which showed the transition to the nematic phase with abundant short rods, the short polymers were segregated from smectic A 4 phase and formed smectic phase in its pure form upon further increase in the amount of short polymers. This smectic phase did not undergo segregation with domains with grain boundaries but aligned along the smectic layer of smectic A4 phase, again showing a more stable smectic phase than the theoretical prediction.

8.4 Smectic Phases of Binary Mixtures of Rod-like Polymers and Spherical Molecules

8.4.1 Theoretical Predictions

At one limit of the length ratio, the binary system of rod-like particles with different lengths becomes a mixture of rod-like particles and spherical particles. The theoretical studies were also conducted on the binary mixtures of rod-like particles and spherical particles in the same manner and have shown the segregation of the spherical particles in the interstitial region of the smectic layers of the rod-like particles.

Koda *et al.* [41] have carried out a theoretical study on the binary mixture of rod-like and spherical particles using isobaric Monte Carlo simulations and shown that the smectic phase was stabilized by segregation of spheres of equal diameter from the rods to be located between the smectic layers of the rods. Dogic *et al.* [42] revealed that the smaller the spheres or the longer the rods are, the greater is the stabilization of the smectic phase, indicating that the entropic depletion force induced between long rods immersed in a fluid of small spheres favors the segregation of the spheres between the smectic layers of rods, although the segregation always competes with macroscopic demixing.

Despite the fact that the various intermolecular interactions are at work between the molecules in real systems (unlike the physical model of the simulations), these theoretical predictions have been largely reproduced in several experimental systems, e.g., silica rods and spheres [43], gold nanorods and nanospheres [44] and rod-like viruses and polystyrene latex [5], in which the diameters of the spheres are smaller than the diameter of the rods. On the contrary, bulk demixing takes place when the diameter of the spheres is sufficiently larger than the diameter of the rods [45,46]. These results clearly indicate that the relative sizes of the particles play a major role in structural formation, however, no systematic experimental study has been reported to clarify this hypothesis because there has been no appropriate experimental system with adjustable sizes of components.

8.4.2 Experimental Verifications

The predicted segregation has been also reproduced using binary mixtures of poly-3 and a methyl-branched alkane, squalane (Figure 8.9A). Figures 8.9B and 8.9C show the small- and wide-angle X-ray

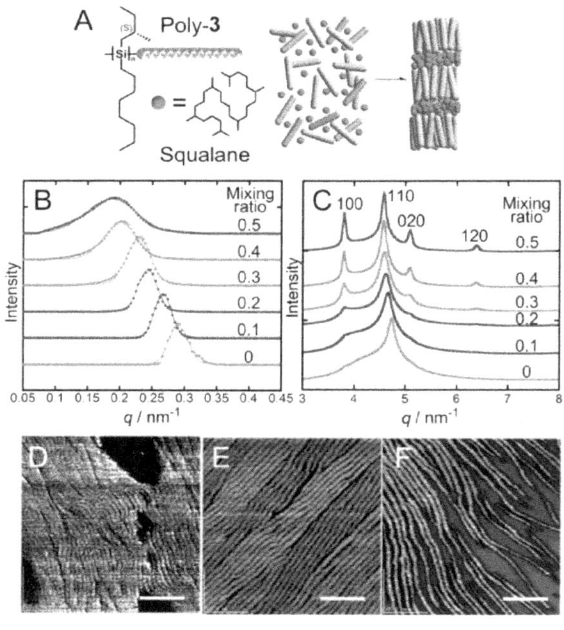

Figure 8.9 (A) Illustration of nano-segregation of squalane between smectic layers of poly-3, (B) small-angle and (C) wide-angle X-ray diffraction profiles of the binary mixtures of poly-3 and squalane with specified mixing ratios, (D-F) AFM images of the binary mixtures of poly-3 and squalane with the mixing ratios (squalane / binary mixture) of (D) 0.0, (E) 0.2, and (F) 0.4. Scale bars: 500 nm. Reproduced from Reference 48 with permission from Lifescience Global.

diffraction profiles of binary mixtures with various mixing ratios. The smectic layer reflection (Figure 8.9B) shifted to the small-angle direction as the mixing ratio of squalane increased while the wide-angle reflections, attributed to the orthorhombic lateral packing (a = 1.57 nm and b = 2.37 nm as indexed in Figure 8.9C) of the polymer, stayed almost the same [47,48]. Thus, the smectic layer spacing increased by 48.8% on the addition of 50% of squalane while the unit cell area of the two-dimensional lattice increased by only 1.02%. These results clearly indicate that the squalane had been preferentially segregated between the smectic layers to expand the layer spacing without penetrating the polymers.

According to theoretical predictions, these behaviors should be size-specific. Therefore, squalane was mixed with poly-3 of different widths, controlled by systematically changing the alkyl side chain length, and the resulting structures were investigated [47,48]. Contrary to the theoretical predictions, the diameter of the polymers showed an optimal value of 5.64 nm, which is smaller than the diameter of squalane, 6.57 nm, as evaluated by molecular dynamics simulations. In other words, the segregation of squalane was suppressed with polymers that had a larger or smaller diameter than the optimal value even though segregation should be promoted if the sphere is smaller in the theoretical prediction. The origin of the discrepancy between theory and experiment remains to be determined by a more detailed experimental study employing size adjustable spheres.

8.5 Conclusions

The structural diversity and functional versatility of organic liquid crystalline compounds are immense. This diversity, thus, might lead to complexity in studying the liquid crystallinity, wherein anisotropically shaped molecules propagate their anisotropy to the entire system by interacting with each other. The molecular theory of liquid crystals is a field of study attempting to determine the origin of the behavior of liquid crystals using simplified physical models of liquid crystals, which has not been complemented by experimental success because of the large deviation of the real systems, wherein miscellaneous diverse interactions at play, from theoretical ones. Thus, a number of organic compounds have been synthesized and their liquid crystallinity has been investigated merely for the purpose of identification and classification without knowing the true origin of the behavior of liquid crystals. We have shown here that rod-like helical polymers are an ideal model for verifying the molecular theory. Even though these polymers are simple homopolymers, these also show conclusive liquid crystalline behavior. In addition, their behavior is clearly based on theoretical prediction because these behave as simple rod-like particles due to their stiffness and structural simplicity. Research on rod-like polymers can, therefore, lead to elucidation of the origin of liquid crystallinity as well as contribute to fabrication of new useful materials for applications away from display applications.

References

1. Livolant, F., and Bouligant, Y. (1986) Liquid crystalline phases given by helical biological polymers (DNA, PBLG and xanthan). Columnar textures. *Journal de Physique*, **47**, 1813-1827.
2. Strzelecka, T. E., Davidson, M. W., and Rill, L. R. (1988) Multiple liquid crystal phases of DNA at high concentrations. *Nature*, **331**, 457-460.
3. Wen, X., Meyer, R. B., and Casper, D. L. D. (1989) Observation of smectic-A ordering in a solution of rigid-rod-like particles. *Physical Review Letters*, **63**, 2760-2763.
4. Dozic, Z., and Fraden, S. (1997) Smectic phase in a colloidal suspension of semiflexible virus particles. *Physical Review Letters*, **78**, 2417-2420.
5. Adams. M., Dozic, Z., Keller, S. L., and Fraden, S. (1998) Entropically driven microphase transition in mixtures of colloidal rods and spheres. *Nature*, **393**, 349-352.
6. Lee, S.-W., Mao, C., Flynn, C. E., and Belcher, A. M. (2002) Ordering of quantum dots using genetically engineered viruses. *Science*, **296**, 892-895.
7. Lee, S.-W., Wood, B. M., and Belcher, A. M. (2003) Chiral smectic C structure of virus-based film. *Langmuir*, **19**, 1592-1598.
8. Li, L., Walda, J., Manna, L., and Alivisatos, A. P. (2002) Semiconductor nanorod liquid crystals. *Nano Letters*, **2**, 557-560.
9. Gabriel, J.-C. P., Camerel, F., Lemarie, B. J., Desvaux, H., Davidson, P., and Batail, P. (2001) Swollen liquid-crystalline lamellar phase based on extended solid-like sheet. *Nature*, **413**, 504-508.
10. Onsager, L. (1949) The effect of shapes on the interaction of colloidal particles. *Annals of the New York Academy of Sciences*, **51**, 627-659.
11. Flory, P. J. (1956) Phase equilibria in solutions of rod-like particles. *Proceedings of the Royal Society A*, **243**, 73-89.

12. Hosino, M., Nakano, N., and Kimura, H. (1979) Nematic-smectic transition in an aligned rod system. *Journal of the Physical Society of Japan*, **46**, 1709-1715.
13. Hosino, M., Nakano, N., and Kimura, H. (1982) Phase transitions in the systems of identical rigid molecules in perfect alignment - Relations of the smetic A and columnar ordering in liquid crystals and the crystalline orderings to the molecular shape. *Journal of the Physical Society of Japan*, **51**, 741-748.
14. Stroobants, A., Lekkerkerker, H. N. W., and Frenkel, D. (1987) Evidence for one-, two-, and three-dimensional order in a system of hard parallel spherocylinders. *Physical Review A*, **36**, 2929-2945.
15. Veerman, J. A. C., and Frenkel, D. (1991) Relative stability of columnar and crystalline phases in a system of parallel hard spherocylinders. *Physical Review A*, **43**, 4334-4343.
16. Bolhuis, P., and Frenkel, D. (1997) Tracing the phase boundaries of hard spherocylinders. *Journal of Chemical Physics*, **106**, 666-687.
17. Veerman, J. A. C., and Frenkel, D. (1992) Phase behavior of disk-like hard-core mesogens. *Physical Review A*, **45**, 5632-5648.
18. Frenkel, D., Moulder, B. M., and McTague, J. P. (1984) Phase diagram of a system of hard ellopsoid. *Physical Review Letters*, **52**, 287-290.
19. Michael, P. A. (1990) Computer simulation of a biaxial liquid crystal. *Liquid Crystals*, **8**, 499-511.
20. Miller, W. G., Wu, C. C., Wee, E L., Santee, G. L., Rai, J. H., and Goebel, K. G. (1974) Thermodynamics and dynamics of polypeptide liquid crystals. *Pure and Applied Chemistry*, **38**, 37-58.
21. Yu, S. M., Conticello, V. P., Zhang, G., Kayser, C., Fournier, M. J., Mason, T. L., and Tirrell, D. A. (1997) Smectic ordering in solutions and films of a rod-like polymer owing to monodispersity of chain length. *Nature*, **389**, 167-170.
22. He, S.-J., Lee, C., Gido, S. P., Yu, S. M., and Tirrell, D. A. (1998) A twist boundary-like twisted smectic phase in monodisperse poly(γ-benzyl α,L-glutamate) produced by recombinant DNA techniques. *Macromolecules*, **31**, 9387-9389.
23. Zhang, G., Fournier, M. J., Mason, T. L., and Tirrell, D. A. (1992) Biological synthesis of monodisperse derivatives of poly(α,L-glutamic acid): Model rod-like polymers. *Macromolecules*, **25**, 3601-3603.
24. Wadati, M., and Isihara, A. (1972) Theory of liquid crystals. *Molecular Crystals and Liquid Crystals*, **17**, 95-108.
25. Kobayashi, K. K. (1970) Theory of translational and orientational melting with application of liquid crystals. I. *Journal of the Physical Society of Japan*, **29**, 101-105.
26. McMillan, W. L. (1971) Simple molecular model for the smectic A phase of liquid crystals. *Physical Review A*, **4**, 1238-1246.
27. Bates, M. A., and Frenkel, D. (1998) Influence of polydispersity on

the phase behavior of colloidal liquid crystals: A Monte Carlo simulation study. *Journal of Chemical Physics*, **109**, 6193-6199.

28. Okoshi, K., Kamee, H., Suzaki, G., Tokita, M., Fujiki, M., and Watanabe, J. (2002) Well-defined phase sequence including cholesteric, smectic A, and columnar phases observed in a thermotropic LC system of simple rigid-rod helical polysilane. *Macromolecules*, **35**, 4556-4559.
29. Fujiki, M. (1996) A correlation between global conformation of polysilane and UV absorption characteristics. *Journal of American Chemical Society*, **118**, 7424-7425.
30. Terao, K., Terao, Y., Teramoto, A., Nakamura, N., Terakawa, I., and Sato, T. (2001) Stiffness of polysilanes depending remarkably on a subtle difference in chiral side chain structure: Poly{n-hexyl-[(S)-2-methylbutyl]silane} and poly{n-hexyl-[(S)-3-methylpentyl]silane}. *Macromolecules*, **34**, 2682-2685.
31. Watanabe, J., and Takashina, Y. (1992) Thermotropic polypeptides VIII. Anomalous phase behavior in low molecular-weight poly(γ-octadecyl L-glutamate). *Polymer Journal*, **24**, 709-713.
32. Okoshi, K., Hagihara, T., Fujiki, M., and Watanabe, J. (2010) Anomalous thermotropic liquid crystalline phase behavior in poly [n-decyl-(RS)-2-methylbutylsilane]s with narrow molecular weight distributions. *Liquid Crystals*, **37**, 1183-1190.
33. Okoshi, K., Saxena, A., Naito, M., Suzaki, G., Tokita, M., Watanabe, J., Fujiki, M. (2004) First observation of a smectic A–cholesteric phase transition in a thermotropic liquid crystal consisting of a rigid-rod helical polysilane. *Liquid Crystals*, **31**, 279-283.
34. Onouchi, H., Okoshi, K., Kajitani, T., Sakurai, S.-i., Nagai, K., Kumaki, J., Onitsuka, K., and Yashima, E. (2008) Two- and three-dimensional smectic ordering of single-handed helical polymers. *Journal of American Chemical Society*, **130**, 229-236.
35. Okoshi, K., Saxena, A., Fujiki, M., Suzaki, G., Watanabe, J., and Tokita, M. (2004) Small-angle X-ray analysis of smectic A cholesteric liquid crystal phase transition in rigid-rod helical polysilane. *Molecular Crystals and Liquid Crystals*, **418**, 57-68.
36. Koda, T., and Kimura, H. (1994) Phase diagram of the nematic-smectic A transition of the binary mixture of parallel hard cylinders of different lengths. *Journal of the Physical Society of Japan*, **62**, 984-994.
37. Varga, S., Velasco, E., Mederos, L., and Vesely, F. J. (2009) Stability of the columnar and smectic phases of length-bidisperse parallel hard cylinders. *Molecular Physics*, **107**, 2481-2492.
38. Varga, S., and Velasco, E. (2010) Modeling and understanding smectic-phase formation in binary mixtures of rodlike polysilane: Comparison of Onsager theory and experiment. *Macromolecules*, **43**, 3956-3963.

39. Okoshi, K., Suzuki, A., Tokita, M., Fujiki, M., and Watanabe, J. (2009) Entropically-driven formation of smectic A1, A2, and A3 phases in binary mixtures of rigid-rod helical polysilanes with different molecular weights. *Macromolecules*, **42**, 3443-3447.
40. Okoshi, K., and Watanabe, J. (2010) Alternating thick and thin layers observed in the smectic phase of binary mixtures of rigid-rod helical polysilanes with different molecular lengths. *Macromolecules*, **43**, 5177-5179.
41. Koda, T., Numajiri, M., and Ikeda, S. (1996) Smectic-A phase of a bidisperse system of parallel hard rods and hard spheres. *Journal of the Physical Society of Japan*, **65**, 3551-3556.
42. Dogic, Z., Frenkel, D., and Fraden, S. (2000) Enhanced stability of layered phases in parallel hard spherocylinders due to addition of hard spheres. *Physical Review E*, **62**, 3925-3933.
43. Bakker, H. E., Dussi, S., Droste, B. L., Besseling, T. H., Kennedy, C. L., Wiegant, E. I., Liu, B., Imhof, A., Dijkstra, M., and van Blaaderen, A. (2016) Phase diagram of binary colloidal rod-sphere mixture from a 3D real-space analysis of sedimentation-diffusion equilibria. *Soft Matter*, **12**, 9238-9245.
44. Ahmad, I., Zandvliet, H. J. W., and Kooij, E. S. (2014) Shape-induced separation of nanospheres and aligned nanorods. *Langmuir*, **30**, 7953-7961.
45. Yasarawan, N., and van Duijneveldt, J. S. (2010) Arrested phase separation of colloidal rod-sphere mixtures. *Soft Matter*, **6**, 353-362.
46. Koenderink, G. H., Vliegenthart, G. A., Kluijtmans, S. G. J. M., van Blaaderen, A., Philipse, A. P., and Lekkerkerker, H. N. W. (1999) Depletion-induced crystallization in colloidal rod-sphere mixtures. *Langmuir*, **15**, 4693-4696.
47. Tanaka, T., Kato, I., and Okoshi, K. (2015) Nano-segregation of squalane between smectic layers of rigid-rod polysilane. *e-Journal of Surface Science and Nanotechnology*, **13**, 121-124.
48. Tanaka, T., Kato, I., and Okoshi, K. (2018) Effect of side chain length on segregation of squalane between smectic layers formed by rod-like polysilane. *Journal of Research Updates in Polymer Science*, **7**, 1-6.

Index

A

B

C

D

E

F

G

M

N

O

P

Q

R

S

T

V

W

www.ingramcontent.com/pod-product-compliance
Lightning Source LLC
LaVergne TN
LVHW011525200926
848484LV00001B/150

* 9 7 8 1 9 2 5 8 2 3 1 7 2 *